RELATED APEX BOOKS

AP U.S. Government & Politics

AP Statistics

RELATED KAPLAN BOOKS

College Admissions and Financial Aid

Guide to the Best Colleges in the U.S.

Kaplan/Newsweek College Catalog

Parent's Guide to College Admissions

Scholarships

What to Study: 101 Fields in a Flash

Yale Daily News Guide to Internships

Yale Daily News Guide to Succeeding in College

You Can Afford College

Test Preparation

ACT

Fast Track ACT

AP Biology

SAT & PSAT

Fast Track SAT & PSAT

SAT Math Mania

SAT Math Workbook

SAT or ACT? Test Your Best\

SAT II: Biology

SAT II: Chemistry

SAT II: Mathematics

SAT II: Writing

SAT Verbal Velocity

SAT Verbal Workbook

Apex
Learning

AP[*]
Calculus AB

An Apex Learning Guide

Simon & Schuster

NEW YORK · LONDON · SINGAPORE · SYDNEY · TORONTO

*AP is a registered trademark of the College Entrance Examination Board, which neither sponsors nor endorses this product.

Kaplan Publishing
Published by Simon & Schuster
1230 Avenue of the Americas
New York, New York 10020

For bulk sales to schools, colleges, and universities, please contact Order Department, Simon & Schuster, 100 Front Street, Riverside, NJ 08075. Phone: (800) 223-2336.
Fax: (800) 943-9831.

The material in this book is up-to-date at the time of publication. The College Entrance Examination Board may have instituted changes in the test after this book was published. Please read all materials you receive regarding the AP Calculus AB Test carefully.

Contributing Editors: Marc Bernstein, Marcy Bullmaster, and Seppy Basili
Mathematics Editor: Robert Reiss
Project Editor: Larissa Shmailo
Cover Design: Cheung Tai
Interior Page Layout: Hugh Haggerty
Production Editor: Maude Spekes
Editorial Coordinator: Dea Alessandro
Executive Editor: Del Franz

Manufactured in the United States of America.
Published simultaneously in Canada.

April 2001
10 9 8 7 6 5 4 3 2 1

ISBN 0-7432-0189-2

Table of Contents

SECTION I:

The Basics

Chapter 1   *Inside the AP Calculus AB Exam*

Before you plunge into studying for the AP Calculus AB exam, let's take a step back and look at the big picture. What's the AP Calculus AB exam all about? How can I prepare for it? How's it scored? This chapter and the next will answer these questions and more.

What is the Advanced Placement (AP) Program?

Through the Advanced Placement Program, you can take college-level courses while you are in high school. Based on your grade on an AP Exam, colleges and universities can grant you placement or college credit or both.

In addition to getting a head start on your college coursework, you can improve your chances of acceptance to competitive schools since colleges know that AP students are better prepared for the demands of college courses. There's also the money you can save on tuition if you receive credit!

About the AP Calculus AB Exam

The AP Calculus AB exam is administered in May by the College Board's AP Services. The exam is 3 hours and 15 minutes long, so developing your stamina is very important. The exam consists of two equally weighted sections: Section I, Multiple-Choice, and Section II, Free-Response.

Section I—Multiple-Choice

There are 45 multiple choice questions. This section is split into two parts: part A consists of 28 questions with no calculator allowed (55 minute time limit) and part B consists of 17 questions for which a calculator is required (50 minute time limit). Each multiple choice question is worth 1 point.

Section II—Free-Response

The second section of the examination consists of 6 free-response questions worth 9 points each. Part A consists of 3 questions for which a calculator is required (45 minutes). Part B consists of 3 questions for which calculators may be not be used (45 minutes). During Part B, you may go back to the first three questions in Part A if you have time, but you will not be permitted to use your calculator.

There's a five minute break between Section I and Section II.

AP Calculus AB at a Glance

	Number of Questions	Time	Calculator?
Section I			
Part A	28	55	No
Part B	17	50	Yes
Section II			
Part A	3	45*	Yes
Part B	3	45	No

*In section II, students may go back to work on questions in Part A after timing for Part B has begun without a calculator. The proctor for the examination will remind you at the beginning of each part of the examination whether or not a calculator is permitted.

What's the AB Stand for?

There are two AP Calculus exams and the AB distinguishes this exam from the AP Calculus BC exam. The AB exam covers one semester of college-level calculus, whereas the BC exam covers a full year of college calculus, although individual colleges may differ in their course offerings and placement/credit policies. You may take only one of the two exams in any given year.

Scoring

The AP Calculus exam is scored on a scale of 1 to 5, with 5 being the highest grade. The scores are defined as follows:

5 Extremely well qualified

4 Well qualified

3 Qualified

2 Possibly qualified

1 No recommendation

In 2000, the average grade of the 137,276 candidates who took the AB test was 3.03. Typically, to obtain a score of 3 or higher, you need to answer about 50 percent of the multi-

Apex Learning

ple choice section correctly, and do acceptable work on the free-response section. Remember that you can miss questions and still get a perfect 5 score.

Raw scores are calculated in points. The multiple-choice section of the AP Calculus exam is worth the same number of points as the free-response section. There is a 1.2 weighting factor used so that the multiple-choice and free-response sections of the exam have equal weight. Each section is worth 54 points, for a total of 108 points.

It would not be correct to say that the AP examination is graded on a curve. How other students do on the same exam will not affect your score. However, some multiple-choice questions are used from one year to the next to allow for a calibration of the scores so that a 4 one year reflects the same statistical strength of performance. For this reason, the cut-offs for each score level do not stay constant from year to year.

Wrong-Answer Penalty

For multiple choice questions, there is a penalty for incorrect answers as opposed to simply leaving answers blank. This is sometimes called a guessing penalty, but it is really a wrong answer penalty. If you guess right, you're in great shape!

Here's how the scoring works: You receive 1 point for a correct answer, 0 points for no answer, and -1/4 point for a wrong answer. For example, getting 30 correct and 15 wrong would give you a score of 30-(15x1/4) = 26.25. We'll talk more about guessing on the AP Calculus AB multiple-choice section in the next chapter, *Taking AP Calculus AB—Strategies for Success*.

What You Need to Bring

- Two calculators (see below) and batteries

- Photo I.D. (driver's license, school I.D., or a valid passport are acceptable)

- Your secondary school code number (see your Guidance Counselor or AP Coordinator)

- Your Social Security number

- Several sharpened No. 2 pencils

- Pencil sharpener

- Eraser

- A watch (in case your exam room doesn't have a clock you can see easily. You need to be able to pace yourself during this long test!)

What NOT to Bring

- DON'T bring scratch paper. You'll make your notes in the test booklet in the spaces provided.

- Don't bring books, compasses, correction fluid, dictionaries, highlighters, notes, or rulers.

- Don't bring beepers or cellular phones, or watches that have beepers or alarms.

- Don't bring food or drinks.

- Don't even wear a T-shirt with math of any kind on it!

Calculators

You can only bring two calculators, both of which must be approved types. To be an approved calculator, it should be able to:

- Produce the graph of a function within an arbitrary viewing window.

- Find the zeros of a function.

- Compute the derivative of a function numerically.

- Compute definite integrals numerically.

You can't use a calculator with a QWERTY keypad, because it makes it too easy to type text (such as test questions) into the calculator. Visit the AP Calculus Website at www.college-board.org/ap/calculus/ or see the AP Bulletin for Students and Parents for a listing of approved calculators.

How Are Exams Graded?

The multiple-choice section of the exam is graded by computer. The free-response booklets are graded by faculty consultants. These are college professors and AP teachers who are specially trained to assess the questions. Usually about six faculty consultants review each free-response booklet.

How Do I Get My Grade?

AP Grade Reports are sent in July to each student's home, high school, and any colleges designated by the student. Students may designate the colleges they would like to receive their grade on the answer sheet at the time of the test. Students may also contact AP Services to forward their grade to other colleges after the exam, or to cancel or withhold a grade.

AP Grades by Phone

AP Grades by phone are available for $13 a call beginning in early July. A touch-tone phone is needed. The toll-free number is (888) 308-0013.

Registration

To register for the AP Calculus AB exam, contact your school guidance counselor or AP Coordinator. If your school does not administer the exam, contact AP Services for a listing of schools in your area that do.

Fees

The fee for each AP Exam is $77. The College Board offers a $22 credit to qualified students with acute financial need. A portion of the exam fee may be refunded if a student does not take the test. There is a $20 late fee for late exam orders. Check with AP Services for applicable deadlines.

Additional Resources

The College Board offers a number of publications about the Advanced Placement Program, including: *Advanced Placement Program Course Description—Calculus, A Guide to the Advanced Placement Program,* and the *AP Bulletin for Students and Parents.* You can order these and other publications online at www.collegeboard.org/ap/ at the Store links, or call AP Services for an order form.

For More Information

For more information about the AP Program and/or the AP Calculus AB exam, contact your school's AP Coordinator or guidance counselor, or contact AP Services at:

AP Services
P.O Box 6671
Princeton, NJ 08541-6671
(609) 771-7300
Toll-free: (888) CALL-4-AP (888-225-5427)
Fax: (609) 530-0482
TTY: (609) 882-4118
Email: apexams@info.collegeboard.org
Website: www. collegeboard.org/ap/calculus/

Chapter 2 Taking AP Calculus AB—Strategies for Success

This chapter will help you tackle the Advanced Placement Calculus AB examination. A passing grade (3, 4, or 5) on the AB examination is accepted for advanced placement or credit equivalent to a semester course in calculus at many colleges and universities. (This guide does not address the additional topics covered in a Calculus BC course which is equivalent to a full year of college calculus.)

What's on the Test?

To be ready for AP Calculus, you should have a good grasp of algebra, geometry, trigonometry, analytic geometry, and elementary functions, including linear, polynomial, rational, exponential, logarithmic, trigonometric, and inverse trigonometric functions. The AP Calculus AB course itself concentrates on extending your knowledge of functions through basic differential and integral calculus. There is much emphasis on *multiple representations*, that is, understanding how to work with functions whether they be presented analytically (by a formula $y = f(x)$), graphically, or as a table of data values.

The breakdown of the AB course description is roughly as follows:

Limits ($\approx 10\%$): calculating limits using algebra, estimating limits from graphs or tables of data, and understanding the key idea of *continuity*, including the Intermediate and Extreme Value Theorems. (Chapter 3)

Differential calculus and its applications ($\approx 40\%$): calculating derivatives using algebra (both at a point and as a function), estimating derivatives from graphs or tables of data, understanding the relationship between continuity and differentiablility, the Mean Value Theorem, higher order derivatives, implicit differentiation, using derivatives to analyze graphs, solving extremum and related rates problems. (Chapters 4, 5, and 7)

Apex Learning   9

Integral calculus and its applications ($\approx 40\%$): calculating indefinite integrals (antiderivatives) and definite integrals using algebra , estimating integrals from graphs or tables of data, Riemann sum approximations, the Fundamental Theorems of Calculus, applications of integrals to finding area, volumes, distances and, in general, accumulated change. (Chapters 6–9)

Separable differential equations ($\approx 10\%$): solving differential equations by separation of variables, estimating solutions graphically using slope fields, exponential and related models arising from differential equations. (Chapter 10)

How to Use This Book

Each chapter of this book lays out the most important objectives of the AP Calculus AB course with examples to illustrate. There are extensive multiple choice and free-response practice problems in each chapter, all with answers immediately following them. Here's a suggestion for using the multiple-choice problems to maximum benefit: for each problem, use a sheet of paper to cover the solution following the problem. Write down your answer on this sheet and then slide it down to reveal the solution and discussion. This method can build your skill in answering these questions.

At the end of the book you'll find two complete Practice Tests with solutions. For these, it is best to try them under conditions as close as possible to those you will face with the real examination. In other words, work through them as best you can under the time constraints and then check your answers at the end.

Estimating Your Score

To estimate your score on either of the Practice Tests, use the following guidelines.

Section I: Multiple-Choice Questions.

Score = (number of correct multiple-choice questions)

$\qquad$ − 1/4 (number of incorrect multiple-choice questions).

Section II: Free-Response Questions.

Questions 1–6 are graded on a nine-point scale.

Determine your score on the free-response section as follows:

Score = # 1 + # 2 + # 3 + # 4 + # 5 + (# 6).

The multiple-choice section of the AP Calculus exam is worth the same number of points as the free-response section. There is a 1.2 weighting factor used so that the multiple-choice and free-response sections of the exam have equal weight.

You can roughly approximate your score on the AP Exam as follows:

First calculate **Overall Score:** = (multiple-choice score $\times$ 6) + (free-response score $\times$ 5) .

Then look up your score in the following table:

Overall Score	AP Approximation
370–540	5
290–369	4
215–289	3
150–214	2
0–149	1

Remember that you can miss questions and still get a 5 score.

AP Calculus AB–Test-Taking Strategies

The AP Calculus AB exam isn't a normal exam. Most normal exams test your memory. This test isn't like that. AP Calculus AB tests problem-solving skills rather than memory. Below are some strategies to put to use on the test.

Guessing on Multiple-Choice Questions

In the multiple-choice part of the examination, you can benefit by using test-smart strategies and techniques.Remember that there is a penalty for incorrect answers versus simply leaving an item blank. You receive 1 point for a correct answer, 0 points for no answer, and $-\frac{1}{4}$ for an incorrect answer. In general, if you can eliminate one or two of the options on a multiple choice item, the odds shift in your favor to go ahead and guess. If you have absolutely no idea, then it may not be wise to guess.

Simplifying Answers on the Free-Response Questions

There are no extra points awarded for simplification of answers, but points may be deducted for *incorrectly* simplified work. Suppose you've been asked to find the derivative of

$$g(x) = x^2 \cos(2 \ln x).$$

The answer

$$g'(x) = 2x \cos(2 \ln x) - x^2 \sin(2 \ln x) \times \left(\frac{2}{x}\right)$$

will receive full credit.

There's no need to simplify your answer unless you need a simplified version. For example, if you wanted to find the maxima and minima of $g(x)$ on the domain $[1, 4]$ you **may** want to simplify:

$$g'(x) = 2x \cos(2 \ln x) - x^2 \sin(2 \ln x) \times \left(\frac{2}{x}\right)$$
$$= 2x \cos(2 \ln x) - 2x \sin(2 \ln x)$$
$$= 2x(\cos(2 \ln x) - \sin(2 \ln x)),$$

but even in this case you could use the calculator to find the zeros of your original form of the derivative.

Be Complete and Show Your Work

Suppose a student wrote this on the free-response section of the exam:

$$\text{If} \quad \frac{dy}{dx} = x^2, \quad \text{then} \quad y = \frac{1}{3}x^3.$$

Oops! The statement would not get full credit because there is *no constant of integration* shown. It should be written:

$$\text{If} \quad \frac{dy}{dx} = x^2, \quad \text{then} \quad y = \frac{1}{3}x^3 + C.$$

You're not required to simplify your numerical answers. For example, if in a certain free-response question, the answer was $h = 2\sqrt[3]{2}$, then you'd still receive full credit for the answer $h = (4)^{2/3}$ or $h = \sqrt[3]{16}$.

Make sure to clearly show your steps in presenting your solution to a problem. For example, suppose you're given this problem:

Find $y = f(x)$ by solving the separable differential equation

$$\frac{dy}{dx} = \frac{2x^2 + 3}{y}$$

with the initial condition $f(1) = 4$.

There are four distinct steps to show here: Separating the variables, integrating to get an implicitly defined family of functions (with a $+C$,), using the initial condition to solve for C, and solving your implicitly defined function for an explicit expression for $f(x)$.

Use Standard Mathematical Notation

For Section II, your work must be expressed in standard mathematical notation and not calculator syntax.

For example, suppose you're given this problem on the part of the free-response section where a calculator is allowed:

Find the volume of the solid generated by revolving the region bounded by $y = x^{2/3}$, $x = 8$ and $y = 0$ about the x-axis.

The solution is obtained by calculating the definite integral

$$\pi \int_0^8 (x^{2/3})^2 \, dx = \pi \int_0^8 x^{4/3} \, dx = \frac{3\pi}{7}x^{7/3} \Big|_0^8 = \frac{384\pi}{7} \approx 172.339$$

If you chose to calculate this integral on your calculator, it would be perfectly acceptable to write simply

$$\pi \int_0^8 (x^{2/3})^2 \, dx \approx 172.339$$

and you would receive full credit for this solution. However, it would **NOT** be acceptable to write something like

$$\pi * \text{fnInt}(x\,\hat{}(4/3),x,0,8) \approx 172.339$$

or some other calculator syntax for the definite integral and you would receive only partial credit in this case.

- If you use your calculator to solve an equation, make sure you write out the equation that you are solving!
- If you use your calculator to calculate a derivative, make sure you write out the function that you are differentiating (and the value at which you are computing it)!
- If you use your calculator to calculate a definite integral, make sure you write out the definite integral!
- If you use your calculator to create a graph that you copy into the test booklet, make sure you indicate the scaling on the axes and label the function that you are graphing!

Avoid Misgridding

A common mistake on multiple-choice questions is marking the right answer–in the wrong place! Be careful about gridding. Also, it's a good idea to grid five or so multiple choice answers at a time to save time and avoid misgridding.

Learn the Directions

Why waste valuable time reading directions when you can have them down pat beforehand? You need every second during the test to answer questions and get points. Become familiar with the directions below before test day. Actual directions for the free-response section of the exam are included in the Practice Tests.

Don't Get Discouraged

It's important to remember that many successful AP Calculus AB test takers miss a number of questions and still get a 5. The test is designed so that the mean will be near 50% in order to provide a full range to base the scores of 1–5 as accurately as possible. Knowing this will stop you from panicking when you hit an impossible question. Relax! You can skip many tough questions on the AP Calculus AB exam and still get a great score! Look at the suggestions in the next section to help you keep your cool.

Managing Stress

The countdown has begun. Your date with the test is looming on the horizon. Anxiety is on the rise. The butterflies in your stomach have gone ballistic and your thinking is getting cloudy. Maybe you think you won't be ready. Maybe you already know your stuff, but you're going into panic mode anyway. Don't freak! It's possible to tame that anxiety and stress-before and during the test. Remember, some stress is normal and good. Anxiety is a motivation to study. The adrenaline that gets pumped into your bloodstream when you're stressed helps you stay alert and think more clearly. But if you feel that the tension is so great that it's preventing you from using your study time effectively, here are some things you can do to get it under control.

Take Control

Lack of control is a prime cause of stress. Research shows that if you don't have a sense of control over what's happening in your life, you can easily end up feeling helpless and hopeless. Try to identify the sources of the stress you feel. Which ones of these can you do something about? Can you find ways to reduce the stress you're feeling about any of these sources?

Focus on Your Strengths

Make a list of areas of strength you have that will help you do well on the test. We all have strengths, and recognizing your own is like having reserves of solid gold at Fort Knox. You'll be able to draw on your reserves as you need them, helping you solve difficult questions, maintain confidence, and keep test stress and anxiety at a distance. And every time you recognize a new area of strength, solve a challenging problem, or score well on a practice test, you'll increase your reserves.

Imagine Yourself Succeeding

Close your eyes and imagine yourself in a relaxing situation. Breathe easily and naturally. Now, think of a real-life situation in which you scored well on a test or did well on an assignment. Focus on this success. Now turn your thoughts to the AP Calculus AB test and keep your thoughts and feelings in line with that successful experience. Don't make comparisons between them; just imagine yourself taking the upcoming test with the same feelings of confidence and relaxed control.

Set Realistic Goals

Facing your problem areas gives you some distinct advantages. What do you want to accomplish in the time remaining? Make a list of realistic goals. You can't help feeling more confident when you know you're actively improving your chances of earning a higher test score.

Exercise Your Frustrations Away

Whether it's jogging, biking, pushups, or a pickup basketball game, physical exercise will stimulate your mind and body, and improve your ability to think and concentrate. A surprising number of students fall out of the habit of regular exercise, ironically because they're spending so much time prepping for exams. A little physical exertion will help you to keep your mind and body in sync and sleep better at night.

Avoid Drugs

Using drugs (prescription or recreational) specifically to prepare for and take a big test is definitely self-defeating. (And if they're illegal drugs, you may end up with a bigger problem than the AP Calculus AB on your hands.) Mild stimulants, such as coffee or cola can sometimes help as you study, since they keep you alert. On the down side, too much of these can also lead to agitation, restlessness, and insomnia. It all depends on your tolerance for caffeine.

Eat Well

Good nutrition will help you focus and think clearly. Eat plenty of fruits and vegetables, low-fat protein such as fish, skinless poultry, beans, and legumes, and whole grains such as brown rice, whole wheat bread, and pastas. Don't eat a lot of sugar and high-fat snacks, or salty foods.

Quick Tips for the Days Just Before the Exam

Do Less

The best test takers do less and less as the test approaches. Taper off your study schedule and take it easy on yourself. You want to be relaxed and ready on the day of the test. Give yourself time off, especially the evening before the exam. By then, if you've studied well, everything you need to know is firmly stored in your memory banks.

Think Positively

Positive self-talk can be extremely liberating and invigorating, especially as the test looms closer. Tell yourself things such as, "I choose to take this test" rather than "I have to"; "I will do well" rather than "I hope things go well"; "I can" rather than "I cannot." Be aware of negative, self-defeating thoughts and images and immediately counter any you become aware of. Replace them with affirming statements that encourage your self-esteem and confidence. Create and practice visualizations that build on your positive statements.

Be Prepared

Get your act together sooner rather than later. Have everything (including choice of clothing) laid out days in advance. Most important, know where the test will be held and the easiest, quickest way to get there. You will gain great peace of mind if you know that all the little details-gas in the car, directions, etcetera-are firmly in your control before the day of the test.

Visit the Test Site

Experience the test site a few days in advance. This is very helpful if you are especially anxious. If at all possible, find out what room your part of the alphabet is assigned to, and try to sit there (by yourself) for a while. Better yet, bring some practice material and do at least a section or two, if not an entire practice test, in that room. In this situation, familiarity doesn't breed contempt, it generates comfort and confidence.

Rest and Relax the Day Before the Test

Forego any practice on the day before the test. It's in your best interest to marshal your physical and psychological resources for 24 hours or so. Even race horses are kept in the paddock and treated like princes the day before a race. Keep the upcoming test out of your consciousness; go to a movie, take a pleasant hike, or just relax. Don't eat junk food or tons of sugar. And-of course-get plenty of rest the night before. Just don't go to bed too early. It's hard to fall asleep earlier than you're used to, and you don't want to lie there thinking about the test.

Handling Stress During the Test

The biggest stress monster will be the test itself. Fear not; there are methods of quelling your stress during the test.

Keep Moving

Keep moving forward instead of getting bogged down in a difficult question. You don't have to get everything right to achieve a fine score. The best test takers skip difficult material temporarily in search of the easier stuff. They mark the ones that require extra time and thought. This strategy buys time and builds confidence so you can handle the tough stuff later.

Work at Your Own Pace

Don't be thrown if other test takers seem to be working more furiously than you are. Continue to spend your time patiently thinking through your answers; it's going to lead to better results. Don't mistake the other people's sheer activity as signs of progress and higher scores.

Keep Breathing

Conscious attention to breathing is an excellent way to manage stress while you're taking the test. Most of the people who get into trouble during tests take shallow breaths: They breathe using only their upper chests and shoulder muscles, and may even hold their breath for long periods of time. Conversely, those test takers who breathe deeply in a slow, relaxed manner are likely to be in better control during the session.

Stretch

If you find yourself getting spaced out or burned out as you're studying or taking the test, stop for a brief moment and stretch. Even though you'll be pausing for a moment, it's a moment well spent. Stretching will help to refresh you and refocus your thoughts.

With what you've just learned here, you're armed and ready to do battle with the test. This book and your studies will give you the information you'll need to answer the questions. It's all firmly planted in your mind. You also know how to deal with any excess tension that might come along, both when you're studying for and taking the exam. You've experienced everything you need to tame your test anxiety and stress. You're going to get a great score.

Good luck with your study of Advanced Placement Calculus AB!

SECTION II:

Calculus Review

Chapter 3   *Limits of Functions*

In this chapter, we'll review objectives for the topic of limits in AP Calculus. Limits provide a very nice language for describing the behavior of functions—both *locally* (near a specific point) and *globally* (over an entire domain). More importantly, the idea of a limit is the cornerstone of calculus. The two main branches of calculus, differential and integral, are based on the mathematical notion of limit.

A. What's a Limit?

▶Objective 1   When given a graph or an algebraic expression, identify when a limit exists.

Example $\lim\limits_{x \to 3} g(x)$ exists in which of the following?

A. $g(x) = \dfrac{x^2 - 9}{x - 3}$

B. $g(x) = |x - 3|$

C. $g(x) = \dfrac{|x - 3|}{x - 3}$

D.

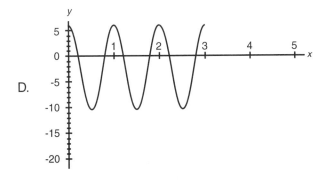

Tip   The right-hand limit must equal the left-hand limit for the overall limit to exist.

Answers   A, B

▶Objective 2   Estimate a limit (approaching from the left or right), from a table of data.

Example

x	2.9	2.99	2.999	3.001	3.01	3.1
$g(x)$	−4.41	−4.9401	−4.994	5.006	5.0601	5.61

From the data above, estimate the value of $\lim\limits_{x \to 3^-} g(x)$, $\lim\limits_{x \to 3^+} g(x)$, and $\lim\limits_{x \to 3} g(x)$.

Tips $\lim\limits_{x \to 3^-}$ means the limit as x is approaching 3 from the left.

$\lim\limits_{x \to 3^+}$ means the limit as x is approaching 3 from the right.

Answers

$$\lim\limits_{x \to 3^-} g(x) = -5. \qquad \lim\limits_{x \to 3^+} g(x) = 5. \qquad \lim\limits_{x \to 3} g(x) \text{ does not exist (DNE).}$$

▶Objective 3   Estimate a limit (approaching from the left or right), from a graph.

Example   Estimate each of the following limits if $y = g(x)$ has a graph given by:

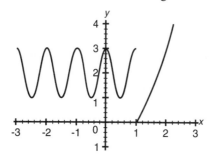

A. $\lim\limits_{x \to -1} g(x)$

B. $\lim\limits_{x \to \frac{3}{2}} g(x)$

C. $\lim\limits_{x \to 1} g(x)$

D. $\lim\limits_{x \to 1^-} g(x)$

E. $\lim\limits_{x \to 1^+} g(x)$

F. $\lim\limits_{x \to 2} g(x)$

Tip   The right-hand limit must equal the left-hand limit in order for the overall limit to exist.

Answers

A. $\lim\limits_{x \to -1} g(x) = 3.$

B. $\lim\limits_{x \to \frac{3}{2}} g(x) = 1.$

C. $\lim\limits_{x \to 1} g(x)$ *DNE*.

D. $\lim\limits_{x \to 1^-} g(x) = 3.$

E. $\lim\limits_{x \to 1^+} g(x) = 0.$

F. $\lim\limits_{x \to 2} g(x) = 3.$

►**Objective 4**   Estimate a limit numerically, using a calculator (including one-sided limits).

Example   Find $\lim\limits_{x \to 1} \dfrac{x - 1}{x^2 - 1}$ and $\lim\limits_{t \to 0} \dfrac{\sqrt{t^2 + 9} - 3}{t^2}$ numerically, using your calculator.

Tip   Warning! If you pick values that are extremely small for t in $\lim\limits_{t \to 0} \dfrac{\sqrt{t^2 + 9} - 3}{t^2}$, you might assume the limit to be 0 (depending on your calculator). Realize that the calculator has only a finite precision. When you square the small values of t in this problem, the numbers become so small that the calculator can't subtract accurately. This is another example of the importance of realizing the limitations of your calculator.

Answers

$$\lim\limits_{x \to 1} \frac{x - 1}{x^2 - 1} = \frac{1}{2} \quad \text{and} \quad \lim\limits_{t \to 0} \frac{\sqrt{t^2 + 9} - 3}{t^2} = \frac{1}{6}.$$

►**Objective 5**   Determine a limit from algebraic formulas, where $\lim\limits_{x \to a} f(x) = f(a)$.

Example   Find the following limits.

A. $\lim\limits_{v \to -1} \sqrt{v^2 - 4v + 2}$

B. $\lim\limits_{x \to 5} (x^2 - 25)$

C. $\lim\limits_{x \to 1.5} [(x - 2)]$

Answers

A. $\lim\limits_{v \to -1} \sqrt{v^2 - 4v + 2} = \sqrt{7}.$

B. $\displaystyle\lim_{x \to 5}(x^2 - 25) = 0.$

C. $\displaystyle\lim_{x \to 1.5}[(x - 2)] = -1.$

▶Objective 6   Determine limits for more complicated expressions, where algebraic manipulation is required, for example, rationalizing, factoring, expanding, finding common denominators, or some combination of all four.

Example   Find the following limits if they exist.

A. $\displaystyle\lim_{h \to 0} \frac{4 - \sqrt{16 + h}}{h}$

B. $\displaystyle\lim_{x \to -2} \frac{x^3 + 8}{x^4 - 16}$

C. $\displaystyle\lim_{x \to 2} \frac{\frac{1}{x} - \frac{1}{2}}{x - 2}$

D. $\displaystyle\lim_{x \to b} \frac{x^2 - b^2}{x - b}$

Tip   Find an algebraic manipulation that changes the form of the expression such that you can simply evaluate the limit.

Answers

A. $\displaystyle\lim_{h \to 0} \frac{4 - \sqrt{16 + h}}{h} = \lim_{h \to 0} \frac{4 - \sqrt{16 + h}}{h} \times \frac{4 + \sqrt{16 + h}}{4 + \sqrt{16 + h}}$

$\displaystyle = \lim_{h \to 0} \frac{16 - (16 + h)}{h(4 + \sqrt{16 + h})} = \lim_{h \to 0} \frac{-1}{4 + \sqrt{16 + h}} = -\frac{1}{8}.$

B. $\displaystyle\lim_{x \to -2} \frac{x^3 + 8}{x^4 - 16} = \lim_{x \to -2} \frac{(x + 2)(x^2 - 2x + 4)}{(x - 2)(x + 2)(x^2 + 4)} = \lim_{x \to -2} \frac{(x^2 - 2x + 4)}{(x - 2)(x^2 + 4)} = -\frac{3}{8}.$

C. $\displaystyle\lim_{x \to 2} \frac{\frac{1}{x} - \frac{1}{2}}{x - 2} = \lim_{x \to 2} \frac{\frac{2 - x}{2x}}{x - 2} = \lim_{x \to 2} \frac{-1}{2x} = -\frac{1}{4}.$

D. $\displaystyle\lim_{x \to b} \frac{x^2 - b^2}{x - b} = \lim_{x \to b} \frac{(x - b)(x + b)}{x - b} = \lim_{x \to b}(x + b) = 2b.$

▶Objective 7   Identify properties of limits.

Example   Find the FALSE statement.

Note: In each statement assume that all the limits involved exist.

$\displaystyle\lim_{x \to a} k = k.$

$\displaystyle\lim_{x \to a} x = a.$

Apex Learning

$$\lim_{x \to a} (f(x) + h(x)) = \lim_{x \to a} f(x) + \lim_{x \to a} h(x).$$

$$\lim_{x \to a} (f(x) \times h(x)) = \lim_{x \to a} f(x) \times \lim_{x \to a} h(x).$$

$$\lim_{x \to a} \frac{f(x)}{h(x)} = \frac{\lim\limits_{x \to a} f(x)}{\lim\limits_{x \to a} h(x)} \quad \text{if} \quad \lim_{x \to a} h(x) \neq 0.$$

$$\lim_{x \to a} |f(x)| = \left| \lim_{x \to a} f(x) \right|.$$

$$\lim_{x \to a} kf(x) = k \lim_{x \to a} f(x).$$

$$\lim_{x \to a} f(x) = f(a) \quad \text{if} \quad f(x) \text{ is a polynomial.}$$

Tip   Do you think $\lim\limits_{x \to a} |f(x)| = \left| \lim\limits_{x \to a} f(x) \right|$ is true? (Yes it is.)

Answer $\lim\limits_{x \to a} |f(x)| = \lim\limits_{x \to a} f(x)$ is false. (Think of $\lim\limits_{x \to -2} |3x| = 6$, where $\lim\limits_{x \to -2} 3x = -6$.)

▶**Objective 8**   Calculate limits involving trigonometric functions.

Example   Find the following limits if they exist.

A. $\lim\limits_{\theta \to \frac{\pi}{6}} \dfrac{\tan \theta}{\cos \theta}$

B. $\lim\limits_{x \to 0} \dfrac{1 - \cos x}{\sin x}$

C. $\lim\limits_{x \to 0} \dfrac{x}{\sin(8x)}$

D. $\lim\limits_{x \to 0} \dfrac{2x + 1 - \cos x}{4x}$

Tip   Remember to use $\lim\limits_{x \to 0} \dfrac{1 - \cos x}{x} = 0$ and $\lim\limits_{x \to 0} \dfrac{\sin x}{x} = 1$ when working these problems.

Answers

A. $\lim\limits_{\theta \to \frac{\pi}{6}} \dfrac{\tan \theta}{\cos \theta} = \dfrac{2}{3}.$

B. $\lim\limits_{x \to 0} \dfrac{1 - \cos x}{\sin x} = \lim\limits_{x \to 0} \dfrac{x(1 - \cos x)}{x \sin x} = \lim\limits_{x \to 0} \dfrac{(1 - \cos x)}{x} \lim\limits_{x \to 0} \dfrac{x}{\sin x} = 0 \times 1 = 0.$

C. $\lim\limits_{x \to 0} \dfrac{x}{\sin(8x)} = \dfrac{1}{8} \lim\limits_{x \to 0} \dfrac{8x}{\sin(8x)} = \dfrac{1}{8}(1) = \dfrac{1}{8}.$

D. $\lim\limits_{x \to 0} \dfrac{2x + 1 - \cos x}{4x} = \lim\limits_{x \to 0} \left(\dfrac{2x}{4x} + \dfrac{1 - \cos x}{4x} \right) = \lim\limits_{x \to 0} \dfrac{2x}{4x} + \lim\limits_{x \to 0} \dfrac{1 - \cos x}{4x}$

$\qquad = \lim\limits_{x \to 0} \dfrac{1}{2} + \dfrac{1}{4} \lim\limits_{x \to 0} \dfrac{1 - \cos x}{x} = \dfrac{1}{2} + \dfrac{1}{4}(0) = \dfrac{1}{2}.$

Exercises: Limits Practice

Questions 1–3 refer to the following graph:

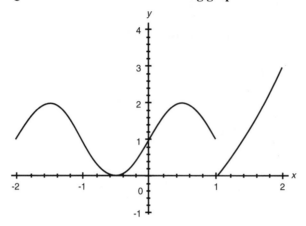

---

1. In this graph of $f(x)$, what is $\lim\limits_{x \to 0} f(x)$?

A. 0

B. 1

C. $-\frac{1}{2}$

D. 2

E. Does not exist

Answer   B. The limit is 1 because the right-hand limit equals the left-hand limit in this case.

---

2. In this graph of $f(x)$, what is $\lim\limits_{x \to 1+} f(x)$?

A. 0

B. 1

C. $-\frac{1}{2}$

D. 2

E. Does not exist

Answer   A. If you approach $x = 1$ from the right, you reach the value $f(x) = 0$.

3. In this graph of $f(x)$, what is $\lim\limits_{x \to 1} f(x)$?

A. 0

B. 1

C. $-\frac{1}{2}$

D. 2

E. Does not exist

Answer E. We say no limit exists, since the right-hand limit does not equal the left-hand limit in this case.

4. Consult the following data. What is $\lim\limits_{x \to 3} g(x)$ likely to be?

x	2.9	2.99	2.999	3.001	3.01	3.1
$g(x)$	4.41	4.9401	4.994	5.006	5.0601	5.61

A. 3

B. 5

C. 5.61

D. 4.41

E. 0

Answer B. It appears the limit will be 5 because $g(x)$ appears to be approaching 5 from both directions.

5. Given that

$$h(x) = \begin{cases} x^2 - x + 6 & \text{if } x < 2 \\ 6 & \text{if } x = 2 \\ x^3 & \text{if } x > 2. \end{cases}$$

which of the following is equal to 8?

I. $\lim\limits_{x \to 2^-} h(x)$

II. $\lim\limits_{x \to 2^+} h(x)$

III. $h(2)$

A. I only

B. II only

C. III only

D. I and II only

E. I, II, and III

Answer   D. I and II only. $h(x)$ approaches 8 from the left and right as $x \to 2$.

Questions 6–10 are True/False.

6. $\lim\limits_{x \to \frac{\pi}{2}} (\sin x) = \dfrac{\sqrt{2}}{2}$.

Answer   False. The limit is 1 because $\sin \frac{\pi}{2} = 1$ from the left and right.

7. $\lim\limits_{x \to 0^-} (\sqrt{x})$ does not exist.

Answer   True. $\lim\limits_{x \to 0^-} (\sqrt{x})$ does not exist, since $\sqrt{x}$ is undefined when $x < 0$.

8. $\lim\limits_{x \to 2} [[x]] = 2$.

(Remember that $[[x]]$ is the greatest integer function of x, or $\text{int}(x)$ on your calculator.)

Answer   False. The limit does not exist, since the left-hand limit $= 1$ and the right-hand limit $= 2$.

9. $\lim\limits_{x \to \frac{5}{2}} [x] = 2$.

(Remember that $[x]$ is the greatest integer function $[[x]]$, or $\text{int}(x)$ on your calculator.)

Answer   True. $\frac{5}{2} = 2.5$, and $[2.5] = 2$. Also $[x] = 2$ for all x in an interval containing 2.5 like $(2.4, 2.6)$.

10. $\lim\limits_{x \to 2} \dfrac{x^2 - 4}{x - 2} = 4$.

Answer   True. It doesn't matter that $\dfrac{x^2 - 4}{x - 2}$ is undefined at $x = 2$, since we only approach 2.

Questions 11–13 refer to the following graph of *y* = *f(x)*:

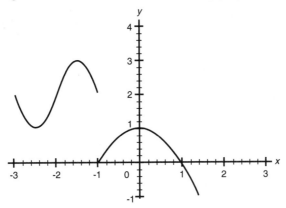

---

11. What is the $\lim\limits_{x \to -1} f(x)$?

A. 0

B. 1

C. $-\dfrac{1}{2}$

D. 2

E. Does not exist

Answer   E. The limit does not exist since the right-hand limit) ≠ the left-hand limit.

---

12. What is $\lim\limits_{x \to 0} f(x)$?

A. 0

B. 1

C. −1

D. 2

E. Does not exist

Answer   B. Notice that $f(x)$ approaches 1 from the right and the left as x goes to zero.

---

13. What is $\lim\limits_{x \to -1^+} f(x)$?

A. 0

B. 1

C. $-\dfrac{1}{2}$

D. 2

E. Does not exist

Answer A. The limit of $f(x)$ as you approach $x = -1$ from the right is 0.

---

14. Consider the table of data for the function $g(x)$, below:

x	2.9	2.99	2.999	3.001	3.01	3.1
$g(x)$	4.41	4.9401	4.994	-5.006	-5.0601	-5.61

From the data given, it would appear that the $\lim\limits_{x \to 3} g(x)$ is likely to be:

A. 3

B. -5

C. 5

D. -5.61

E. Does not exist

Answer E. Notice that the left-hand limit appears to be 5, while the right-hand limit appears to be -5. So left-hand limit $\neq$ right-hand limit, and thus the limit does not exist.

---

15. Suppose that

$$g(x) = \begin{cases} x^2 - x + 5 & \text{if } x < 2 \\ 5 & \text{if } x = 2 \\ x^3 - 1 & \text{if } x > 2. \end{cases}$$

Which of the following is equal to 7?

I. $\lim\limits_{x \to 2^-} h(x)$

II. $\lim\limits_{x \to 2^+} h(x)$

III. $\lim\limits_{x \to 2} h(x)$

A. I only

B. II only

C. III only

D. I and II only

E. I, II, and III

Answer E. First, convince yourself that both the left-hand limit and the right-hand limit are equal to 7. Then, since left-hand limit = right-hand limit = 7, the overall limit is also equal to 7. (Remember, it doesn't matter what the value of the function is at $x = 2$!)

---

16. $\lim\limits_{x \to \frac{\pi}{3}} (\tan x) =$

A. $-\sqrt{3}$

B. 1

C. $\sqrt{3}$

D. -1

E. Does not exist

Answer C. As x approaches $\frac{\pi}{3}$ from either the left or the right, $\tan(x)$ approaches $\sqrt{3}$.

---

17. $\lim\limits_{x \to -2} |x| =$

A. -2

B. 2

C. 0

D. -1

E. Does not exist

Answer B. We have $\lim\limits_{x \to -2} |x| = \lim\limits_{x \to 2} x = 2$.

18. $\lim\limits_{x\to\frac{9}{2}} [[x]] =$

(Remember that $[[x]]$ represents the greatest integer function of x.)

A. 4

B. 5

C. 4.5

D. -4

E. Does not exist

Answer A. Notice that for all x close enough to $\frac{9}{2}$ ($= 4.5$), we have $[[x]] = 4$.

19. $\lim\limits_{x\to 9} [[x]] =$

A. 9

B. 8

C. 7

D. 10

E. Does not exist

Answer E. As x goes to 9 from the left (so that x is a little less than 9), the value of $[[x]]$ is 8. As x goes to 9 from the right (so that x is a little more than 9), the value of $[[x]]$ is 9. So we have left-hand limit $\neq$ right-hand limit so the limit does not exist.

20. $\lim\limits_{x\to 2} \sqrt{16}$ is:

A. $\sqrt{2}$

B. 2

C. 16

D. 4

E. None of these

Answer D. $\lim\limits_{x\to 2} \sqrt{16} = 4$, since $\sqrt{16}$ does not vary as x changes.

21. Which of the following properties is true?

A. $\lim\limits_{x \to a} (f(x) + g(x)) = \lim\limits_{x \to a} f(x) + \lim\limits_{x \to a} g(x)$

B. $\lim\limits_{x \to a} (f(x) \times g(x)) = \lim\limits_{x \to a} f(x) \times \lim\limits_{x \to a} g(x)$

C. $\lim\limits_{x \to a} (mx + b) = ma + b$

D. $\lim\limits_{x \to a} (kf(x)) = k \lim\limits_{x \to a} f(x)$

E. All of these are correct

Answer   E. All of these are limit properties, so all are true.

22. What is $\lim\limits_{x \to 6} \sqrt[4]{x}$?

A. $\sqrt[6]{4}$

B. $\sqrt[4]{6}$

C. $\sqrt[4]{x}$

D. $\sqrt[6]{x}$

E. None of these

Answer   B. By simple substitution, $\lim\limits_{x \to 6} \sqrt[4]{x} = \sqrt[4]{6}$.

23. What is $\lim\limits_{x \to 2} (2x^2 + 4x - 5)$?

A. 8

B. 7

C. 10

D. 11

E. This limit does not exist

Answer   D. By substitution, $\lim\limits_{x \to 2} (2x^2 + 4x - 5) = 2(2)^2 + 2(2) - 5 = 11$.

24. What is $\displaystyle\lim_{x \to 1} \frac{4x^2}{7x - 4}$?

A. $\frac{16}{3}$

B. $\frac{4}{7}$

C. $-\frac{4}{3}$

D. $\frac{4}{3}$

E. This limit does not exist

Answer   D. By substitution, $\displaystyle\lim_{x \to 1} \frac{4x^2}{7x - 4} = \frac{4}{7 - 4} = \frac{4}{3}$.

---

25. What is $\displaystyle\lim_{x \to 2} \frac{x^2 - 4}{x - 2}$?

A. ∞

B. 4

C. 0

D. 2

E. This limit does not exist

Answer   B.

$$\lim_{x \to 2} \left(\frac{x^2 - 4}{x - 2} \right)$$
$$= \lim_{x \to 2} \left(\frac{(x - 2)(x + 2)}{(x - 2)} \right)$$
$$= \lim_{x \to 2} (x + 2) = 2(2) = 4.$$

---

26. What is $\sin(u)$ essentially equal to for small values of u?

A. 0

B. u

C. $\cos(u)$

D. 1

E. -1

Answer   B. Looking at graphs of $\sin(u)$ and u show they are close at small values.

27. What is $\lim\limits_{x \to \pi} \sin(\frac{x}{4})$?

A. 0

B. 1

C. $\frac{\sqrt{2}}{2}$

D. $\frac{\sqrt{3}}{2}$

E. $\frac{1}{2}$

Answer C. By simple substitution we learn that $\lim\limits_{x \to \pi} \sin(\frac{x}{4}) = \sin(\frac{\pi}{4}) = \frac{1}{2}\sqrt{2}$.

28. What is $\lim\limits_{x \to \frac{\pi}{2}} \tan(\frac{x}{2})$?

A. $\sqrt{3}$

B. 1

C. $\frac{\sqrt{3}}{3}$

D. 0

E. Does not exist

Answer B. $\lim\limits_{x \to \frac{\pi}{2}} \tan\left(\frac{x}{2}\right) = \tan\left(\frac{\pi}{4}\right) = 1$.

29. What is $\lim\limits_{x \to 0} \frac{\sin x}{x}$?

A. 1

B. 0

C. $\frac{\sqrt{2}}{2}$

D. -1

E. Does not exist

Answer A. Since $\sin x \approx x$ for small x, then $\lim\limits_{x \to 0} \frac{\sin x}{x} = \lim\limits_{x \to 0} \frac{x}{x} = 1$.

30. What is $\lim\limits_{x \to 0} \dfrac{\sin(4x)}{\sin(x)}$?

A. 1

B. 0

C. 4

D. $\dfrac{1}{4}$

E. Does not exist

Answer C.

$$\begin{aligned}
\lim_{x \to 0} \frac{\sin(4x)}{\sin(x)} &= \lim_{x \to 0} \left(\frac{\sin(4x)}{\sin(x)} \times \frac{4x}{4x} \right) \\
&= \lim_{x \to 0} \left(\frac{\sin(4x)}{4x} \times \frac{x}{\sin x} \times \frac{4}{1} \right) \\
&= 4 \lim_{x \to 0} \left(\frac{\sin(4x)}{4x} \right) \times \lim_{x \to 0} \left(\frac{x}{\sin(x)} \right) \\
&= 4 \times 1 \times 1 = 4
\end{aligned}$$

where we used limit identities and the fact that $4x \to 0$ as $x \to 0$.

31. What is $\lim\limits_{x \to 0} \left(\dfrac{x + 1 - \cos x}{4x} \right)$?

A. 1

B. 0

C. 4

D. $\dfrac{1}{4}$

E. Does not exist

Answer D.

$$\lim_{x \to 0} \left(\frac{x + 1 - \cos x}{4x} \right) = \lim_{x \to 0} \frac{x}{4x} + \frac{1}{4} \lim_{x \to 0} \frac{1 - \cos x}{x} = \frac{1}{4} + \frac{1}{4}(0) = \frac{1}{4}.$$

32. What is $\lim\limits_{x \to 0} \left(\dfrac{1 - \cos(3x)}{3x} \right)$?

  A. 1

  B. 0

  C. 3

  D. $\dfrac{1}{3}$

  E. Does not exist

Answer B. Let $u = 3x$, so we have $\lim\limits_{u \to 0} \dfrac{1 - \cos u}{u} = 0$.

More Limit Practice

1. What is $\lim\limits_{x \to 7} \sqrt{36}$?

2. What is $\lim\limits_{x \to 1} \dfrac{5x^2}{3x - 8}$?

3. What is $\lim\limits_{x \to 3} \dfrac{x^2 - 9}{x - 3}$?

4. What is $\lim\limits_{x \to 4} \dfrac{x^2 - 4^2}{\sqrt{x} - 2}$?

5. What is $\lim\limits_{x \to 0} \dfrac{\sqrt{3 + x} - \sqrt{3}}{x}$?

6. What is $\lim\limits_{x \to 5} \dfrac{x^2 - 4x - 5}{x^2 - 8x + 15}$?

7. For what values of a does $\lim\limits_{x \to a} [[\tfrac{x}{2}]]$ exist?

In each of the next six problems, you'll be given a function f and a value of a. Then you'll need to find:

A) $\lim\limits_{x \to a^-} f(x)$ **B)** $\lim\limits_{x \to a^+} f(x)$ **C)** $\lim\limits_{x \to a} f(x)$

Example If you were given $f(x) = \begin{cases} \sin x & \text{if } x < \frac{\pi}{2} \\ \cos x & \text{if } x \geq \frac{\pi}{2}. \end{cases}$ and $a = \frac{\pi}{2}$, then your answer should be something like:

A. $\lim\limits_{x \to \frac{\pi}{2}^-} \begin{cases} \sin x & \text{if } x < \frac{\pi}{2} \\ \cos x & \text{if } x \geq \frac{\pi}{2}. \end{cases} = \lim\limits_{x \to \frac{\pi}{2}^-} \sin x = \sin \frac{\pi}{2} = 1$ since when coming from the left of $\frac{\pi}{2}$, the function is simply $\sin x$.

B. $\lim\limits_{x \to \frac{\pi}{2}^+} \begin{cases} \sin x & \text{if } x < \frac{\pi}{2} \\ \cos x & \text{if } x \geq \frac{\pi}{2}. \end{cases} = \lim\limits_{x \to \frac{\pi}{2}^+} \cos x = \cos \frac{\pi}{2} = 0$, since when coming from the right of $\frac{\pi}{2}$, the function is simply $\cos x$.

C. $\lim\limits_{x \to \frac{\pi}{2}} \begin{cases} \sin x & \text{if } x < \frac{\pi}{2} \\ \cos x & \text{if } x \geq \frac{\pi}{2}. \end{cases}$ is undefined, since the right-hand limit doesn't equal the left-hand limit.

---

8. $f(x) = \dfrac{\sin x}{x} \qquad a = \pi$

---

9. $f(x) = \sin \dfrac{\pi}{[[x]]} \qquad a = 3$

---

10. $f(x) = \begin{cases} \dfrac{\sin x}{x} & \text{if } x < 0 \\ \cos x & \text{if } x \geq 0 \end{cases}$ and $a = 0$

---

11. $f(x) = \dfrac{\sin^2(x) - 1}{\sin(x) - 1} \qquad a = \frac{\pi}{2}$

---

12. $f(x) = \begin{cases} \sin x & \text{if } x < \frac{\pi}{6} \\ \tan x & \text{if } x = \frac{\pi}{6} \\ \cos x & \text{if } x > \frac{\pi}{6} \end{cases}$ and $a = \frac{\pi}{6}$

---

13. $f(x) = \begin{cases} \sin x & \text{if } x < \frac{\pi}{4} \\ \tan x & \text{if } x = \frac{\pi}{4} \\ \cos x & \text{if } x > \frac{\pi}{4} \end{cases}$ and $a = \frac{\pi}{4}$

Answers to More Limit Practice

---

1. $\lim\limits_{x \to 7} \sqrt{36} = 6$, since $\sqrt{36}$ doesn't vary as x changes.

2. $\lim\limits_{x \to 1} \left(\dfrac{5x^2}{3x - 8} \right) = \dfrac{5}{3 - 8} = -1.$

3. $\lim\limits_{x \to 3} \left(\dfrac{x^2 - 9}{x - 3} \right) = \lim\limits_{x \to 3} \left(\dfrac{(x - 3)(x + 3)}{x - 3} \right) = \lim\limits_{x \to 3} (x + 3) = 2(3) = 6.$

4.

$$\begin{aligned}
\lim\limits_{x \to 4} \left(\dfrac{x^2 - (4)^2}{\sqrt{x} - 2} \right) &= \lim\limits_{x \to 4} \left(\dfrac{(x - 4)(x + 4)}{\sqrt{x} - 2} \right) \\
&= \lim\limits_{x \to 4} \left(\dfrac{(\sqrt{x} - 2)(\sqrt{x} + 2)(x + 4)}{\sqrt{x} - 2} \right) \\
&= \lim\limits_{x \to 4} \left(\dfrac{(\sqrt{x} + 2)(x + 4)}{1} \right) \\
&= (\sqrt{4} + 2)(4 + 4) = 32.
\end{aligned}$$

5.

$$\lim\limits_{x \to 0} \left(\dfrac{\sqrt{3 + x} - \sqrt{3}}{x} \right)$$

$$\begin{aligned}
&= \lim\limits_{x \to 0} \left(\dfrac{\sqrt{3 + x} - \sqrt{3}}{x} \right) \times \dfrac{\sqrt{3 + x} + \sqrt{3}}{\sqrt{3 + x} + \sqrt{3}} \\
&= \lim\limits_{x \to 0} \dfrac{3 + x - 3}{x(\sqrt{3 + x} + \sqrt{3})} \\
&= \lim\limits_{x \to 0} \dfrac{1}{(\sqrt{3 + x} + \sqrt{3})} = \dfrac{1}{2\sqrt{3}} = \dfrac{1}{6}\sqrt{3}.
\end{aligned}$$

6. $\lim\limits_{x \to 5} \left(\dfrac{x^2 - 4x - 5}{x^2 - 8x + 15} \right) = \lim\limits_{x \to 5} \dfrac{(x + 1)(x - 5)}{(x - 3)(x - 5)} = \lim\limits_{x \to 5} \dfrac{x + 1}{x - 3} = \dfrac{5 + 1}{5 - 3} = 3.$

7.

$\lim\limits_{u \to n} [[u]]$ doesn't exist for integers n because the right-hand limit doesn't equal the left. Thus $\lim\limits_{x \to a} [[\frac{x}{2}]]$ will exist everywhere except when $\frac{a}{2}$ is an integer, which is when a is an even integer. So $\lim\limits_{x \to a} [[\frac{x}{2}]]$ exists for all a except when a is an even integer.

8.

A. $\lim\limits_{x \to \pi^-} \dfrac{\sin x}{x} = \dfrac{\sin \pi}{\pi} = 0.$

B. $\displaystyle \lim_{x \to \pi^+} \frac{\sin x}{x} = \frac{\sin \pi}{\pi} = 0.$

C. $\displaystyle \lim_{x \to \pi} \frac{\sin x}{x} = \frac{\sin \pi}{\pi} = 0.$

9.

A. $\displaystyle \lim_{x \to 3^-} \sin \frac{\pi}{[[x]]} = \sin\left(\frac{\pi}{2}\right) = 1.$

B. $\displaystyle \lim_{x \to 3^+} \sin \frac{\pi}{[[x]]} = \sin\left(\frac{\pi}{3}\right) = \frac{1}{2}\sqrt{3}.$

C. $\displaystyle \lim_{x \to 3} \sin \frac{\pi}{[[x]]}$ does not exist.

10.

A. $\displaystyle \lim_{x \to 0^-} \begin{cases} \dfrac{\sin x}{x} & \text{if } x < 0 \\ \cos x & \text{if } x \geq 0 \end{cases} = \lim_{x \to 0^-} \frac{\sin x}{x} = 1.$

B. $\displaystyle \lim_{x \to 0^+} \begin{cases} \dfrac{\sin x}{x} & \text{if } x < 0 \\ \cos x & \text{if } x \geq 0 \end{cases} = \lim_{x \to 0^+} \cos x = 1.$

C. $\displaystyle \lim_{x \to 0} \begin{cases} \dfrac{\sin x}{x} & \text{if } x < 0 \\ \cos x & \text{if } x \geq 0 \end{cases} = 1.$

11.

A. $\displaystyle \lim_{x \to \frac{\pi}{2}^-} \frac{\sin^2(x) - 1}{\sin(x) - 1} = \lim_{x \to \frac{\pi}{2}^-} \frac{(\sin(x) - 1)(\sin(x) + 1)}{\sin(x) - 1} = \lim_{x \to \frac{\pi}{2}^-} (\sin(x) + 1) = 2.$

B. $\displaystyle \lim_{x \to \frac{\pi}{2}^+} \frac{\sin^2(x) - 1}{\sin(x) - 1} = \lim_{x \to \frac{\pi}{2}^+} \frac{(\sin(x) - 1)(\sin(x) + 1)}{\sin(x) - 1} = \lim_{x \to \frac{\pi}{2}^+} (\sin(x) + 1) = 2.$

C. $\displaystyle \lim_{x \to \frac{\pi}{2}} \frac{\sin^2(x) - 1}{\sin(x) - 1} = 2.$

12.

A. $\displaystyle \lim_{x \to \frac{\pi}{6}^-} \begin{cases} \sin x & \text{if } x < \frac{\pi}{6} \\ \tan x & \text{if } x = \frac{\pi}{6} \\ \cos x & \text{if } x > \frac{\pi}{6} \end{cases} = \lim_{x \to \frac{\pi}{6}^-} \sin x = \frac{1}{2}.$

B. $\lim\limits_{x \to \frac{\pi}{6}^+} \begin{cases} \sin x & \text{if } x < \frac{\pi}{6} \\ \tan x & \text{if } x = \frac{\pi}{6} \\ \cos x & \text{if } x > \frac{\pi}{6} \end{cases} = \lim\limits_{x \to \frac{\pi}{6}^+} \cos x = \frac{1}{2}\sqrt{3}.$

C. $\lim\limits_{x \to \frac{\pi}{6}} \begin{cases} \sin x & \text{if } x < \frac{\pi}{6} \\ \tan x & \text{if } x = \frac{\pi}{6} \\ \cos x & \text{if } x > \frac{\pi}{6} \end{cases}$ does not exist.

13.

A. $\lim\limits_{x \to \frac{\pi}{4}^-} \begin{cases} \sin x & \text{if } x < \frac{\pi}{4} \\ \tan x & \text{if } x = \frac{\pi}{4} \\ \cos x & \text{if } x > \frac{\pi}{4} \end{cases} = \lim\limits_{x \to \frac{\pi}{4}^-} (\sin x) = \frac{1}{2}\sqrt{2}.$

B. $\lim\limits_{x \to \frac{\pi}{4}^+} \begin{cases} \sin x & \text{if } x < \frac{\pi}{4} \\ \tan x & \text{if } x = \frac{\pi}{4} \\ \cos x & \text{if } x > \frac{\pi}{4} \end{cases} = \lim\limits_{x \to \frac{\pi}{4}^+} (\cos x) = \frac{1}{2}\sqrt{2}.$

C. $\lim\limits_{x \to \frac{\pi}{4}} \begin{cases} \sin x & \text{if } x < \frac{\pi}{4} \\ \tan x & \text{if } x = \frac{\pi}{4} \\ \cos x & \text{if } x > \frac{\pi}{4} \end{cases} = \frac{1}{2}\sqrt{2}.$

B. Asymptotic and Unbounded Behavior

▶Objective 1   Estimate limits at infinity, using numerical or graphical techniques.

Example 1   For the graph of $f(x)$ below find all the following:

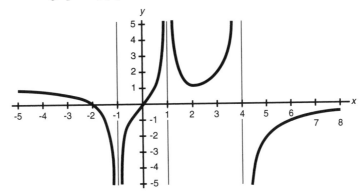

A. $\lim\limits_{x \to -1^-} f(x)$

B. $\lim\limits_{x \to -1^+} f(x)$

C. $\lim\limits_{x \to 1^-} f(x)$

D. $\lim\limits_{x \to 1^+} f(x)$

E. $\lim\limits_{x \to 4^-} f(x)$

F. $\lim\limits_{x \to 4^+} f(x)$

Tip Vertical asymptotes occur at $x = a$ if $\lim\limits_{x \to a^+} f(x) = +\infty$ or $-\infty$, or if $\lim\limits_{x \to a^-} f(x) = +\infty$ or $-\infty$.

Answers

A. $\lim\limits_{x \to -1^-} f(x) = -\infty.$

B. $\lim\limits_{x \to -1^+} f(x) = -\infty.$

C. $\lim\limits_{x \to 1^-} f(x) = \infty.$

D. $\lim\limits_{x \to 1^+} f(x) = \infty.$

E. $\lim\limits_{x \to 4^-} f(x) = \infty.$

F. $\lim\limits_{x \to 4^+} f(x) = -\infty.$

Example 2 Find $\lim\limits_{x \to \infty} \dfrac{\sin x}{x}$.

Tip While $\sin(x)$ oscillates between -1 and 1, x keeps getting bigger, and $\dfrac{\sin x}{x}$ goes to zero.

Answer $\lim\limits_{x \to \infty} \dfrac{\sin x}{x} = 0.$

▶Objective 2 Solve limits involving infinity, using algebraic manipulation.

Example 1 Find:

A. $\lim\limits_{x \to \infty} \dfrac{3x^2 - x + 4}{x - x^2}$

B. $\lim\limits_{x \to 2^-} \dfrac{x + 2}{x^2 - 4}$

C. $\lim\limits_{x \to 2^+} \dfrac{x + 2}{x^2 - 4}$

D. $\lim\limits_{x\to 2} \dfrac{x+2}{x^2-4}$

Tip   Stating $-\infty$ or ∞ communicates more information than simply saying DNE (does not exist).

Answers

A. $\lim\limits_{x\to\infty} \dfrac{3x^2-x+4}{x-x^2} = \lim\limits_{x\to\infty} \dfrac{3x^2-x+4}{x-x^2} \times \dfrac{\frac{1}{x^2}}{\frac{1}{x^2}} = \lim\limits_{x\to\infty} \dfrac{3-\frac{1}{x}+\frac{4}{x^2}}{\frac{1}{x}-1} = -3.$

B. $\lim\limits_{x\to 2^-} \dfrac{x+2}{x^2-4} = \lim\limits_{x\to 2^-} \dfrac{x+2}{(x-2)(x+2)} = \lim\limits_{x\to 2^-} \dfrac{1}{(x-2)} = -\infty.$

C. $\lim\limits_{x\to 2^+} \dfrac{x+2}{x^2-4} = \lim\limits_{x\to 2^+} \dfrac{1}{(x-2)} = \infty.$

D. $\lim\limits_{x\to 2} \dfrac{x+2}{x^2-4}$ is undefined.

(The left-hand side shoots off to $-\infty$, and the right-hand side shoots off to ∞.)

Example 2   Given that $\lim\limits_{x\to\infty} \dfrac{6x^n-3x+4}{ax^3-4x^2+9} = 3$, what are the values of a and n?

Answer $n=3$, since the limit would be 0 or ∞ in any other case. This forces a to have a value of 2.

▶Objective 3   Use limits to find and describe asymptotes.

Example   Use limits (one-sided, if necessary) to describe all the asymptotes in this graph.

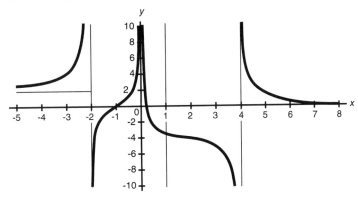

Tip $\lim\limits_{x\to\infty} f(x) = a$ is generally used for horizontal asymptotes, and $\lim\limits_{x\to\pm a} f(x) = \pm\infty$ is generally used for vertical asymptotes.

Answers

$\lim\limits_{x\to-\infty} f(x) = 2$; $\lim\limits_{x\to\infty} f(x) = 0$; $\lim\limits_{x\to-2^-} f(x) = \infty$; $\lim\limits_{x\to-2^+} f(x) = -\infty$;

$\lim\limits_{x\to 0^-} f(x) = \infty$; $\lim\limits_{x\to 0^+} f(x) = \infty$; $\lim\limits_{x\to 4^-} f(x) = -\infty$; $\lim\limits_{x\to 4^+} f(x) = \infty$.

▶Objective 4   Reconstruct the graph of a function when given limits that describe the function.

Example   Draw a rough sketch of a function that satisfies all the following:

$$\lim_{x \to -\infty} f(x) = 3; \qquad \lim_{x \to \infty} f(x) = 0;$$

$$\lim_{x \to 0^-} f(x) = -\infty; \qquad \lim_{x \to 0^+} f(x) = \infty;$$

$$\lim_{x \to 3^-} f(x) = \infty; \qquad \lim_{x \to 3^+} f(x) = -\infty.$$

Tip   Draw each asymptote first and then sketch the graph.

Answers

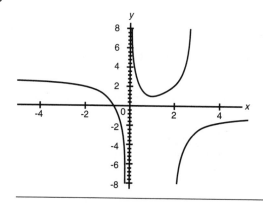

▶Objective 5   Determine limits that compare growth rates of functions (by evaluating limits at infinity of the ratio of the two functions).

Example   Find each of the following limits:

A. $\lim\limits_{x \to \infty} \dfrac{x^4 - 7x + 9}{4 + 5x + 3x^4}$

B. $\lim\limits_{x \to \infty} \dfrac{2^x + x^3}{x^2 + 3^x}$

C. $\lim\limits_{x \to 0} \dfrac{x^3 - 7x + 9}{7^x + x^3}$

D. $\lim\limits_{x \to \infty} \dfrac{2^x + x^3}{x^2 + 2^x}$

E. $\lim\limits_{x \to \infty} \dfrac{\sqrt{x^3}}{x^2 + x}$

F. $\lim\limits_{x \to \infty} (x^5 - 5^x)$

Tip   Remember which functions will outgrow others as x goes to infinity.

Answers

A. $\lim\limits_{x\to\infty} \dfrac{x^4 - 7x + 9}{4 + 5x + 3x^4} = \dfrac{1}{3}$.

B. $\lim\limits_{x\to\infty} \dfrac{2^x + x^3}{x^2 + 3^x} = 0$.

C. $\lim\limits_{x\to 0} \dfrac{x^3 - 7x + 9}{7^x + x^3} = 9$.

If you put 0 here, you didn't notice that the limit goes to 0 and not ∞. Read the question carefully on exams.

D. $\lim\limits_{x\to\infty} \dfrac{2^x + x^3}{x^2 + 2^x} = 1$.

E. $\lim\limits_{x\to\infty} \dfrac{\sqrt{x^3}}{x^2 + x} = 0$.

F. $\lim\limits_{x\to\infty} (x^5 - 5^x) = -\infty$.

►Objective 6   Compare relative magnitudes of functions, including algebraic and exponential functions.

Example   1. Place all of the following functions in order of growth rate from fastest to slowest.

A. $a(x) = x^4 - 3x + 9$

B. $b(x) = 4x^{\frac{5}{2}}$

C. $c(x) = \sqrt{x^9 - 3x}$

D. $d(x) = 100^x$

E. $e(x) = e$

F. $f(x) = e^x$

G. $g(x) = 2^x - x^3$

H. $h(x) = 9x^2$

Tip   Exponential functions a^x $(a > 1)$ grow faster than power functions x^n.

Answers

D. $d(x) = 100^x$

F. $f(x) = e^x$

G. $g(x) = 2^x - x^3$

C. $c(x) = \sqrt{x^9 - 3x}$

Apex Learning   45

A. $a(x) = x^4 - 3x + 9$

B. $b(x) = 4x^{\frac{5}{2}}$

H. $h(x) = 9x^2$

E. $e(x) = e$

▶Objective 7   Solve problems by comparing relative rates of growth.

Example   The following numerical data comes from the function $h(t) = 2t - 1.1^t + t^3$.

$h(t)$	1.9	133.39	1017.4	1.2498×10^5	9.8642×10^5
t	1	5	10	50	100

Would you agree that $\lim_{t \to \infty} h(t) = \infty$? Why or why not?

Tip   Use what you know about growth of functions to predict end behavior (for example, behavior as $x \to \infty$).

Answer $\lim_{t \to \infty} h(t) = -\infty$, since the -1.1^t term will eventually dominate and make the function values decrease without bound.

▶Objective 8   Find when a limit does not exist and identify why the limit does not exist (for example, right-hand-limit $\neq$ left-hand-limit, vertical asymptote, or oscillations).

Example   Determine each limit; if the limit does not exist, explain why not.

A. $\lim_{x \to 2} \dfrac{\sqrt{x^2 - 4x + 4}}{x - 2}$

B. $\lim_{x \to 2} \dfrac{x^2 - 4x + 4}{x - 2}$

C. $\lim_{x \to 0} \cos\left(\dfrac{3}{2x}\right)$

D. $\lim_{x \to 2} \dfrac{x^2 + 4x + 4}{x - 2}$

E. $\lim_{x \to 0} x \cos\left(\dfrac{3}{2x}\right)$

Tip   Review the questions on limits that do not exist.

Answers

A. $\lim\limits_{x \to 2} \dfrac{\sqrt{x^2 - 4x + 4}}{x - 2}$ is undefined. This one has a hidden absolute value:

$$\lim\limits_{x \to 2} \frac{\sqrt{x^2 - 4x + 4}}{x - 2} = \lim\limits_{x \to 2} \frac{\sqrt{(x - 2)(x - 2)}}{x - 2} = \lim\limits_{x \to 2} \frac{|x - 2|}{x - 2},$$

so that expression $\dfrac{|x - 2|}{x - 2}$ changes from a value of -1 for $x < 2$ to a value of 1 for $x > 2$.

Thus the right-hand limit $\neq$ the left-hand limit, and the limit does not exist.

B. $\lim\limits_{x \to 2} \dfrac{x^2 - 4x + 4}{x - 2} = 0$, since $\lim\limits_{x \to 2} \dfrac{x^2 - 4x + 4}{x - 2} = \lim\limits_{x \to 2} \dfrac{(x - 2)(x - 2)}{x - 2} = \lim\limits_{x \to 2} (x - 2) = 0.$

C. $\lim\limits_{x \to 0} \cos(\dfrac{3}{2x})$ is undefined, since the function oscillates as $x \to 0$.

D. $\lim\limits_{x \to 2} \dfrac{x^2 + 4x + 4}{x - 2}$ is undefined. We have a vertical asymptote here as shown in this graph.

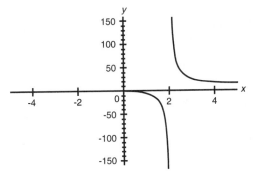

E. $\lim\limits_{x \to 0} x \cos(\dfrac{3}{2x}) = 0.$

Though this one oscillates around the origin, the function does eventually go to zero as $x \to 0$, since the x in front makes the amplitude of the oscillations smaller and smaller.

▶Objective 9   Analyze situations that can be described in terms of limits of functions.

Example   Salt water with a concentration of 4.5 grams of salt per gallon flows into a large trough that initially contains 40 gallons of pure water. The salt water is flowing into the trough at a rate of 5 gallons per minute.

A. Find the volume of water, $V(t)$, in the tank as a function of time.

B. Find the amount of salt, $A(t)$, in the tank as a function of time.

C. Find the concentration of salt, $C(t)$, (in grams per gallon) as a function of time.

D. Find $\lim\limits_{t \to \infty} C(t)$. What does this limit represent?

Tip   Always take your time and translate into math symbols what is given. Draw a picture if one will help you visualize what's happening.

Answers

A.   Find the volume of water in the tank as a function of time.

$$V(t) = 40 + 5t$$

B.   Find the amount of salt in the tank as a function of time. $A(t) = 4.5(5)t = 22.5t$, since 5 gallons per minute are coming into the trough and there are 4.5 grams per gallon.

C.   Find the concentration of salt (in grams per gallon) as a function of time.

$$C(t) = \frac{A(t)}{V(t)} = \frac{22.5t}{40 + 5t}$$

D.   Find $\lim_{t \to \infty} C(t)$. What does this limit represent?

$$\lim_{t \to \infty} C(t) = \lim_{t \to \infty} \frac{22.5t}{40 + 5t} = \lim_{t \to \infty} \frac{22.5t}{40 + 5t} \times \frac{\frac{1}{t}}{\frac{1}{t}} = \lim_{t \to \infty} \frac{22.5}{\frac{40}{t} + 5}$$
$$= \frac{22.5}{5} = 4.5 \text{ grams per gallon.}$$

The initial 40 gallons of salt become less and less important as more and more salt water is added to the tank. The concentration in the tank will always be less than the salt water coming in, but its concentration will approach 4.5 grams/gallon.

Here is a graph of $C(t) = \dfrac{A(t)}{V(t)} = \dfrac{22.5t}{40 + 5t}$. You can see the horizontal asymptote.

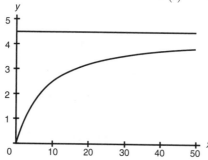

Exercises on Asymptotic and Unbounded Behavior

1.   Which of the following is a vertical asymptote for $f(x) = \dfrac{x + 1}{x - 2}$?

A. $y = 1$

B. $x = \dfrac{1}{2}$

C. $x = 2$

D. $x = 0$

E.   There are no vertical asymptotes.

Answer   C. $\lim\limits_{x \to 2^-} f(x) = -\infty$ and $\lim\limits_{x \to 2^+} f(x) = \infty$.

---

2.   For $f(x) = \dfrac{1}{x}$, which of the following statements are true:

I. $\lim\limits_{x \to 0^+} f(x) = \infty$

II. $\lim\limits_{x \to -\infty} f(x) = 0$

III. $\lim\limits_{x \to \infty} f(x) = 0$

A.   I only

B.   I and III only

C.   II and III only

D.   III only

E.   I, II, and III

Answer   E.

---

3.   At what values of a does $\lim\limits_{x \to a^+} f(x) = -\infty$ for $f(x) = \dfrac{x^2 - x - 6}{x^2 + 3x - 4}$?
(Use your calculator if you need it.)

A. -2 and 2

B. -2 and 3

C. 1 and -4

D. 4

E. 0

Answer   C. 1 and -4 are the roots of the expression in the denominator (but are not also roots of the numerator).

4. Which of these statements properly implies a vertical asymptote?

A. $\lim_{x \to a} f(x) = \infty$

B. $\lim_{x \to \infty} f(x) = a$

C. $f(a) = \infty$

D. $f(x) = a$

E. $f(x) = 4$

Answer   A. $\lim_{x \to a} f(x) = \infty$ will give a vertical asymptote of the form $x = a$.

5. $g(x) = \dfrac{4x^2 - 30x - 16}{x^2 + x}$ will have vertical asymptotes at:

A. $x = 0$

B. $x = -\dfrac{1}{2}$ and 8

C. $y = 4$

D. $x = 0$ and -1

E. $g(x) = 4$

Answer   D. The vertical asymptotes will be at $x = 0$ and -1.

6. $g(x) = \dfrac{4x^2 - 30x - 16}{x^2 + x}$ will have horizontal asymptote(s) at:

A. $x = 0$

B. $x = \dfrac{-1}{2}$ and 8

C. $y = 4$

D. $x = 0$ and -1

E. $y = \dfrac{1}{4}$

Answer   C. $y = 4$ is a horizontal asymptote, since

$$\lim_{x \to \infty} \frac{4x^2 - 30x - 16}{x^2 + x} = \lim_{x \to \infty} \frac{(4x^2 - 30x - 16)/x^2}{(x^2 + x)/x^2} = \lim_{x \to \infty} \frac{4 - \dfrac{30}{x} - \dfrac{16}{x^2}}{1 + \dfrac{1}{x}} = \frac{4}{1} = 4.$$

7. Which of the graphs below satisfy $\lim\limits_{x \to 1} f(x) = \infty$ and $\lim\limits_{x \to -\infty} f(x) = 1$?

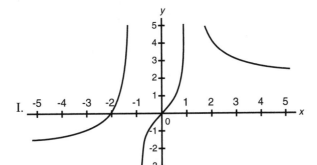

I.

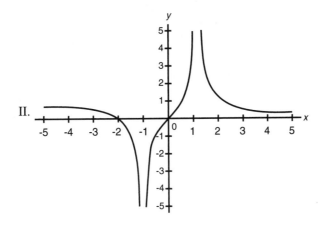

II.

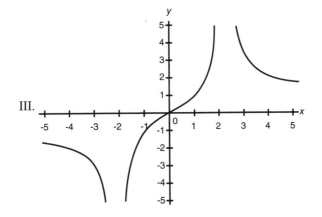

III.

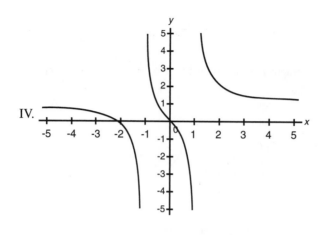

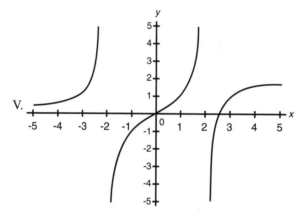

A. I

B. II

C. III

D. IV

E. V

Answer   B. It is graph II because the function value appears to go to ∞ as $x \to 1$ from either direction, and the function value appears to go to 1 as $x \to -\infty$.

8. Which of the following graphs satisfy $\lim\limits_{x \to 2^-} f(x) = \infty$ and $\lim\limits_{x \to -\infty} f(x) = 0$?

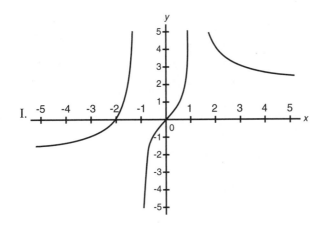

I.

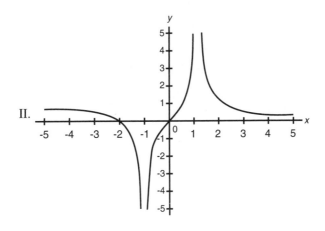

II.

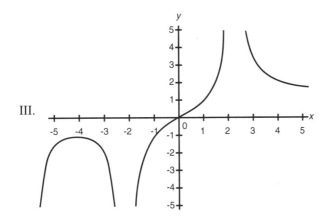

III.

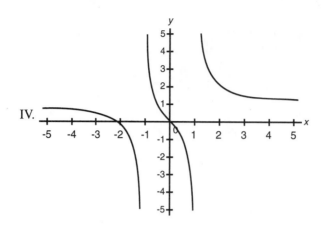

IV.

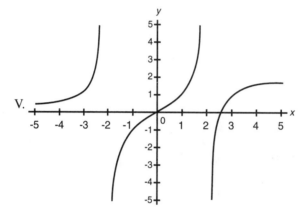

V.

A. I

B. II

C. III

D. IV

E. V

Answer   E. It is graph V because the function value appears to go to ∞ as x approaches 2 from the left, and the function value appears to go to 0 as $x \to -\infty$.

9.   Which of the following graphs satisfy $\lim\limits_{x \to -\infty} f(x) = 1$ and $\lim\limits_{x \to 1} f(x) = \infty$?

Apex Learning

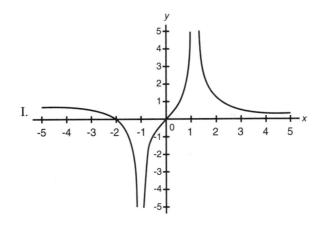

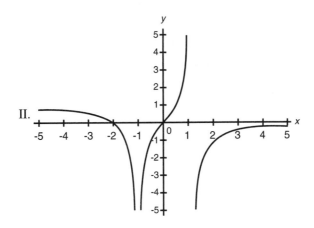

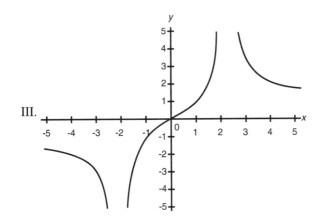

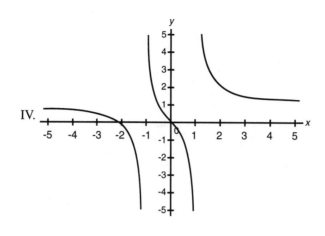

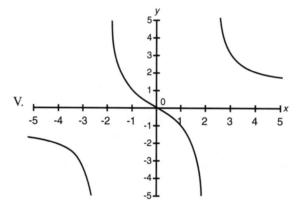

A. I

B. II

C. III

D. IV

E. V

Answer   A. The function value appears to go to ∞ as x approaches 1 from either direction, and the function value appears to go to 1 as $x \to -\infty$.

10.   Which of the following graphs satisfy $\lim\limits_{x \to 1^+} f(x) = \infty$ and $\lim\limits_{x \to 1^-} f(x) = -\infty$?

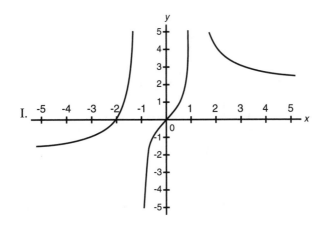

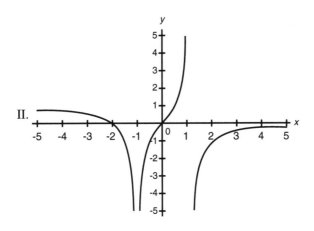

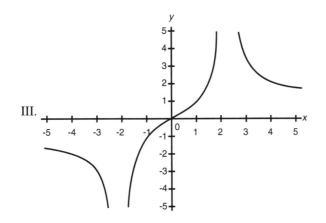

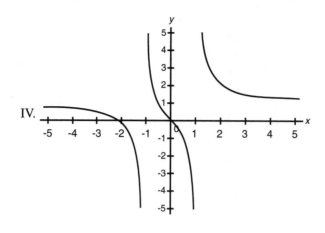

IV.

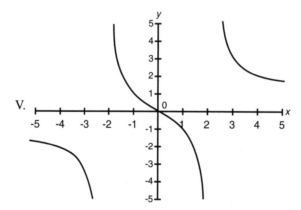

V.

A. I

B. II

C. III

D. IV

E. V

Answer   D. The function value appears to go to ∞ as x approaches 1 from the right, and the function value appears to go to $-\infty$ as x approaches 1 from the left.

11.   Which of the following graphs satisfy $\lim_{x \to \infty} f(x) = 1$ and $\lim_{x \to 2^-} f(x) = -\infty$?

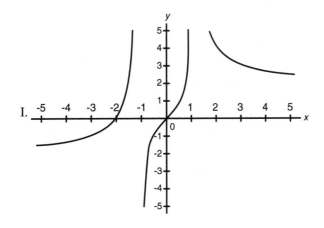

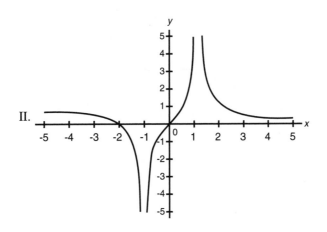

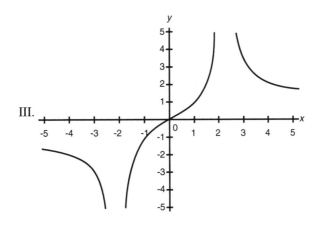

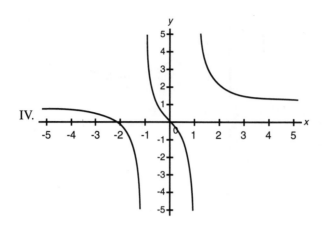

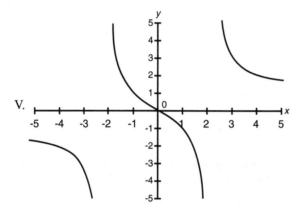

A. I

B. II

C. III

D. IV

E. V

Answer   E. The function value appears to go to 1 as x approaches ∞, and the function value appears to go to $-\infty$ as x approaches 2 from the left.

_____

12.   Which of the following graphs satisfy $\lim\limits_{x \to \infty} f(x) = 1$ and $\lim\limits_{x \to -2} f(x) = -\infty$?

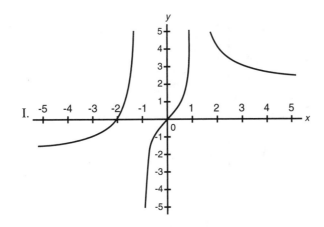

I.

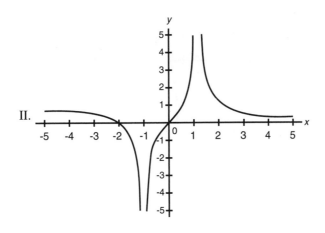

II.

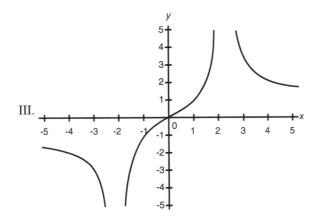

III.

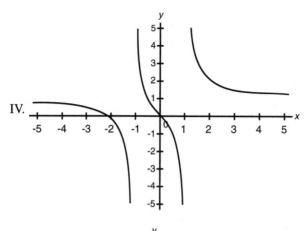

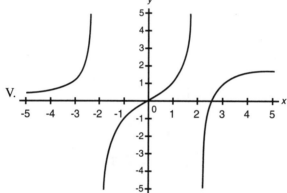

A. I

B. II

C. III

D. IV

E. V

Answer   C. The function value appears to go to 1 as x approaches ∞, and the function value appears to go to $-\infty$ as x approaches -2 from either direction.

---

13. $\lim\limits_{x \to \infty} \dfrac{2x^4 + 7x^2 + 9}{3 - 4x^2 - 5x^4} =$

A. $-\dfrac{2}{5}$

B. -3

C. $\dfrac{2}{5}$

D. 0

E. $-\infty$

Answer   A. Look at the coefficients of the highest power terms.

---

14. $\lim\limits_{x \to \infty} \dfrac{5x^4 + 7x^2 + 9}{x^4} =$

A.   22

B. -5

C.   0

D.   5

E. ∞

Answer   D. Look at the coefficients of the highest power terms.

---

15. $\lim\limits_{x \to 1} \dfrac{5x^3 + 7x^2 + 2}{2x^3} = ?$

A. $\dfrac{5}{2}$

B. -7

C.   0

D. ∞

E.   7

Answer   E. Simply plug in $x = 1$ and calculate the value.

---

16. $\lim\limits_{x \to \infty} \dfrac{5x^4 + 7x^2 + 9}{x^5 + x^2 - 9} =$

A.   5

B.   0

C. $-\dfrac{21}{9}$

D. -5

E. $-\infty$

Answer   B. When the denominator grows faster than numerator, the quotient goes to zero as $x \to \infty$.

17. $\lim\limits_{x \to \infty} \dfrac{x^3 + 7x^2 + 9}{\sqrt{9x^6 + x^2 - 9}} =$

A. ∞

B. $\dfrac{1}{3}$

C. 0

D. 1

E. -1

Answer   B. The denominator $\sqrt{9x^6 + x^2 - 9}$ will act like $\sqrt{9x^6} = 3x^3$ when $x \to \infty$.

18. Which of the following functions grows the fastest as $x \to \infty$?

A. $a(x) = 3x^2$

B. $b(x) = x^{\frac{3}{2}}$

C. $c(x) = \sqrt{x^6 - 3x}$

D. $d(x) = 100^x$

E. $f(x) = e^x$

Answer   D. $d(x) = 100^x$ grows fastest because this is the exponential function with the largest base.

19. Which of the following functions grows the fastest as $x \to \infty$?

A. $f(x) = 4$

B. $g(x) = 2^x - x^3$

C. $c(x) = \sqrt{x^6 - 3^x}$

D. $h(x) = x^4 - 3x + 9$

E. $k(x) = e^x$

Answer   E. $k(x) = e^x$ grows fastest because this is the exponential function with the largest base.

20. Which of the following functions grows the fastest as $x \to \infty$?

A. $f(x) = 4$

B. $b(x) = x^{\frac{3}{2}}$

C. $c(x) = \sqrt{x^6 - 3x}$

D. $g(x) = 2^x - x^3$

E. $h(x) = x^4 - 3x + 9$

Answer   D. $g(x) = 2^x - x^3$ grows fastest because exponential functions always grow faster than polynomials.

___

21.   Which of the following functions grows the fastest as $x \to \infty$?

A. $f(x) = 4$

B. $b(x) = x^{\frac{3}{2}}$

C. $c(x) = \sqrt{x^6 - 3x}$

D. $a(x) = 3x^2$

E. $h(x) = x^4 - 3x + 9$

Answer   E. $h(x) = x^4 - 3x + 9$ grows fastest because the term x^4 will grow faster than all the other choices.

___

22.   Which of the following functions grows the fastest as $x \to \infty$?

A. $f(x) = 4$

B. $b(x) = x^{\frac{3}{2}}$

C. $c(x) = \sqrt{x^6 - 3x}$

D. $a(x) = 3x^2$

E. None

Answer   C. $c(x) = \sqrt{x^6 - 3x}$ grows fastest because $\sqrt{x^6 - 3x}$ will act like $\sqrt{x^6} = x^3$, which is of higher degree than the other choices.

___

23. $\displaystyle \lim_{x \to \infty} \frac{x^4 - 7x + 9}{4 + 5x + x^3} =$

A. 0

B. $\dfrac{1}{4}$

C. 1

D. 4

E. Does not exist

Answer   E. Notice that there is an x^4 term in the numerator and an x^3 term in the denominator. Thus, $\displaystyle\lim_{x\to\infty}\dfrac{x^4-7x+9}{4+5x+x^3}=\infty$, or does not exist.

---

24. $\displaystyle\lim_{x\to\infty}\dfrac{2^x+x^3}{x^2+3^x}=$

A. 0

B. 1

C. $\dfrac{3}{2}$

D. $\dfrac{2}{3}$

E. Does not exist

Answer   A. Since exponentials will outgrow power functions, we only need to consider the 2^x and the 3^x terms. Since 3^x has a larger base, it will outgrow 2^x and $\displaystyle\lim_{x\to\infty}\dfrac{2^x+x^3}{x^2+3^x}=0$.

---

25. $\displaystyle\lim_{x\to0}\dfrac{x^3-7x+9}{4^x+x^3}=$

A. 0

B. $\dfrac{1}{4}$

C. 1

D. 9

E. Does not exist

Answer   D. We're not asking about relative growth here (which would involve $\displaystyle\lim_{x\to\infty}$ or $\displaystyle\lim_{x\to-\infty}$). In this problem we are asked to find $\displaystyle\lim_{x\to0}$, which we can find by simply plugging zero in for x:
$\displaystyle\lim_{x\to0}\dfrac{x^3-7x+9}{4^x+x^3}=9$.

26. For the function $g(t) = 4t^4 - 4^t$, which of the following statements are true?

I. $\lim_{t \to 0} g(t) = -1$

II. $\lim_{t \to \infty} g(t) = -\infty$

III. $g(t)$ has 2 roots.

A. I only

B. II only

C. III only

D. I and II only

E. I, II, and III

Answer   D. I ($\lim_{t \to 0} g(t) = -1$) and II ($\lim_{t \to \infty} g(t) = -\infty$) are true, but III is not true.

You can plug in $x = 0$ to see that I is true, and II is true since the exponential function 4^t outgrows the power function $4t^4$. As for III, it would first appear that $g(t) = 4t^4 - 4^t$ has only 2 roots (see the graph below).

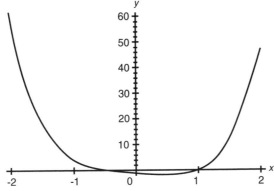

But $g(t)$ has 3 roots! Since we know that the -4^t term will eventually be more important than the $4t^4$ term, we know that $\lim_{t \to \infty} g(t) = -\infty$, and the graph will eventually come back down far to the right (as the graph shows at a different scaling):

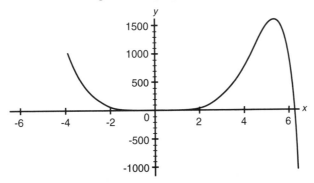

27. $\lim\limits_{x \to \frac{\pi}{3}} \cot(3x) =$

A. $\sqrt{3}$

B. 1

C. $\frac{\sqrt{3}}{3}$

D. 0

E. Does not exist

Answer E. Try graphing this one and notice that the graph has an asymptote at $x = \frac{\pi}{3}$. The limit in fact does not exist.

28. $\lim\limits_{x \to 0} \dfrac{\cos(x) - 1}{x} =$

A. 1

B. 0

C. $\frac{\sqrt{2}}{2}$

D. -1

E. Does not exist

Answer B. $\cos(x) - 1$ approaches 0 much faster than x and $\lim\limits_{x \to 0} \dfrac{\cos(x) - 1}{x} = 0$. This is one of the basic limits that you should memorize.

29. $\lim\limits_{x \to 0} [\cos(x) - x] =$

A. 1

B. 0

C. $\frac{\sqrt{3}}{2}$

D. $\frac{1}{2}$

E. Does not exist

Answer A. We have $\lim\limits_{x \to 0} [\cos(x) - x] = \cos(0) - 0 = 1 - 0 = 1$

30. Which of the following functions grows the fastest?

A. $b(t) = t^4 - 3t + 9$

B. $f(t) = 2^t - t^3$

C. $h(t) = 5^t + t^5$

D. $c(t) = \sqrt{t^2 - 5t}$

E. $d(t) = (1.1)^t$

Answer C. $h(t) = 5^t + t^5$ grows the fastest because its dominant term is the exponential function with the largest base.

---

31. Which of the following functions grows the fastest?

A. $f(t) = 2^t - t^3$

B. $a(t) = t^{\frac{5}{2}}$

C. $e(t) = e$

D. $g(t) = 3t^2 - t$

E. $b(t) = t^4 - 3t + 9$

Answer A. $f(t) = 2^t - t^3$ is the only exponential function listed (and it has a base which is greater than 1), so it grows the fastest.

---

32. Which of the following functions grows the fastest?

A. $g(t) = 3t^2 - t$

B. $i(t) = \ln(t^{100})$

C. $e(t) = e$

D. $c(t) = \sqrt{t^2 - 5t}$

E. $a(t) = t^{\frac{5}{2}}$

Answer E. The power function with the largest exponent is $a(t) = t^{\frac{5}{2}}$, and it outgrows any logarithmic function. So $a(t) = t^{\frac{5}{2}}$ grows the fastest.

---

33. Which of the following functions grows the *slowest*?

A. $b(t) = t^4 - 3t + 9$

B. $f(t) = 2^t - t^3$

C. $h(t) = 5^t + t^5$

D. $c(t) = \sqrt{t^2 - 5t}$

E. $d(t) = (1.1)^t$

Answer   D. The exponential functions all grow faster than the power functions. Of the power functions listed, $c(t)$ grows like t, while $b(t)$ grows like t^4. So $c(t)$ grows the *slowest*.

---

34.   Which of the following functions grows the *slowest*?

A. $g(t) = 3t^2 - t$

B. $i(t) = \ln(t^{100})$

C. $e(t) = e$

D. $c(t) = \sqrt{t^2 - 5t}$

E. $a(t) = t^{\frac{5}{2}}$

Answer   C. The constant function $e(t) = e$ doesn't grow at all!

---

35.   Which of the following functions grows the *slowest*?

A. $j(t) = \frac{1}{4}\ln(t^{200})$

B. $a(t) = t^{\frac{5}{2}}$

C. $i(t) = \ln(t^{100})$

D. $g(t) = 3t^2 - t$

E. $b(t) = t^4 - 3t + 9$

Answer   A. The logarithmic functions grow the slowest here. But $j(t) = \frac{1}{4}\ln(t^{200}) = 200 \times \frac{1}{4}\ln t = 50\ln t$ while $i(t) = \ln(t^{100}) = 100\ln t$. Thus, $j(t)$ grows the *slowest* (twice as slow as $i(t)$).

Which Limits Do Not Exist?

---

1. $\lim\limits_{x \to 0} \sin(x)$

Answer   The limit exists. At $x = 0$, $\sin(x)$ approaches 0 from the left and the right.

2. $\lim\limits_{x \to 1} \dfrac{1}{x-1}$

Answer The limit does not exist, because $\lim\limits_{x \to 1+} \dfrac{1}{x-1} = \infty$ and $\lim\limits_{x \to 1-} \dfrac{1}{x-1} = -\infty$.

3. $\lim\limits_{x \to 1} \dfrac{x^2 - 2x + 1}{x - 1}$

Answer The limit exists. $\lim\limits_{x \to 1} \dfrac{x^2 - 2x + 1}{x - 1} = \lim\limits_{x \to 1} (x - 1) = 0$.

Even though the denominator is zero, we have a "hole" in the graph, not an asymptote.

4. $\lim\limits_{x \to 0} \dfrac{\sqrt{x^2 - 2x + 1}}{x - 1}$

Answer The limit exists. $\lim\limits_{x \to 0} \dfrac{\sqrt{x^2 - 2x + 1}}{x - 1} = -1$.

(When x goes to zero, we can simply plug in values.)

5. $\lim\limits_{x \to 1} \dfrac{\sqrt{x^2 - 2x + 1}}{x - 1}$

Answer The limit does not exist. $\lim\limits_{x \to 1} \dfrac{\sqrt{x^2 - 2x + 1}}{x - 1} = \lim\limits_{x \to 1} \dfrac{\sqrt{(x-1)^2}}{x - 1} = \lim\limits_{x \to 1} \dfrac{\sqrt{(x-1)^2}}{x - 1}$

$= \lim\limits_{x \to 1} \dfrac{|x - 1|}{x - 1} =$ undefined. This is because $\lim\limits_{x \to 1^-} \dfrac{|x - 1|}{x - 1} = -1$ while $\lim\limits_{x \to 1^+} \dfrac{|x - 1|}{x - 1}$
$= 1$.

6. $\lim\limits_{x \to 2} \dfrac{(2 - x)^2}{x - 2}$.

Answer The limit does exist.
$\lim\limits_{x \to 2} \dfrac{(2 - x)^2}{x - 2} = \lim\limits_{x \to 2} \dfrac{(2 - x)(2 - x)}{x - 2} = \lim\limits_{x \to 2} \dfrac{-(x - 2)(2 - x)}{x - 2} = \lim\limits_{x \to 2} -(2 - x) = 0$.

7. $\lim\limits_{x \to \infty} \cos x$

Answer The limit does not exist. $\cos(x)$ oscillates back and forth and does not approach any limit as $x \to \infty$.

8. $\lim\limits_{x \to \infty} \dfrac{\sin x}{x^2}$

Answer   The limit does exist. $\lim\limits_{x \to \infty} \dfrac{\sin x}{x^2} = 0$.

9. $\lim\limits_{x \to 0} |x|$

Answer   The limit does exist. $\lim\limits_{x \to 0} |x| = 0$ because $|x|$ approaches 0 from the left and the right.

10. $\lim\limits_{x \to 0} \cos \left(\dfrac{1}{x^2} \right)$

Answer   The limit does not exist. Since $\frac{1}{x^2}$ gets big as $x \to 0$, this limit is similar to $\lim\limits_{u \to \infty} \cos(u)$. However, for large u, $\cos(u)$ oscillates between -1 and $+1$, so the limit doesn't exist. Therefore, $\lim\limits_{x \to 0} \cos \left(\dfrac{1}{x^2} \right)$ does not exist either.

More Limit Problems

1. $\lim\limits_{x \to \infty} 2 \cos (2x)$

Answer $\lim\limits_{x \to \infty} 2 \cos (2x)$ does not exist. The function values oscillate between -2 and 2.

2. $\lim\limits_{x \to \infty} \dfrac{2 \cos (2x)}{x^2}$

Answer $\lim\limits_{x \to \infty} \dfrac{2 \cos (2x)}{x^2} = 0$, since the x^2 grows without bound, which squeezes the limit to 0.

3. $\lim\limits_{x \to 0} \cos \left(\dfrac{1}{x^2} \right)$

Answer $\lim\limits_{x \to 0} \cos \left(\dfrac{1}{x^2} \right)$ does not exist The function oscillate between -1 and 1, since as x goes to zero, $\dfrac{1}{x^2}$ goes to infinity. (This is one of the "wild" ones.)

4. $\lim\limits_{x\to\infty}\cos\left(\dfrac{1}{x^2}\right)$

Answer $\lim\limits_{x\to\infty}\cos\left(\dfrac{1}{x^2}\right) = 1$, since $\lim\limits_{x\to\infty}\dfrac{1}{x^2} = 0$ and $\cos(0) = 1$.

5. $\lim\limits_{x\to\infty}\cos\left(\dfrac{4x^2 - x + 9}{x^3 - 3x^2 - 7}\right)$

Answer $\lim\limits_{x\to\infty}\cos\left(\dfrac{4x^2 - x + 9}{x^3 - 3x^2 - 7}\right) = 1$, since $\lim\limits_{x\to\infty}\dfrac{4x^2 - x + 9}{x^3 - 3x^2 - 7} = 0$ and $\cos(0) = 1$.

6. a) Find $\lim\limits_{x\to n^-}(x - [[x]])$, where n is any integer.

Answer $\lim\limits_{x\to n^-}(x - [[x]]) = n - (n-1) = 1$. Notice that $\lim\limits_{x\to n^-}(x - [[x]]) = 1$, since $\lim\limits_{x\to n^-}(x) = n$ and $\lim\limits_{x\to n^-}([[x]]) = n - 1$.

   b) Find $\lim\limits_{x\to n^+}(x - [[x]])$, where n is any integer.

Answer $\lim\limits_{x\to n^+}(x - [[x]]) = n - n = 0$.

   c) Find $\lim\limits_{x\to n}(x - [[x]])$, where n is any integer.

Answer $\lim\limits_{x\to n}(x - [[x]])$ does not exist, since the left-hand limit doesn't equal the right.

7. a) Find $\lim\limits_{x\to 0^-}\left(\dfrac{|x|}{x}\right)$.

Answer $\lim\limits_{x\to 0^-}\left(\dfrac{|x|}{x}\right) = \lim\limits_{x\to 0^-}\left(\dfrac{-x}{x}\right) = -1$.

   b) Find $\lim\limits_{x\to 0^+}\left(\dfrac{|x|}{x}\right)$.

Answer $\lim\limits_{x\to 0^+}\left(\dfrac{|x|}{x}\right) = \lim\limits_{x\to 0^+}\left(\dfrac{x}{x}\right) = 1$.

   c) Find $\lim\limits_{x\to 0}\left(\dfrac{|x|}{x}\right)$.

Answer This limit does not exist, since the left- and right-hand limits aren't equal.

8. a) Find $\displaystyle\lim_{x\to-5^-} \frac{x^2+5x}{\sqrt{x^2+10x+25}}$.

Answer

$$\lim_{x\to-5^-} \frac{x^2+5x}{\sqrt{x^2+10x+25}} = \lim_{x\to-5^-} \frac{x(5+x)}{\sqrt{(5+x)(5+x)}}$$
$$= \lim_{x\to-5^-} \frac{x(5+x)}{|(5+x)|}$$
$$= \lim_{x\to-5^-} \frac{x(5+x)}{-1(5+x)}$$
$$= \lim_{x\to-5^-} \left(\frac{x}{-1}\right) = 5.$$

b) Find $\displaystyle\lim_{x\to-5^+} \frac{x^2+5x}{\sqrt{x^2+10x+25}}$.

Answer

$$\lim_{x\to-5^+} \frac{x^2+5x}{\sqrt{x^2+10x+25}} = \lim_{x\to-5^+} \frac{x(5+x)}{\sqrt{(5+x)(5+x)}}$$
$$= \lim_{x\to-5^+} \frac{x(5+x)}{|(5+x)|}$$
$$= \lim_{x\to-5^+} \frac{x(5+x)}{(5+x)}$$
$$= \lim_{x\to-5^+} \left(\frac{x}{1}\right) = -5.$$

c) Find $\displaystyle\lim_{x\to-5} \frac{x^2+5x}{\sqrt{x^2+10x+25}}$.

Answer   This limit does not exist, since the left- and right-hand limits aren't equal.

9. a) Find $\displaystyle\lim_{x\to1^+} \left(\frac{x^2-1}{|x-1|}\right)$.

Answer $\displaystyle\lim_{x\to1^+} \left(\frac{x^2-1}{|x-1|}\right) = \lim_{x\to1^+} \left(\frac{x^2-1}{x-1}\right) = \lim_{x\to1^+} (x+1) = 2.$

b) Find $\displaystyle\lim_{x\to1^-} \left(\frac{x^2-1}{|x-1|}\right)$.

Answer

$$\lim_{x\to1^-} \left(\frac{x^2-1}{|x-1|}\right) = \lim_{x\to1^-} \left(\frac{x^2-1}{-1(x-1)}\right)$$
$$= \lim_{x\to1^-} \left(\frac{(x-1)(x+1)}{-1(x-1)}\right)$$
$$= \lim_{x\to1^-} \left(\frac{(x+1)}{-1}\right) = -2.$$

c) Find $\lim\limits_{x \to 1} \left(\dfrac{x^2 - 1}{|x - 1|} \right)$.

Answer This limit does not exist, since the left- and right-hand limits aren't equal.

Applications of Limits

Read each question carefully, making sure to show all your work and reasoning.

1. A local baseball stadium is free for children under five years old. They charge $5.00 for children 5–10, $6.00 for adults over 65, and $9.00 for everyone else.

A. Write a piecewise function that gives the price $p(x)$ for attending a game for someone who is x years old.

Answer $p(x) = \begin{cases} 0 & \text{if } 0 \le x < 5 \\ 5 & \text{if } 5 \le x \le 10 \\ 9 & \text{if } 10 < x \le 65 \\ 6 & \text{if } 65 < x. \end{cases}$

Don't worry if you have your $<$'s and your $\le$'s in different places; there are different ways to interpret the English. But do be sure that every single number is covered exactly once in your intervals. For example, does your function give a correct price for someone who is age $9\frac{1}{2}$? 10? $10\frac{1}{2}$?

B. Find all values of a where $\lim\limits_{x \to a} p(x)$ exists and all values where $\lim\limits_{x \to a} p(x)$ fails to exist. Describe why the limit fails to exist in those situations.

Answer $\lim\limits_{x \to a} p(x)$ exists for $x \ge 0, x \ne 5, 10, 65$. The limit doesn't exist at 5, 10, 65, since the right-hand limit $\ne$ left-hand limit.

2. The focal length for a lens is given by the equation $\dfrac{1}{f} = \dfrac{1}{D_O} + \dfrac{1}{D_I}$, where D_O is the distance from the lens to the object and D_I is the distance from the lens to the image.

A. Write a function that gives D_I as a function of D_O, assuming that a lens has a constant focal length f.

Answer

$$\frac{1}{f} = \frac{1}{D_O} + \frac{1}{D_I}$$
$$\Rightarrow \frac{1}{D_I} = \frac{1}{f} - \frac{1}{D_O} = \frac{D_O - f}{f D_O}$$
$$\Rightarrow D_I = \frac{f D_O}{D_O - f}.$$

B. Find $\lim\limits_{D_O \to \infty} D_I(D_O)$.

Answer $\lim\limits_{D_O \to \infty} D_I(D_O) = \lim\limits_{D_O \to \infty} \dfrac{fD_O}{D_O - f} = f \lim\limits_{D_O \to \infty} \dfrac{D_O}{D_O - f} = f.$

C. Describe in your own words what this limit represents.

Answer

As the object being focused is moves farther and farther away, the image becomes focused at a focal length away from the lens. (This is one way to find the focal length of a lens.)

---

3. Special relativity states that the relativistic mass (the mass of a moving object) of an object m is given by

$$m = \frac{m_0}{\sqrt{1 - \dfrac{v^2}{c^2}}},$$

where m_0 is the mass of the object if it isn't moving; c is the speed of light ($3 \times 10^8 \ m/s$); and v is the speed of the object.

A. Find $\lim\limits_{v \to 0} m(v)$ and explain what your result implies for the mass of a moving object.

Answer $\lim\limits_{v \to 0} m(v) = m_0.$

If the object is not moving, then it has a mass equal to its mass when it's not moving.

B. Find $\lim\limits_{v \to c-} m(v)$ and explain what your result implies for the mass of a moving object.

Answer $\lim\limits_{v \to c^-} m(v) = \infty.$

The mass of an object increases without bound as the speed of the object approaches the speed of light. (Which is why no object with mass can go the speed of light.)

C. Why did we use a left-hand limit in part b?

Answer Mathematically, if $v = c$, then we get zero in the denominator. If $v > c$, then we get a negative number under the $\sqrt{\ }$, which is not a real number.

Physically, no object with mass can go the speed of light (or faster).

C. Continuity

▶Objective 1   State the definition of continuity at a point.

Example   State the definition of continuity at a point.

Tip   A function is continuous if you can draw the graph without lifting your pencil off of the paper. (But this isn't the formal definition!)

Answer $f(x)$ is continuous at $x = a$ if $\lim_{x \to a} f(x)$ exists and $f(a)$ exists and

$$\lim_{x \to a} f(x) = f(a).$$

▶**Objective 2**   Determine if a function is continuous at a certain point, using the limit definition.

Example   Describe why $f(x) = [[x]]$ does not meet the definition of continuity at $x = 3$. Explain why it does at $x = 2.5$.

Tip $f(x)$ is continuous at $x = a$ if $\lim_{x \to a} f(x)$ and $f(a)$ both exist and $\lim_{x \to a} f(x) = f(a)$.

Answers   At $x = 3$. In this case $\lim_{x \to a} f(x)$ does not exist, so that the conditions of continuity can't be satisfied. At $x = 2.5$, $\lim_{x \to a} f(x)$ does exists and is equal to 2. $f(a) = f(2.5)$ exists and is also equal to 2, therefore $\lim_{x \to a} f(x) = f(a) = 2$, and $f(x) = [[x]]$ is continuous at $x = 2.5$.

▶**Objective 3**   Determine the type of discontinuity that may exist, graphically and analytically.

Example   Label all the discontinuities in each function as either a jump, infinite, or removable discontinuity and state the x value where each discontinuity takes place.

A. $g(x) = \begin{cases} \dfrac{x^2 - 16}{x + 4} & \text{if } x < 1 \\ \dfrac{4}{x - 7} & \text{if } x \geq 1 \end{cases}$

B.

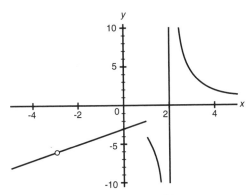

(Note the open hole at $x = -3$.)

C. $h(x) = [[\frac{x}{2}]]$, where $[[u]]$ is the greatest integer function.

Tip Remember to think of these in terms of limits, since the definition of continuity is that $\lim\limits_{x \to a} f(x) = f(a)$. Look for places where the functions might be undefined, as well as places where different pieces glue together.

Answers

A. $g(x) = \begin{cases} \dfrac{x^2 - 16}{x + 4} & \text{if } x < 1 \\ \dfrac{4}{x - 7} & \text{if } x \geq 1 \end{cases}$

At $x = -4$ we have a removable discontinuity, since all we would have to do is define the function to be -8 at $x = -4$ in order to make it continuous.

At $x = 1$ we have a jump discontinuity, since the right-hand limit $\neq$ the left-hand limit.

At $x = 7$ we have an infinite discontinuity, since we have a vertical asymptote.

B.

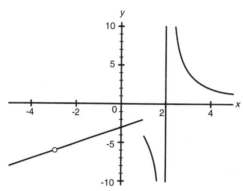

(Note the open hole at $x = -3$.)

At $x = -3$ we have a removable discontinuity, since all we would have to do is define the function to be -6 at $x = -3$ in order to make it continuous.

At $x = 1$ we have a jump discontinuity, since the right-hand limit $\neq$ the left-hand limit.

At $x = 2$ we have an infinite discontinuity, since we have a vertical asymptote.

C. $h(x) = [[\frac{x}{2}]]$ where $[[u]]$ is the greatest integer function.

This function has jump discontinuities whenever $\dfrac{x}{2}$ is equal to an integer, which is for all even integers. The graph looks like:

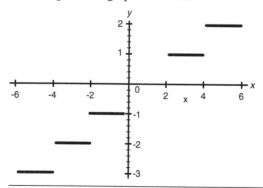

▶Objective 4   Describe discontinuities in terms of limits.

Example   Label each of the following as indicating a jump, infinite, or removable discontinuity.

A. $\lim\limits_{x \to a^-} f(x) = L_1 \neq L_2 = \lim\limits_{x \to a^+} f(x).$

B. $\lim\limits_{x \to a^-} f(x) = \infty.$

C. $\lim\limits_{x \to a} f(x) = L_1 \neq f(a).$

D. $\lim\limits_{x \to a^+} f(x) = -\infty.$

E. $\lim\limits_{x \to a} f(x) = L,$ but $f(a)$ does not exist.

Tip   Visualize a graph for each case.

Answers

A. $\lim\limits_{x \to a^-} f(x) = L_1 \neq L_2 = \lim\limits_{x \to a^+} f(x) :$ jump

B. $\lim\limits_{x \to a^-} f(x) = \infty :$ infinite

C. $\lim\limits_{x \to a} f(x) = L_1 \neq f(a) :$ removable

D. $\lim\limits_{x \to a^+} f(x) = -\infty :$ infinite

E. $\lim\limits_{x \to a} f(x) = L,$ but $f(a)$ does not exist: removable

---

▶Objective 5   Solve for parameters in equations that represent continuous functions.

Example 1   Suppose $h(t) = \begin{cases} ct + 2 & \text{if } t < -2 \\ \sin(\frac{\pi t}{2}) & \text{if } -2 \leq t \leq 3. \\ t^2 - d & \text{if } t > 3 \end{cases}$

Given that $h(t)$ is continuous, find the values of c and d.

Tip   Each piece of the function must match each of the other pieces at the junctions.

Answers $c = 1$ and $d = 10.$

Since $\lim\limits_{t \to -2^-} h(t) = \lim\limits_{t \to -2^+} h(t),$

$c(-2) + 2 = \sin(\frac{\pi(-2)}{2}) = 0,$ so that $c = 1;$

since $\lim\limits_{t \to 3^-} h(t) = \lim\limits_{t \to 3^+} h(t),$

$\sin(\frac{3\pi}{2}) = 3^2 - d;$

$-1 = 9 - d,$ so that $d = 10.$

Example 2   Let $f(x) = \dfrac{\sqrt{x + c^2} - c}{x}$, where $c > 0$.

A.   What is the domain of f?

B.   How can you define f at $x = 0$ for f to be continuous there?

Tip   Although putting in actual values in for c can help you visualize the problem, you must **not** plug in values for c when you present your solution. The idea is to keep the function general for any values of c greater than zero.

Answers

A.   The domain of x is $\{x \; : \; x \geq -c^2, \; x \neq 0\}$.

B.   In order to be continuous at $x = 0$, $\lim\limits_{x \to a} f(x) = f(a)$, so we should find $\lim\limits_{x \to 0} f(x)$.

$$\lim_{x \to 0} \frac{\sqrt{x + c^2} - c}{x} = \lim_{x \to 0} \frac{\sqrt{x + c^2} - c}{x} \times \frac{\sqrt{x + c^2} + c}{\sqrt{x + c^2} + c} = \lim_{x \to 0} \frac{x + c^2 - c^2}{x(\sqrt{x + c^2} + c)}$$

$$= \lim_{x \to 0} \frac{x}{x(\sqrt{x + c^2} + c)}$$

$$= \lim_{x \to 0} \frac{1}{\sqrt{x + c^2} + c} = \frac{1}{2c}$$

Therefore we should define $f(0) = \dfrac{1}{2c}$ to make the function f continuous at $x = 0$.

▶ **Objective 6**   Know the statements of the Intermediate Value Theorem and the Extreme Value Theorem, including their hypotheses.

The **Intermediate Value Theorem** states that if f is a function continuous for all points in the closed interval $[a, b]$, for every value C between $f(a)$ and $f(b)$ there is at least one c between a and b for which $f(c) = C$.

The **Extreme Value Theorem** states that if f is a function continuous for all points in the closed interval $[a, b]$, f takes on both a maximum value M and a minimum value m over that interval.

▶ **Objective 7**   State why each hypothesis is needed in the Intermediate Value Theorem and the Extreme Value Theorem.

Example   Consider the function $f(x) = \begin{cases} 1 + x, & -\frac{1}{2} \leq x < 0 \\ \frac{1}{2}, & 0 \leq x \leq \frac{1}{2}. \end{cases}$

Explain why the Extreme Value Theorem fails to hold for $f(x)$ on $[-\frac{1}{2}, \frac{1}{2}]$.

Also, consider the function $g(x) = \begin{cases} -1, & -1 \leq x < 0 \\ 1, & 0 \leq x \leq 1. \end{cases}$

Explain why the Intermediate Value Theorem fails to hold for $g(x)$ on $[-1, 1]$.

Tip Review the statements of the Intermediate Value Theorem and the Extreme Value Theorem.

Answer In order for the EVT and IVT to hold on an interval, you need a function to be continuous on that interval.

Note that $f(x)$ is not continuous on $[-\frac{1}{2}, \frac{1}{2}]$ and there is no c in $[-\frac{1}{2}, \frac{1}{2}]$ such that $f(c) \geq f(x)$ for all x in $[-\frac{1}{2}, \frac{1}{2}]$.

Note that $g(x)$ is also not continuous on $[-1, 1]$ and there is no c in $[-1, 1]$ such that $g(c) = 0$.

▶Objective 8 Use the Intermediate Value Theorem and the Extreme Value Theorem to predict some of the behavior of a continuous function over a closed interval.

Example Show that there's a root of the equation in the given interval.

 A. $2x^3 + x^2 + 2 = 0$ on interval $(-2, -1)$

 B. $x^4 + 1 = \dfrac{1}{x}$ on interval $(\frac{1}{2}, 1)$

Answers

 A. First, let $f(x) = 2x^3 + x^2 + 2$. We know that all polynomials are continuous, so $f(x)$ is continuous on the closed interval $[-2, -1]$. Then since $f(-2) = -16 + 4 + 2 = -10$ and $f(-1) = -2 + 1 + 2 = 1$, we can say that there exists at least one value of x on interval $(-2, -1)$ where $f(x) = 0$ (Intermediate Value Theorem).

 B. First, let $f(x) = x^4 + 1 - \dfrac{1}{x}$. We know that all polynomials are continuous, and $\dfrac{1}{x}$ is continuous except at $x = 0$, so $f(x)$ is continuous on the closed interval $[\frac{1}{2}, 1]$. Then since $f(\frac{1}{2}) = \dfrac{1}{16} + 1 - 2 = -\dfrac{15}{16}$ and $f(1) = 1 + 1 - 1 = 1$, we can say that there exists at least one value of x on interval $(\frac{1}{2}, 1)$ where $f(x) = 0$ (Intermediate Value Theorem).

Exercises on Continuity

---

1. If $f(x)$ is continuous for all real numbers and $\lim\limits_{x \to a^-} f(x) = b$, then which of the following statements necessarily is true?

 I. $\lim\limits_{x \to a^+} f(x) = b$

 II. $\lim\limits_{x \to a} f(x) = b$

 III. $f(a) = b$

 A. I only

 B. I and II only

C. II only

D. III only

E. I, II, and III.

Answer   E. These are all properties of continuous functions.

---

2. If $g(x)$ is equal to $\dfrac{x^2 - 9}{x - 3}$ when $x \neq 3$, and $g(x)$ is continuous for **all** real numbers, then what is the value of $g(3)$?

A. 6

B. Undefined

C. 0

D. 3

E. ∞

Answer   A. Factoring leads you to $\displaystyle\lim_{x \to 3} \dfrac{x^2 - 9}{x - 3} = \lim_{x \to 3}(x + 3) = 6$, and you know this is equal to the value of g at 3.

Question 3–5 refer to the following graph:

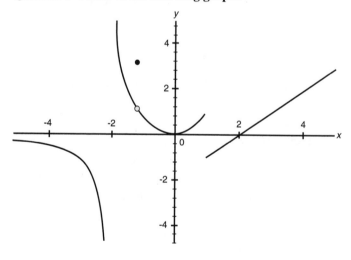

---

3. An infinite discontinuity appears to occur at:

A. $x = -2$

B. $x = -1$

C. $x = 0$

D. $x = 1$

E. Nowhere

Answer A. That's where the function value approaches ∞ from the right and $-\infty$ from the left.

---

4. A jump discontinuity appears to occur at:

A. $x = -2$

B. $x = -1$

C. $x = 0$

D. $x = 1$

E. Nowhere

Answer D. That's where the function value approaches -1 from the right and 1 from the left.

---

5. A removable discontinuity appears to occur at:

A. $x = -2$

B. $x = -1$

C. $x = 0$

D. $x = 1$

E. Nowhere

Answer B. At $x = -1$ the function value approaches 1 from the right and left, but $f(-1) \neq 1$.

---

6. If $g(x) = \begin{cases} \dfrac{x-4}{\sqrt{x}-2} & \text{if } x \neq 4 \\ a & \text{if } x = 4 \end{cases}$ is continuous, what is the value of a?

A. 2

B. 4

C. $\sqrt{2}$

D. $2\sqrt{2}$

E. None of these

Answer B. $\lim\limits_{x \to 4} \dfrac{x-4}{\sqrt{x}-2} = 4$.

7. If $g(x) = \begin{cases} 2x + b & \text{if } x < 0 \\ 5 & \text{if } 0 \le x \le 2 \\ ax^2 - bx + 3 & \text{if } x > 2 \end{cases}$ is continuous, what is the value of a?

A. 0

B. $\dfrac{1}{2}$

C. 2

D. 3

E. None of these

Answer   D. $\lim\limits_{x \to 0^-} 2x + b = 5 \Longrightarrow b = 5$, then $\lim\limits_{x \to 2^+} ax^2 - bx + 3 = 5 \Longrightarrow a = 3$.

Questions 8–11 refer to the following graph of $g(x)$:

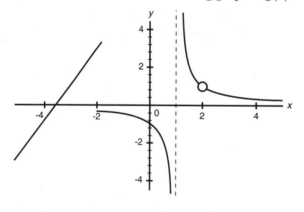

8. An infinite discontinuity appears to occur at $x =$

A. -2

B. 1

C. 2

D. 0

E. $2, -2$

Answer   B. Vertical asymptotes are infinite discontinuities. There is a vertical asymptote at $x = 1$.

9. A jump discontinuity appears to occur at $x =$

A. -2

B. 1

C. 2

D. 0

E. $2, -2$

Answer A. The function jumps from 3 to $-\dfrac{2}{5}$ at $x = -2$.

10. A removable discontinuity appears to occur at $x =$

A. -2

B. 1

C. 2

D. 0

E. $2, -2$

Answer C. The graph has a "hole" at $x = 2$, so there is a removable discontinuity there.

11. If $\lim\limits_{x \to a^+} g(x) = \infty$, then it appears from the graph that $a =$

A. 2

B. -2

C. 0

D. 1

E. 4

Answer D. The function goes to positive infinity as x get closer and closer to 1 from the right.

12. Suppose $g(x) = \begin{cases} \dfrac{1}{x+1} & \text{if } x < 1 \\ 2x - 1 & \text{if } x \geq 1 \end{cases}$

The **best** description concerning the continuity of $g(x)$ is that the function:

A.   is continuous.

B.   has a jump discontinuity.

C.   has an infinite discontinuity.

D.   has a removable discontinuity.

E.   has both jump and infinite discontinuity.

Answer   E. The function has a vertical asymptote at $x = -1$, so it has an infinite discontinuity. Furthermore, the function has a jump discontinuity at $x = 1$, since $\lim\limits_{x \to 1^-} g(x) = .5$, while $\lim\limits_{x \to 1^+} g(x) = 1$.

---

13.   Suppose $g(x) = \begin{cases} \dfrac{x^2 + 2x + 1}{x + 1} & \text{if } x < 1 \\ 2x & \text{if } x \geq 1 \end{cases}$

The **best** description concerning the continuity of $g(x)$ is that the function:

A.   is continuous.

B.   has a jump discontinuity.

C.   has an infinite discontinuity.

D.   has a removable discontinuity.

E.   has both infinite and removable discontinuities.

Answer   D. The function has a removable discontinuity at $x = -1$.

Notice that $\lim\limits_{x \to -1} \dfrac{x^2 + 2x + 1}{x + 1} = \lim\limits_{x \to -1} \dfrac{(x + 1)(x + 1)}{x + 1} = \lim\limits_{x \to -1} x + 1 = 0$.

So we could make the function continuous by redefining $g(-1) = 0$.

---

14.   Suppose $g(x) = \begin{cases} \dfrac{1}{x - 2} & \text{if } x < 1 \\ 2x - 3 & \text{if } x \geq 1. \end{cases}$

The **best** description concerning the continuity of $g(x)$ is that the function:

A. is continuous.

B. has a jump discontinuity.

C. has an infinite discontinuity.

D. has a removable discontinuity.

E. None of the above

Answer A. The function is continuous. If you are not certain, graph it and see!

---

15. Suppose $g(x) = \begin{cases} \dfrac{1}{x-2} & \text{if } x < 1 \\ 2x - 4 & \text{if } x \geq 1. \end{cases}$

The **best** description concerning the continuity of $g(x)$ is that the function:

A. is continuous.

B. has a jump discontinuity.

C. has an infinite discontinuity.

D. has a removable discontinuity.

E. None of the above

Answer B. The function has a jump discontinuity at $x = 1$.

We have $\displaystyle \lim_{x \to 1^-} g(x) = \lim_{x \to 1^-} \frac{1}{x-2} = -1$, while $\displaystyle \lim_{x \to 1^+} g(x) = \lim_{x \to 1^+} 2x - 4 = -2$.

---

16. If $h(x)$ is equal to $\dfrac{x^2 - 4}{x + 2}$ when $x \neq -2$, and $h(x)$ is continuous for all real numbers, then what is $h(-2)$?

A. 0

B. -2

C. -4

D. 2

E. This is impossible. There is an infinite discontinuity at $x = -2$.

Answer C. $\displaystyle \lim_{x \to -2} \frac{x^2 - 4}{x + 2} = -4$, so $h(-2) = -4$.

17. If the following function is continuous, then what is the value of b?

$$g(t) = \begin{cases} \dfrac{2t^2 + 2t - 24}{t - 3} & \text{if } t \neq 3 \\ b & \text{if } t = 3 \end{cases}$$

A. 0

B. 3

C. 7

D. 14

E. None of these

Answer D. The function g can be made continuous by choosing $b = 14$, since

$$\lim_{t \to 3} \frac{2t^2 + 2t - 24}{t - 3} = \lim_{t \to 3} \frac{2(t + 4)(t - 3)}{(t - 3)} = \lim_{t \to 3} 2(t + 4) = 14.$$

18. If the following function is continuous, then what is the value of a?

$$h(t) = \begin{cases} 2t + b & \text{if } t < 0 \\ 2\cos(t) - 3 & \text{if } 0 \leq t \leq \frac{\pi}{2} \\ a\sin(t) + 5b & \text{if } t > \frac{\pi}{2} \end{cases}$$

A. 0

B. 1

C. $\dfrac{\pi}{2}$

D. 2

E. 4

Answer D. Find b first, then find a:

Since $2\cos(t) - 3$ equals -1 when $t = 0$, b is equal to -1.

Now $2\cos(t) - 3$ equals -3 when $t = \frac{\pi}{2}$, and so $a\sin(t) + 5b$ must also equal -3 when $t = \frac{\pi}{2}$, which leads us to the conclusion that $a = 2$.

19. $f(x)$ is a continuous function and $f(1) = 2$, $f(3) = 4$. Which of the following values must $f(x)$ **necessarily** attain on the domain $(1, 3)$?

I. $\frac{3}{2}$

II. $\frac{5}{2}$

III. 3

A.   I only

B.   II only

C.   II and III only

D.   I, II, and III

E.   None of these

Answer   C. Both $\frac{5}{2}$ and 3 must be attained as function values on the domain since the Intermediate Value Theorem guarantees that $f(x)$ takes on each value between 2 and 4 over the interval $1 \le x \le 3$.

---

20. $g(x)$ is a continuous function and $g(-1) = 2$, $g(-2) = 0$, $g(-3) = 4$. Which of the following $g(x)$ values must **necessarily** exist on the domain $(-3, -1)$?

I. $\frac{3}{2}$

II. $\frac{5}{2}$

III. 3

A.   I only

B.   II only

C.   II and III only

D.   I, II, and III

E.   None of these

Answer   D. If the values exist on the domain $(-3, -2)$, then they exist on the domain $(-3, -1)$.

---

21.   Using your calculator to calculate the values and the Intermediate Value Theorem, determine the minimum number of roots of $g(x) = x^5 + 2x^4 - 5x^3 + 2x - 1$ on the domain $[-5, 5]$.

A. 2

B. 3

C. 4

D. 5

E. None

Answer   B. $g(-5) < 0$, $g(-3) > 0$, $g(1) < 0$, and $g(2) > 0$. Since g is continous on $[-5, 5]$, the Intermediate Value Theorem says there's a root between -5 and -3, between -3 and 1, and between 1 and 2.

---

22.   What is the minimum value of $y = 4 - x^2$ on the domain $[-1, 2)$?

A. $x = 2$

B. $y = 0$

C. $y = -1$

D. $y = 3$

E. Does not exist

Answer   E. Because the domain is open at 2, the minimum is not guaranteed. In this case, the minimum would be at $x = 2$, which is not included in this interval. So, the minimum on this domain does not exist.

---

23. $h(x)$ is a continuous function where $h(-2) = -1$, $h(1) = -2$, and $h(3) = 2$. Which of the following statements are necessarily true?

I. The maximum value of h on the domain $[-2, 3]$ is greater than or equal to 2.

II. The minimum value of h on the domain $[-2, 3]$ is equal to -2.

III. $h(x)$ must have values strictly between -2 and 2.

A. I only

B. III only

C. I and II only

D. I, II, and III

E. None of these

Answer   A. The maximum value of h must be ≥ 2; since we know $h(3) = 2$, the maximum on this closed interval cannot be any less.

Chapter 4   *Derivatives*

In this chapter we'll review the definition of derivative and the objectives for understanding how it is used. Imagine being able to attach a speedometer to a function—a device that would actually measure how fast the outputs to the function changed relative to changes in the inputs. This is exactly what a *derivative* supplies us. When real world quantities change, measuring the *rate* of change can give us insights and be applied to solving problems. If the quantity can be modeled with a function, then the derivative provides us with that measurement.

A. Derivative at a Point

▶Objective 1   Calculate average rates of change in various situations where one quantity changes in relation to another quantity.

Example   A piece of chocolate is pulled from a refrigerator (6°C) and placed on a counter (22°C). The temperature of the chocolate is given by:

min	0	4	8	12	16	20	24	28	32	36
temp	6.00	9.87	12.81	15.04	16.72	18.00	18.97	19.70	20.26	20.68

What is the average rate of change in the temperature of the chocolate from 8 to 20 minutes?

Tip   Notice how this relates to $m = \dfrac{\Delta y}{\Delta x}$.

Answer

$$\frac{\Delta Temp}{\Delta time} = \frac{18.00 - 12.81}{20 - 8} = .4325 \; \frac{°C}{min}.$$

▶Objective 2   Estimate instantaneous rates of change using data and graphs.

Example   A piece of chocolate is pulled from a refrigerator (6°C) and placed on a counter (22°C). The temperature of the chocolate is given by:

min	0	4	8	12	16	20	24	28	32	36
temp	6.00	9.87	12.81	15.04	16.72	18.00	18.97	19.70	20.26	20.68

Estimate the instantaneous rate of change of temperature at time equals 22 minutes.

Tip   The smaller the interval you use, the better the estimate for the instantaneous rate of change.

Answer

$$\frac{\Delta Temp}{\Delta time} = \frac{18.97 - 18.00}{24 - 20} = .2425 \; \frac{°C}{min}.$$

▶Objective 3   Define instantaneous rate of changes as a limit of an average rate of change.

Graphically you can look at this as the slope of a secant line as the secant line becomes a tangent line through the limiting process. You should understand that this is just the derivative:

$$\lim_{h \to 0} \frac{f(x + h) - f(x)}{h} \quad \text{or} \quad \lim_{x \to a} \frac{f(x) - f(a)}{x - a}.$$

▶Objective 4   Calculate an instantaneous rate of change using the limit definition of the derivative.

Example   A bug's position (in inches) along a straight path is given by the equation $s(t) = 2t + t^2$, on the interval [0, 4] minutes. Using the limit definition of the derivative, find its instantaneous velocity at $t = 3$ min.

Tip   It's fine to take the derivative using the formula as a check for this kind of a problem, but you should be able to use the limit definition when asked to.

Answer   The rate of change of the bug's position is the velocity, so you just need the instantaneous slope of this function:

$$\lim_{t \to 3} \frac{s(t) - f(3)}{t - 3} = \lim_{t \to 3} \frac{2t + t^2 - (2(3) + 3^2)}{t - 3} = \lim_{t \to 3} \frac{t^2 + 2t - 15}{t - 3}$$
$$= \lim_{t \to 3} \frac{(t + 5)(t - 3)}{t - 3} = \lim_{t \to 3} (t + 5) = 8 \; \frac{\text{inches}}{\text{min}}.$$

►Objective 5   Using the concept of the limit, explain how the slope of a tangent line is related to the slopes of secant lines.

The slope of a secant line is given by

$$\frac{f(b) - f(a)}{b - a},$$

where $(a, f(a))$ and $(b, f(b))$ are the two points on the graph of $y = f(x)$.

The slope of the tangent line is the limit of this difference quotient as b approaches a:

$$\lim_{b \to a} \frac{f(b) - f(a)}{b - a}.$$

In other words, as the intervals get smaller, the secant lines approach the tangent line, so the limit of the slopes of the secant lines as the length of the interval goes to zero is the slope of the tangent line.

►Objective 6   Find derivatives of functions using the definition of the derivative.

Example   Using limits, find the derivative of $y = \sqrt{x}$.

Answer

$$\begin{aligned}
f'(x) &= \lim_{h \to 0} \frac{f(x+h) - f(x)}{h} = \lim_{h \to 0} \frac{\sqrt{(x+h)} - \sqrt{(x)}}{h} \\
&= \lim_{h \to 0} \frac{\sqrt{(x+h)} - \sqrt{(x)}}{h} \times \frac{\sqrt{(x+h)} + \sqrt{(x)}}{\sqrt{(x+h)} + \sqrt{(x)}} \\
&= \lim_{h \to 0} \frac{(x+h) - (x)}{h\left(\sqrt{(x+h)} + \sqrt{(x)}\right)} = \lim_{h \to 0} \frac{h}{h\left(\sqrt{(x+h)} + \sqrt{(x)}\right)} \\
&= \lim_{h \to 0} \frac{1}{\left(\sqrt{(x+h)} + \sqrt{(x)}\right)} = \frac{1}{2\sqrt{x}}.
\end{aligned}$$

►Objective 7   Estimate the graph of a derivative function from the graph of its original function.

Example   Draw a sketch of the derivative of the function given by the following graph:

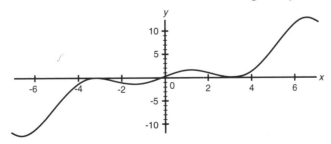

Tip   The derivative is positive when the original function is increasing and negative when the original function is decreasing. The derivative is zero when the original function is flat. Pay attention to where the steepest parts are, where there are positive or negative slopes, and where the flat parts are.

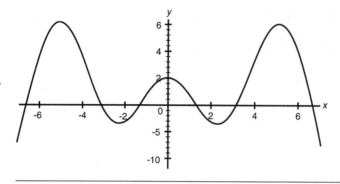

Answer

▶Objective 8   Estimate the graph of the original function when given the graph of the derivative.

Example   Draw a rough draft of an original function that has the following graph as its derivative:

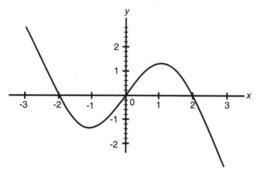

Tips   The key to this one is to see that when the derivative crosses the x-axis, the original function will be flat. Thus at $x = -2$, $x = 0$, $x = 2$ the graph is flat. Also, when the derivative is positive (which occurs on the intervals $(-3, 2)$ and $(0, 2)$), the original function is increasing. Between -2 and 0 it's decreasing as well as to the right of 2, since the derivative is negative.

Also, remember that the derivative tells you the shape of the original function, but nothing about its actual y-values. So, any graph with this basic shape, no matter whether it was shifted up or down from the one given, would be correct.

Answer

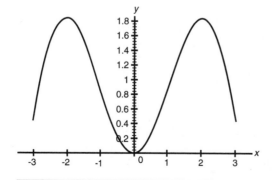

►Objective 9   Be familiar with the Leibniz (differential or fractional) and prime notations for derivatives.

Leibniz notation: $\dfrac{dy}{dx}$

Leibniz notiation is used when it's important to state the independent variable (x in this case). In $\dfrac{d\theta}{dt}$ we say we're taking the derivative of θ "with respect to" t.

Prime notation: y' or f'

Prime notation can be used when it's clear what the independent variable is; y' is typically equivalent to $\dfrac{dy}{dx}$.

►Objective 10   Be able to use the numerical derivative functionality of your calculator to compute (approximate) values of derivatives.

Tip   The syntax for **nDeriv** to approximate $f'(a)$ when you know the expression for $f(x)$ is **nDeriv**(*expression*, x, a) where *expression* is $f(x)$, x is the variable, and a is the point.

Example 1   Suppose a ball is thrown straight up into the air, and the height of the ball above the ground is given by the function $h(t) = 6 + 37t - 16t^2$, where h is in feet and t is in seconds. What is the velocity of the ball at time $t = 3.2$?

A.   53 ft/sec

B. -102.4 ft/sec

C. -65.4 ft/sec

D. -32.3 ft/sec

E.   32.4 ft/sec

Answer   C. -65.4 ft/sec. Try the different answer choices on your calculator until you find the correct choice, which is choice D. Here is what happens with Choice D. On your calculator, enter **nDeriv** $(6 + 37x - 16x^2, x, 3.2)$, and press ENTER . This gives you the derivative of the height function with respect to time (the velocity) at time $t = 3.2$.

Example 2   At what time t does the ball stop going up and start returning to Earth?

A.   1.42341 seconds

B.   .04684 seconds

C.   2.46551 seconds

D.   1.15625 seconds

E.   4.33232 seconds

Answer   D. $t = 1.15625$ seconds. Try the $d\|$ event answer choices on your calculator until you find the correct choice, which is choice D. Here is what happens with choice D. On your calculator, enter **nDeriv**$(6 + 37x - 16x^2, x, 1.15625)$, and press $\boxed{\text{ENTER}}$. You will see that the derivative of the function is 0 at time $t = 1.15625$, which means that the ball has velocity 0 at that time. Practically speaking, the ball has stopped moving upward and is about to start falling back to earth. You can also see this by graphing the parabola $6 + 37x - 16x^2$ and finding its highest point.

Multiple-Choice Practice: Rates of Change and Slopes

A car races around a one-mile racetrack. The data in the table below tells us the total time elapsed during the race as the car completes each lap. The first two problems refer to this data.

Completed Laps	0	1	2	3	4	5
Elapsed time (min.: sec.)	0:00	0:40	1:18	1:50	2:25	3:01

Completed Laps	6	7	8	9	10
Elapsed time (min.: sec.)	3:37	4:10	4:48	5:32	6:05

---

1. What is the average velocity (in miles per hour) of the car over the first 10 laps?

A. 36.5 mph

B. 98.63 mph

C. 1.64 mph

D. 99.17 mph

E. 36.3 mph

Answer   B. The average velocity is given by

$$\frac{\text{miles through lap 10} - \text{miles through lap 0}}{\text{time through lap 10} - \text{time through lap 0}}$$

$$= \frac{10}{6:05} = \frac{10}{6 + \dfrac{5}{60}} = \frac{10}{\dfrac{73}{12}} = \frac{120}{73} \text{ miles per minute}$$

$$= \frac{120}{73} \times 60 \text{ mph} \approx 98.63 \text{ mph}.$$

Always be careful about units!

---

2. What is your best local linear approximation of the instantaneous velocity (in miles per hour) of the car at 4 minutes into the race?

A.   100 mph

B.   94.74 mph

C.   98.63 mph

D.   109.09 mph

E.   105.25 mph

Answer   D. The best local linear approximation of the instantaneous velocity of the car at 4:00 into the race is gotten by taking the data points slightly before 4:00 and slightly after 4:00 (3:37 and 4:10), and implementing the expression for average velocity over an interval:

$$\frac{7-6}{4:10-3:37} = \frac{1}{(4+\frac{10}{60})-(3+\frac{37}{60})} = \frac{1}{\frac{33}{60}} = \frac{60}{33} \text{ miles per minute}$$

$$= \frac{60}{33} \times 60 \text{ mph} \approx 109.9 \text{ mph}.$$

3.   Following is the graph of $y = f(x)$ along with various line segments connecting points on the graph:

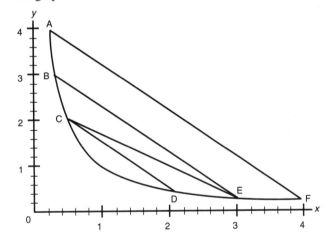

The average rate of change of f over the interval $[.5, 2]$ is the slope of which line segment?

A.   *AF*

B.   *BE*

C.   *CD*

D. CE

E. None of the above

Answer   C. Line segment CD has slope $\dfrac{f(2) - f(.5)}{2 - .5}$, which is precisely the average rate of change of f on the interval $[.5, 2]$.

---

4. Suppose $f(x) = x^4 - 4x^2 + 2$.

   The average rate of change of the function on the interval $[1.9, 2.1]$ is:

A. 16.08

B. -108.32

C. 12.72

D. -31.92

E. 31.92

Answer   A. Average rate of change of the function on the interval $[1.9, 2.1]$ is $\dfrac{f(2.1) - f(1.9)}{2.1 - 1.9}$.

---

5. Below is the graph of $f(x) = x^4 - 4x^2 + 2$ along with the graph of a line.

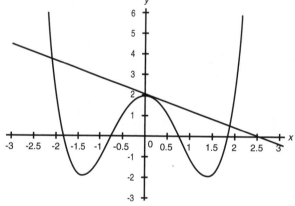

   The slope of the line is equal to which of the following:

A. the average rate of change of the function $f(x)$ over the interval $[-2.1, 1.9]$.

B. the average rate of change of the function $f(x)$ over the interval $[-2.1, .1]$.

C. the average rate of change of the function $f(x)$ over the interval $[.1, 1.9]$.

D. the instantaneous rate of change of the function $f(x)$ at the point $x = .1$.

E. All of the above

Answer E. The slope of the line is equal to all the following expressions:

$$\frac{f(1.9) - f(-2.1)}{1.9 - (-2.1)}, \frac{f(.1) - f(-2.1)}{.1 - (-2.1)}, \frac{f(1.9) - f(.1)}{1.9 - .1}, \text{ and } \lim_{x \to .1} \frac{f(x) - f(.1)}{x - .1}.$$

---

6. Suppose $f(x) = x^4 - 4x^2 + 2$. The instantaneous rate of change of the function at 2 is:

A. -16

B. $x^2(x + 2)$

C. 0

D. 16

E. the limit doesn't exist

Answer D. You need to compute the limit,

$$\lim_{x \to 2} \frac{f(x) - f(2)}{x - 2} = \lim_{x \to 2} \frac{x^4 - 4x^2 + 2 - (2^4 - 4 \times 2^2 + 2)}{x - 2}$$

$$= \lim_{x \to 2} \frac{x^4 - 4x^2}{x - 2} = \lim_{x \to 2} \frac{x^2(x^2 - 4)}{x - 2}$$

$$= \lim_{x \to 2} x^2(x + 2) = 16.$$

---

7. Suppose $f(x) = \sin(200\pi x)$.

The instantaneous rate of change of $f(x)$ at 0 is given by the formula $\lim_{x \to 0} \frac{f(x) - f(0)}{x - 0}$.

Approximating this instantaneous rate of change by taking $x = .01$ you get:

A. 10.94

B. -100

C. 0

D. 100

E. 7.54

Answer C. To approximate $\lim_{x \to 0} \frac{f(x) - f(0)}{x - 0}$, you can choose a very small value of x:

$$\frac{\sin(200\pi(.01)) - \sin(200\pi(0))}{.01 - 0} = \frac{0 - 0}{.01} = 0.$$

---

8. The cost of producing x units of a certain commodity is $C(x) = 1000 + 5.70x + .7x^2$. What is the average rate of change of C with respect to x when the production level is raised from $x = 100$ to $x = 120$?

A. 3,194

B. 859.7

C. 959.7

D. 159.7

E. 5.7

Answer   D. The average rate of change of C with respect to x, when the production level is raised from $x = 100$ to $x = 120$, is given by

$$\frac{C(120) - C(100)}{120 - 100}$$
$$= \frac{1,000 + 5.70(120) + .7(120^2) - 1,000 - 5.70(100) - .7(100^2)}{20}$$
$$= 159.7$$

9. The cost of producing x units of a certain commodity is $C(x) = 1,000 + 5.70x + .7x^2$. What is the average rate of change of C with respect to x when the production level is raised from $x = 100$ to $x = 101$?

A. 146.4

B. 145.7

C. 140.7

D. 5.7

E. 7.32

Answer   A. The average rate of change of C with respect to x, when the production level is raised from $x = 100$ to $x = 101$, is given by

$$\frac{C(101) - C(100)}{101 - 100}$$
$$= \frac{1,000 + 5.70(101) + .7(101^2) - 1,000 - 5.70(100) - .7(100^2)}{1}$$
$$= 146.4$$

10. The cost of producing x units of a certain commodity is $C(x) = 1,000 + 5.70x + .7x^2$. what is the instantaneous rate of change of C with respect to x when $x = 100$?

A. 146.4

B. 140.7

C. 146.33

D.   145.7

E.   Not defined

Answer   D. The instantaneous rate of change of C with respect to x, when $x = 100$ is given by

$$\lim_{x \to 100} \frac{C(x) - C(100)}{x - 100}$$
$$= \lim_{x \to 100} \frac{1,000 + 5.70x + .7x^2 - (1,000 + 5.70(100) + .7(100^2))}{x - 100}$$
$$= \lim_{x \to 100} \frac{.7x^2 + 5.7x - 7570}{x - 100}.$$

You can approximate this limit numerically by substituting values of x closer and closer to $x = 100$, noticing that they converge toward 145.7.

Alternatively (and more rigorously), you can try to evaluate the limit algebraically.

The numerator $.7x^2 + 5.7x - 7570$ factors (by quadratic formula or long division) into $(x - 100)(.7x + 75.7)$. Thus, the limit becomes

$$\lim_{x \to 100} \frac{.7x^2 + 5.7x - 7570}{x - 100}$$
$$= \lim_{x \to 100} \frac{(x - 100)(.7x + 75.7)}{x - 100}$$
$$= \lim_{x \to 100} (.7x + 75.7) = 145.7.$$

The displacement s (in meters) of a particle moving in a straight line is a function of time t (in seconds). The next two problems refer to the table below that gives the particle's displacement for various times:

t	$s(t)$
0	29
.1	27.3756
.2	25.7665
.3	24.1724
.4	22.5936
.5	21.03
.6	19.4816
.7	17.9484
.8	16.4304
.9	14.9276
1.0	13.44
1.1	11.9676
1.2	10.5104
1.3	9.0684
1.4	7.6416
1.5	6.23
1.6	4.8336
1.7	3.4524
1.8	2.0864
1.9	.7356

2.0   −.6

---

11. The average velocity of the particle between time $t = 1$ and time $t = 1.8$ seconds is:

A. 11.3536

B. 14.192

C. 12.7534

D. −11.3536

E. −14.192

Answer E. The average velocity of the particle between time $t = 1$ and time $t = 1.8$ seconds is given by:

$$\frac{s(1.8) - s(1.0)}{1.8 - 1.0} = \frac{2.0864 - 13.44}{.8} = -14.192.$$

---

12. Which of the following is the *best* approximation of the instantaneous velocity of the particle at time $t = 1$ second?

A. −14.8

B. −14.724

C. −15.0

D. −14.876

E. −15.236

Answer A. This one's a little tricky. You could approximate the instantaneous velocity of the particle at time $t = 1$ second by either the average velocity from $t = 1$ to $t = 1.1$ (you get -14.724) OR the average velocity from $t = .9$ to $t = 1.0$ (you get -14.876), but this particle appears to be increasing its speed around $t = 1$. (Remember, speed is the absolute value of velocity.)

The average velocity from $t = 1$ to $t = 1.1$ (AFTER $t = 1$) is when the particle is moving more quickly than the instantaneous velocity at $t = 1$. The average velocity from $t = .9$ to $t = 1.0$ (BEFORE $t = 1$) is when the particle is moving more slowly than the instantaneous velocity at $t = 1$.

A better approximation then would be to take the average of these, in which case you get

$$\frac{-14.876 + (-14.724)}{2} = -14.8.$$

13. A cylindrical tank holds 20,000 gallons of water, which can be drained from the bottom of the tank in 20 minutes. Below is a table of data giving the volume V of water (in gallons) in the tank at time t in minutes, where time $t = 0$ corresponds to the instant the tank starts to drain.

t	$V(t)$
0	20,000
1	18,050
2	16,200
3	14,450
4	12,800
5	11,250
6	9,800
7	8,450
8	7,200
9	6,050
10	5,000
11	4,050
12	3,200
13	2,450
14	1,800
15	1,250
16	800
17	450
18	200
19	50
20	0

The average rate of water draining from the tank is greatest during which of the following time intervals?

A. From time $t = 4$ to $t = 5$

B. From time $t = 4$ to $t = 7$

C. From time $t = 2$ to $t = 4$

D. From time $t = 14$ to $t = 20$

E. From time $t = 7$ to $t = 14$

Answer   C. From time $t = 2$ to time $t = 4$ the water drains at an average rate of $1,700$ gallons per minute, since

$$\frac{V(4) - V(2)}{4 - 2} = \frac{12,800 - 16,200}{4 - 2} = -1,700 \text{ gallons/minute.}$$

The other average rates are not as high.

14. A cylindrical tank holds 20,000 gallons of water, which can be drained from the bottom of the tank in 20 minutes. The function

$$V(t) = 20,000(1 - \frac{t}{20})^2$$

gives the volume V of water (in gallons) in the tank at time t in minutes, where time $t = 0$ corresponds to the instant the tank starts to drain. At time $t = 3$ minutes, the water is draining out of the tank at an instantaneous rate of:

A. $-2,200$ gallons/minute

B. $1,700$ gallons/minute

C. $2,200$ gallons/minute

D. $-1,800$ gallons/minute

E. $1,800$ gallons/minute

Answer   B. The instantaneous rate of change of V with respect to t at $t = 3$ is given by

$$\lim_{t \to 3} \frac{V(t) - V(3)}{t - 3} = \lim_{t \to 3} \frac{20,000(1 - \frac{t}{20})^2 - 14,450}{t - 3}$$

$$= \lim_{t \to 3} \frac{20,000(1 - \frac{t}{10} + \frac{t^2}{400}) - 14,450}{t - 3}$$

$$= \lim_{t \to 3} \frac{50t^2 - 2,000t + 5,550}{t - 3}$$

$$= \lim_{t \to 3} \frac{50(t - 3)(t - 37)}{t - 3} = \lim_{t \to 3} \frac{50(t - 37)}{1}$$

$$= 50(3 - 37) = -1,700.$$

So the water is draining out at the rate of 1,700 gallons/minute.

15. Which of the following is the definition of the derivative of a function $y = f(x)$ at a point $(x, f(x))$?

A. $\frac{dy}{dx} = f'(x) = \lim_{x \to 0} \frac{f(x + h) - f(x)}{h}$

B. $\frac{dy}{dx} = f'(x) = \lim_{h \to 0} \frac{f(x + h) - f(x)}{h}$

C. $\frac{dy}{dx} = f'(x) = \lim_{x \to 0} \frac{f(h) - f(x + h)}{h}$

D. $\frac{dy}{dx} = f'(x) = \lim_{h \to 0} \frac{f(h) - f(x + h)}{h}$

E. $\dfrac{dy}{dx} = f'(x) = \lim\limits_{x \to a} \dfrac{f(x + a) - f(x)}{a}$

Answer   B. $\dfrac{dy}{dx} = f'(x) = \lim\limits_{h \to 0} \dfrac{f(x + h) - f(x)}{h}$.

---

16.   Below is the graph of $y = f(x)$ (the curve), along with a straight line.

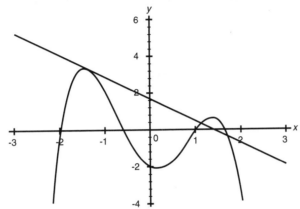

Which of the following statements is **not** true?

A.   The line is the tangent line to the graph of $y = f(x)$ at $x = -1.4$.

B.   The slope of the line is the derivative of the function $f(x)$ at $x = -1.4$.

C.   The line is a secant line.

D.   The derivative of the function $f(x)$ at $x = -1.4$ is positive.

E.   The line is not the tangent line to the graph of $y = f(x)$ at $x = 1.1$.

Answer   D. The derivative of the function $f(x)$ at $x = -1.4$ is the slope of the tangent line to the graph $y = f(x)$ at $x = -1.4$, and the slope of the line is obviously negative.

---

17.   The difference quotient for the function $f(x) = 3x^2 + 2x + 5$ is:

A.   4

B. $\lim\limits_{h \to 0} \dfrac{3(x + h)^2 + 2(x + h) + 5 - 3x^2 - 2x - 5}{h}$

C. $\dfrac{3(h)^2 + 2(h) + 5 - 3x^2 + 2x + 5}{h}$

D. $\lim\limits_{h \to 0} \dfrac{3(h)^2 + 2(h) + 5 - 3x^2 - 2x - 5}{h}$

E. $\dfrac{3(x + h)^2 + 2(x + h) + 5 - 3x^2 - 2x - 5}{h}$

Answer   E. The difference quotient is the slope of the secant line passing through points $(x, f(x))$ and $(x + h, f(x + h))$:

$$\frac{f(x + h) - f(x)}{(x + h) - x} = \frac{3(x + h)^2 + 2(x + h) + 5 - 3x^2 - 2x - 5}{h}.$$

---

18.   Which of the following guarantees that the function $f(x)$ does **not** have a derivative at the point $(x, f(x))$?

A.   The function is not continuous at the point $(x, f(x))$.

B.   The graph of the function has a horizontal tangent line at the point $(x, f(x))$.

C.   The graph of the function is "smooth" (no "corner") at the point $(x, f(x))$.

D. $\lim\limits_{h \to 0} \dfrac{f(x + h) - f(x)}{h} = x$

E. $\lim\limits_{h \to 0} f(x + h) = f(x)$

Answer   A. Differentiable functions are automatically continuous, so if a function is NOT continuous at a point, it CANNOT be differentiable there.

---

19.   The derivative of the function $f(x) = 3x^2 + 2x + 5$ at $x = -3$ is:

A. -16

B. 24

C. $\dfrac{3(x + h)^2 + 2(x + h) + 5 - 3x^2 - 2x - 5}{h}$

D. 14

E.   nonexistent

Answer   A. Using the definition of the derivative:

$$\lim_{h \to 0} \frac{f(x + h) - f(x)}{h}$$
$$= \lim_{h \to 0} \frac{3(-3 + h)^2 + 2(-3 + h) + 5 - 3(-3)^2 - 2 \times (-3) - 5}{h}$$
$$= \lim_{h \to 0} \frac{3(9 - 6h + h^2) - 6 + 2h - 3 \times 9 + 6}{h}$$
$$= \lim_{h \to 0} \frac{-18h + 2h + 3h^2}{h}$$
$$= \lim_{h \to 0} (-18 + 2 + 3h) = -16.$$

20. The distance (in feet) above the earth of a falling anvil is given by the equation

$$f(t) = -16t^2 + 1,000.$$

What is the anvil's instantaneous vertical velocity at time $t = 10$?

A. 160

B. −160

C. 240

D. −240

E. −320

Answer   E. Just start with the definition of the derivative at $t = 10$, simplify the difference quotient, and then take the limit:

$$\lim_{h \to 0} \frac{f(t+h) - f(t)}{h}$$
$$= \lim_{h \to 0} \frac{-16(10+h)^2 + 1,000 - (-16(10^2) + 1,000)}{h}$$
$$= \lim_{h \to 0} \frac{-16(100 + 20h + h^2) + 16 \times 100}{h}$$
$$= \lim_{h \to 0} \frac{-320h - 16h^2}{h}$$
$$= \lim_{h \to 0} (-320 - 16h) = -320.$$

The next two problems refer to the following situation:

A certain cylindrical tank holds 20,000 gallons of water, which can be drained from the bottom of the tank in 20 minutes. The volume V of water remaining in the tank after t minutes is given by the function $V(t) = 20,000(1 - \frac{t}{20})^2$, where V is in gallons, $0 \le t \le 20$ is in minutes, and $t = 0$ represents the instant the tank starts draining.

21. How fast is the water draining four and a half minutes after it begins?

A. 1,550 gallons/minute

B. 1,000 gallons/minute

C. 9,000 gallons/minute

D. 3,050 gallons/minute

E. 800 gallons/minute

Answer   A. At time $t = 4.5$, the tank is draining at the rate of 1,550 gallons/minute. We just need to find the derivative of the volume with respect to time, at time $t = 4.5$. On the calculator, enter **nDeriv**$(20,000(1 - \frac{x}{20})^2, x, 4.5)$, and press [**ENTER**]. You get $-1,550$. The negative sign just indicates that the volume is decreasing at the rate of $1,550$ gallons/minute.

---

22.   The average rate of change in volume of water in the tank from time $t = 0$ to $t = 20$ is

$$\frac{V(20) - V(0)}{20 - 0} = -1,000 \text{ gallons/minute.}$$

At what time t is the instantaneous rate of the water draining from the tank at 1,000 gallons/minute?

A. $t = 8$ minutes

B. $t = 9$ minutes

C. $t = 10$ minutes

D. $t = 11$ minutes

E. $t = 12$ minutes

Answer   C. Water is draining from the tank at the rate of 1,000 gallons/minute at time $t = 10$. Simply check the derivative at each of the values of t; you see that the derivative is $-1,000$ when $t = 10$. This means that at time $t = 10$, the water is draining at the rate of $1,000$ gallons/minute.

The next two problems refer to the following situation:

A tray of lasagna comes out of the oven at $200°F$ and is placed on a table where the surrounding room temperature is $70°F$. The temperature T (in $°F$) of the lasagna is given by the function $T(t) = e^{(4.86753-t)} + 70, 0 \le t$, where t is time (in hours) after taking the lasagna out of the oven.

---

23.   What is the rate of change in the temperature of the lasagna exactly 2 hours after taking it out of the oven?

A. -15.36 degrees/hour

B. -17.59 degrees/hour

C. -19.22 degrees/hour

D. -20.21 degrees/hour

E. -22.37 degrees/hour

Answer   B. The rate of change in the temperature of the lasagna exactly 2 hours after taking it out of the oven is -17.59 degrees/hour.

24. At which of the following times is the lasagna cooling SLOWEST?

A. 3 hours after being taken out of the oven

B. 5 hours after being taken out of the oven

C. 6 hours after being taken out of the oven

D. 7 hours after being taken out of the oven

E. 9 hours after being taken out of the oven

Answer   E. Check the derivatives of each choice above by using the **nDeriv** function on your calculator. You will see that the derivative is smallest (and negative) for $x = 9$. Never eat lasagna that's been sitting on a table for 9 hours!

(You can also see this by graphing $T(t)$ and noticing that it gets less and less steep as time goes by.)

25. Two particles are moving in straight lines. The displacement (in meters) of particle 1 is given by the function $S_1(t) = e^{4\cos(t)}$, where t is in seconds. The displacement (in meters) of particle 2 is given by the function $S_2(t) = -\dfrac{t^3}{3} - \dfrac{t^2}{2} + 2$, where t is in seconds.

Find the first positive time at which the particles have (approximately) the same velocity.

A. $t = 1.569$ seconds

B. $t = 0$ seconds

C. $t = 2.366$ seconds

D. $t = 0.588$ seconds

E. $t = 1.011$ seconds

Answer   A. The first positive time at which the two particles have (approximately) the same velocity is at $t = 1.569$ seconds. Check this with **nDeriv** on your calculator. It's easiest to look at the difference $y(t) = (s_1 - s_2)(t)$ to find where its derivative is approximately zero.

The next two problems refer to the following situation:

The population of a colony of bacteria is modeled by the function $P(x) = 50(e^{-x} - e^{-x^2}) + 10$, for $0 \le x$, where population P is in thousands, x is in hours, and $x = 0$ corresponds to the moment of introduction of a certain chemical into the colony's environment.

26. At which of the times below is the rate of population growth the greatest?

A. $x = 0.2$

B. $x = 0.8$

C. $x = 1.2$

D. $x = 2.2$

E. $x = 3$

Answer B. The time at which the rate of population growth is greatest is time $x = 0.8$ hours. You can check this using the **nDeriv** function on your calculator.

---

27. At which time(s) is the bacteria population growing at an instantaneous rate of 0, going from positive growth just prior, to negative growth right after?

A. $x = 0.39382$

B. $x = 0.58746$

C. $x = 1.71842$

D. $x = 2.23498$

E. $x = .39382$ and $x = 2.23498$

Answer C. Only at time $x = 1.71842$ is the bacteria population at an instant of 0 rate of population growth, going from positive growth just prior, to negative growth right after. Use **nDeriv** on your calculator to check that the derivative of P with respect to x at $x = 1.71842$ is approximately 0. You can graph the function $P(x)$ to see that just prior to this time the population is increasing, and right after this time the population is decreasing.

Free-Response Practice Questions

---

1. Here is the graph of $y = f(x)$:

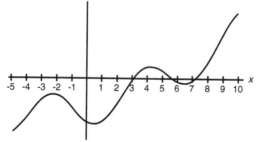

Arrange the following numbers in increasing order and explain your reasoning.

0    $f'(0)$ $f'(3)$ $f'(-1)$ $f'(4)$ $f'(-2)$

---

2. Sketch the graph of a function f for which
$f(-4) = -4$; $f(0) = -2$; $f(4) = 8$; $f'(-2) = 2$; $f'(0) = 0$; and $f'(2) = 4$.

3. Below is the graph of $y = f(x)$:

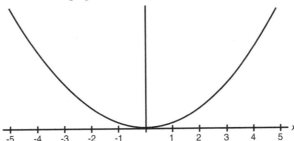

Notice that the scale is missing from the y-axis—it's not needed for this problem.

A. Arrange the following in increasing order, and explain your reasoning.

$$\frac{f(4) - f(2)}{2}; \quad \frac{f(-2) - f(-3)}{1}; \quad f'(-2); \quad f'(-3); \quad f'(2); \ f'(4)$$

B. For which values of x is $\dfrac{f(x) - f(-2)}{x - (-2)} > f'(-2)$?

For which values of x is $\dfrac{f(x) - f(-2)}{x - (-2)} < f'(-2)$?

For which values of x is $\dfrac{f(x) - f(-2)}{x - (-2)} = f'(-2)$?

In each case, explain your answers.

4. Suppose that the tangent line to $y = g(x)$ at $(-3, 5)$ passes through the point $(2, -4)$.

Find $g(-3)$, $g'(-3)$, and the equation of the tangent line to $y = g(x)$ at $(-3, 5)$.

5. Consider the function $f(x) = e^x$. You're interested in approximating $f'(x)$ for $x = 2, 4$, and 10 by using the slopes of secant lines over small intervals. Calculate each of the following (to four decimal places) for $x = 2$, $x = 4$, and $x = 10$ and then use this information to approximate $f'(2)$, $f'(4)$ and $f'(10)$.

$f(x)$ $\qquad\qquad\qquad$ $\dfrac{f(x + .1) - f(x)}{.1}$ $\qquad\qquad$ $\dfrac{f(x + .01) - f(x)}{.01}$

$\dfrac{f(x + .001) - f(x)}{.001}$ $\qquad$ $\dfrac{f(x + .0001) - f(x)}{.0001}$ $\qquad$ $f'(x) \approx?$

Compare the computed values of $f(x)$ and $f'(x)$, and make a supposition concerning $f(x)$ and $f'(x)$ that you think might be true for ALL x. (You don't have to prove the supposition.)

6. Suppose $f(x) = 3000 \sin(3000x)$. Your friend notices that

$$\frac{f(.1) - f(0)}{.1} = -29992.68, \qquad \frac{f(.01) - f(0)}{.01} = -296409.49$$

$$\frac{f(.001) - f(0)}{.001} = 423360.02, \qquad \frac{f(.0001) - f(0)}{.0001} = 8865606.20$$

Your friend then concludes that $f'(0)$ does not exist. Explain why your friend might conclude this, and discuss whether this is the correct conclusion.

7. Suppose $g(x) = \cot(x)$. Estimate $g'(\frac{\pi}{4})$ in two ways.

A. First, calculate the difference quotient for successively smaller values of h, and use your calculations to estimate the derivative.

B. Second, graph the function with your calculator. Zoom into a small enough window so that the slope of the curve at $x = \frac{\pi}{4}$ is obvious.

8. If $g(x) = 1 + x + x^2$, find $g'(1)$ and the equation of the tangent line to $y = g(x)$ at the point $(1, 3)$. (Evaluate the limit here; do not use approximation.)

9. Suppose $g(x) = \dfrac{x}{1 - 2x}$.

A. At what x values does $g'(x)$ not exist and why?

B. Find $g'(a)$ for arbitrary a (assuming $g'(a)$ exists). You must use the limit definition and show all work.

10. The cost of extracting x gallons of oil from a new oil well is given by the function $C = f(x)$, where C is in dollars.

A. What is the (practical) meaning of the mathematical expression $\dfrac{df}{dx}$, and what units does it have?

B. What is the practical meaning of the expression $f'(1200) = 3.50$?

Answers to Free-Response Practice

1. $f'(-1), \quad f'(0), \quad f'(-2), \quad 0, \quad f'(4), \quad f'(3)$

These are just slopes of the curve. The question is really asking you to look at the curve at the various locations and compare its steepness. The slopes at -2 and 0 are close; you may have had them switched. But if you draw the tangent lines with a ruler, you can see that the slope is greater at $x = -2$.

2. There are lots of correct answers, of course. Here's one answer:

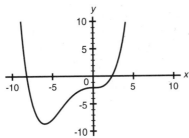

The important features to look for are whether your graph goes through the three named points $((-4, -4), (0, -2),$ and $(4, 8))$, and that it's flat at $x = 0$, increasing at $x = -2$, and increasing more steeply at $x = 2$. (Your graph definitely does not have to have a valley at $x = 6$.)

3.

A. $f'(-3); \dfrac{f(-2) - f(-3)}{1}; f'(-2); f'(2); \dfrac{f(4) - f(2)}{2};$ and $f'(4)$

For the first group, $f'(-3); \dfrac{f(-2) - f(-3)}{1}; f'(-2); f'(-3)$ and $f'(-2)$ are the slopes of the tangent lines to the graph at those points.

The value of $\dfrac{f(-2) - f(-3)}{1}$ is the slope of the secant connecting $x = -2$ and $x = -3$.

If you draw these lines with a ruler, you can see that $f'(-3) < \dfrac{f(-2) - f(-3)}{1} < f'(-2)$.

The same steps work for the second group, $f'(2); \dfrac{f(4) - f(2)}{2};$ and $f'(4)$.

B. $\dfrac{f(x) - f(-2)}{x - (-2)} > f'(-2)$ for all $x > -2$, since the slope of the secant connecting $(-2, f(-2))$ to $(x, f(x))$ is greater than the slope of the tangent to the graph at $(-2, f(-2))$.

$\dfrac{f(x) - f(-2)}{x - (-2)} < f'(-2)$ for all $x < -2$, since the slope of the secant connecting $(-2, f(-2))$ to $(x, f(x))$ is less than the slope of the tangent to the graph at $(-2, f(-2))$.

$\dfrac{f(x) - f(-2)}{x - (-2)} = f'(-2)$ for NO values of x, because that difference quotient is not defined at $x = 2$. The limit of the secant line and its slope at $x = -2$ are the tangent line and the derivative, but you can't plug $x = -2$ into the left-hand side to get equality. (Did we fool you? Be careful.)

---

4. $g(-3) = 5$ and $g'(-3) = -\frac{9}{5}$ (the slope of the line between $(-3, 5)$ and $(2, -4)$)).

The equation of the tangent line is $y = -\frac{9}{5}(x + 3) + 5$.

5.

For $x = 2$:

$f(x) = 7.3891$ $\dfrac{f(x + .1) - f(x)}{.1} = 7.7711$ $\dfrac{f(x + .01) - f(x)}{.01} = 7.4261$

$\dfrac{f(x + .001) - f(x)}{.001} = 7.3928$ $\dfrac{f(x + .0001) - f(x)}{.0001} = 7.3894$

Therefore, $f'(x) \approx 7.3894$

For $x = 4$:

$f(x) = 54.5982$ $\dfrac{f(x + .1) - f(x)}{.1} = 57.4214$ $\dfrac{f(x + .01) - f(x)}{.01} = 54.8721$

$\dfrac{f(x + .001) - f(x)}{.001} = 54.6255$ $\dfrac{f(x + .0001) - f(x)}{.0001} = 54.6009$

Therefore, $f'(x) \approx 54.6009$

For $x = 10$:

$f(x) = 22,026.4658$ $\dfrac{f(x + .1) - f(x)}{.1} = 23,165.4363$ $\dfrac{f(x + .01) - f(x)}{.01} = 22,136.9662$

$\dfrac{f(x + .001) - f(x)}{.001} = 22,037.4827$ $\dfrac{f(x + .0001) - f(x)}{.0001} = 22,027.5672$

Therefore, $f'(x) \approx 22,027.5672$

The best approximation should come from the difference quotient with the tiniest interval. You could use an even better approximation (for instance, use an increment of 0.000001), if you like.

If you compare the computed values of $f(x)$ and $f'(x)$, you'll notice that they're very similar. This suggests that $f(x) = f'(x)$ for all x.

6. The friend is looking at difference quotients with $x = 0$, for successively smaller values of h. The difference quotients don't seem to be converging, so it appears that the limit does not exist, and therefore the derivative does not exist.

Your friend's conclusion is wrong, however. The friend isn't taking small enough values of h. The period of this sine curve is $\dfrac{2\pi}{3,000} \approx 0.002$, so her h's cover too much variation in the graph. The curve has a well-defined tangent at $x = 0$ (with a very large slope!).

7.

A.   You should have difference quotients and calculations for at least three decreasing values of h to demonstrate that they appear to be approaching a limit. The difference quotient here is $\dfrac{\cot(\frac{\pi}{4} + h) - \cot(\frac{\pi}{4})}{h}$, and the derivative is in fact -2.

B.   Your graph should contain a small enough window so that the curve approximates a straight line. The scale should be such that the curve appears to has slope -2. For example, using this window, you could calculate the slope of this curve (which looks nearly straight) using the points $(.74, 1.1)$ and $(.84, .9)$ to get -2.

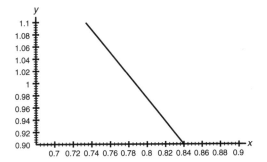

8.

$$g'(1) = \lim_{h \to 0} \frac{g(1 + h) - g(1)}{h}$$
$$= \lim_{h \to 0} \frac{1 + (1 + h) + (1 + h)^2 - 3}{h}$$
$$= \lim_{h \to 0} \frac{h^2 + 3h}{h}$$
$$= 3.$$

Then, the equation of the tangent line is $y = 3(x - 1) + 3 = 3x$.

9.

A.   At $x = \frac{1}{2}$, since the function has a vertical asymptote there.

B. $g'(a) = \dfrac{1}{(1 - 2a)^2}.$

Steps:

$$g'(a) = \lim_{h \to 0} \frac{g(a+h) - g(a)}{h}$$

$$= \lim_{h \to 0} \left\{ \left(\frac{a+h}{1 - 2(a+h)} - \frac{a}{1 - 2a} \right) \Big/ h \right\}$$

$$= \lim_{h \to 0} \left\{ \left(\frac{(1 - 2a)(a+h)}{(1 - 2a)(1 - 2(a+h))} - \frac{a(1 - 2(a+h))}{(1 - 2a)(1 - 2(a+h))} \right) \Big/ h \right\}$$

$$= \lim_{h \to 0} \left\{ \left(\frac{a + h - 2a^2 - 2ah - a + 2a^2 + 2ah}{(1 - 2a)(1 - 2(a+h))} \right) \Big/ h \right\}$$

$$= \lim_{h \to 0} \left\{ \left(\frac{h}{(1 - 2a)(1 - 2(a+h))} \right) \Big/ h \right\}$$

$$= \lim_{h \to 0} \frac{1}{(1 - 2a)(1 - 2(a+h))}$$

$$= \frac{1}{(1 - 2a)^2}.$$

10.

A. $\frac{df}{dx}$ represents the instantaneous rate of change of cost of extracting x gallons of oil from the well with respect to x. More practically, $\frac{df}{dx}$ tells how fast the cost will change as the amount of oil extracted increases. The units are dollars per gallon.

B. With the level of oil extraction at $1{,}200$ gallons, the rate of cost increase for extracting more oil is $\$3.50$ per gallon.

B. Computing and Interpreting Derivatives

▶Objective 1   Determine the derivative of basic power functions and polynomials.

Example   Find $\frac{dy}{dx}$ when $y = f(x) = \sqrt[4]{x}$.

Answer

$$y = \sqrt[4]{x} = x^{\frac{1}{4}}, \text{ so } \frac{dy}{dx} = \frac{1}{4} x^{-\frac{3}{4}}.$$

▶Objective 2   Determine derivatives of functions defined as a sum of other functions.

Example   Find $g'(x)$ when $g(x) = \sin x + x^4 - 6x + 4$.

Tip   Don't forget that you're really finding the derivative given by

$$g'(x) = \lim_{h \to 0} \frac{g(x+h) - g(x)}{h}$$

when you do this. It's easier to use the shortcuts than it is to calculate the limit directly!

Answer $g'(x) = \cos x + 4x^3 - 6$.

►Objective 3   Determine derivatives of functions defined as a product of other functions.

Example   If $h(\theta) = \theta^2 \sin \theta$, find $h'(\theta)$.

Tip   Memorize the product rule: $(uv)' = u'v + uv'$.

Answer $h'(\theta) = 2\theta \sin \theta + \theta^2 \cos \theta$.

►Objective 4   Determine derivatives of functions defined as a quotient of other functions.

Example   If $h(\theta) = \dfrac{\theta^2}{\sin \theta}$, find $h'(\theta)$.

Tip   Memorize the quotient rule: $\left(\dfrac{u}{v}\right)' = \dfrac{vu' - v'u}{v^2}$.

Answer $h'(\theta) = \dfrac{2\theta \sin \theta - \theta^2 \cos \theta}{(\sin \theta)^2} = \dfrac{2\theta \sin \theta - \theta^2 \cos \theta}{\sin^2 \theta}$.

►Objective 5   Know the product and quotient rules from memory.

> **Product rule** : $(uv)' = u'v + uv'$.

> **Quotient rule** : $\left(\dfrac{u}{v}\right)' = \dfrac{vu' - v'u}{v^2}$.

▶Objective 6   Know the derivatives of each of the six basic trigonometric functions:

$$\sin, \quad \cos, \quad \tan, \quad \csc, \quad \sec, \quad \cot$$

While you might opt to simply memorize all six of these, you could also determine the derivatives whenever you need them by simply using the derivative of sin and cosine combined with the quotient rule.

$$\frac{d}{dx}\sin x = \cos x \qquad \frac{d}{dx}\cos x = -\sin x \qquad \frac{d}{dx}\tan x = \sec^2 x$$

$$\frac{d}{dx}\csc x = -\csc x \cot x \qquad \frac{d}{dx}\sec x = \sec x \tan x \qquad \frac{d}{dx}\cot x = -\csc^2 x$$

▶Objective 7   Determine derivatives that may require a combination of the sum, product, and quotient rules for functions that are algebraic, trigonometric, or combinations of both.

Example   If $y = x^2 - \dfrac{\sqrt{x}\tan x}{x^2 - 3x + 1}$, find $\dfrac{dy}{dx}$.

Tip   Don't worry about simplifying answers like these on the AP Exam.

Answer

$$\frac{dy}{dx} = 2x - \frac{(x^2 - 3x + 1)\left(\frac{1}{2\sqrt{x}}\tan x + \sqrt{x}\sec^2 x\right) - (2x - 3)(\sqrt{x}\tan x)}{(x^2 - 3x + 1)^2}.$$

▶Objective 8   Use the derivative to solve problems where calculating the slope of a function will help you to determine the solution.

Example   If the profit from selling x number of items each month is given by

$$P(x) = 40\sqrt{x} - 0.4x - 400,$$

find the rate of change of P with respect to x if $x = 700, 1,400, 2,000,$ and $3,500$.

Tip   Remember that the derivative is just the rate of change of a function.

Answer

$$P'(x) = \frac{20}{\sqrt{x}} - 0.4$$

$$P'(700) = \frac{20}{\sqrt{700}} - 0.4 = 0.35593$$

$$P'(1400) = \frac{20}{\sqrt{1,400}} - 0.4 = 0.13452$$

$$P'(2000) = \frac{20}{\sqrt{2,000}} - 0.4 = 0.04721$$

$$P'(3500) = \frac{20}{\sqrt{3,500}} - 0.4 = -0.06194$$

Note that the profit per item sold decreases (and eventually goes negative), because the company must lower the price in order to sell that many items.

▶Objective 9   Looking at a graph of a function, predict features about the graph of its derivative.

Example   Given the following graph of a differentiable function $g(x)$:

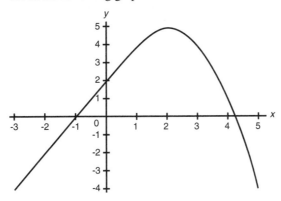

A.   What is the value of $g'(-1)$?

B.   If $g'(c) = 0$, what is the value of c?

C.   Estimate the value of $g'(4)$

Tip   Recognize that the derivative represents the slope of the original function for any x value.

Answers

A. $g'(-1) = slope = \dfrac{\Delta y}{\Delta x} = 2.$

B.   The curve is flat and thus has a slope of 0 when $c = 2$.

C. $g'(4) = slope = \dfrac{\Delta y}{\Delta x} \approx \dfrac{4 - (-4)}{3 - 5} = -4.$

(We used a secant slope to estimate the slope of the tangent line—you could also draw a tangent line and compute its slope. Many different answers could be correct. Be sure to show all your work.)

▶Objective 10   Looking at a graph of a derivative, predict features about the graph of the original function.

Example   The following graph is of f', the **derivative of** f:

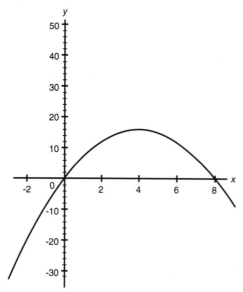

On the interval shown, answer the following questions about the graph of f:

A. At what x values is the graph of $y = f(x)$ horizontal?

B. What is the slope of $y = f(x)$ at $x = 2$?

C. At what x value is $y = f(x)$ increasing the most?

Tip   Read the question carefully, so that you notice whether you are given a graph of the derivative or of the original function.

Answer

A. The graph of $y = f(x)$ is horizontal when $f'(x) = 0$, which is when $x = 0$ and 8.

B. From the graph, $f'(2) = 12$, which is the slope of $y = f(x)$ at $x = 2$.

C. $y = f(x)$ is increasing the most when $f'(x)$ has the largest value. In this case that occurs at $x = 4$.

▶Objective 11   Determine when a function is differentiable.

Example   State all of the x values where the following function is not differentiable.

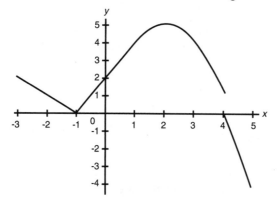

Tip Points of discontinuity are always points of nondifferentiability.

Answer The function isn't differentiable at $x = -1$, since there is a corner there.

It's also not differentiable at $x = 4$, because the function is discontinuous there.

►Objective 12 Determine the derivatives of piecewise functions.

Example Find the derivative of

$$g(x) = \begin{cases} \sin(\pi x) + 1 & \text{if } x \leq 1 \\ x^2 & \text{if } x > 1. \end{cases}$$

Tip In order for the derivative to exist at the junction, the function must be continuous and the values of the derivative must match at that value as well.

Answer

$$g'(x) = \begin{cases} \pi \cos(\pi x) & \text{if } x < 1 \\ \text{undefined} & \text{if } x = 1. \\ 2x & \text{if } 1 < x. \end{cases}$$

Although the function is continuous at $x = 1$, the derivative still doesn't exist at $x = 1$ because the slope as you approach $x = 1$ from the right doesn't equal the slope as you approach $x = 1$ from the left. Be careful with your $<$'s and $\leq$'s

---

►Objective 13 Use Rolle's Theorem and the Mean Value Theorem to relate average rate of change to instantaneous rate of change for a differentiable function over a closed interval.

Rolle's Theorem: Suppose f is a function, differentiable for all points in the open interval (a, b), and continuous for all points in the closed interval $[a, b]$. If $f(a) = f(b) = 0$, then there is at least one number c between a and b for which $f'(c) = 0$.

The Mean Value Theorem: Suppose f is a function, differentiable for all points in the open interval (a, b), and continuous for all points in the closed interval $[a, b]$. Then there is at least one point c between a and b for which $f'(c) = \dfrac{f(b) - f(a)}{b - a}$.

Example 1 Find the value of x that satisfies the conclusion of the Mean Value Theorem for

$f(x) = x^2 \sin x$ on the domain $[0, \frac{\pi}{2}]$ (using a calculator).

Answer

$$f(0) = 0$$

$$f\left(\frac{\pi}{2}\right) = \left(\frac{\pi}{2}\right)^2 \sin\left(\frac{\pi}{2}\right) = \frac{1}{4}\pi^2$$

$$\frac{f\left(\frac{\pi}{2}\right) - f(0)}{\frac{\pi}{2} - 0} = \frac{\frac{1}{4}\pi^2 - 0}{\frac{\pi}{2} - 0} = \frac{1}{2}\pi$$

$$f'(x) = 2x\sin x + x^2\cos x$$

$$f'(x) = 2x\sin x + x^2\cos x = \frac{1}{2}\pi$$

$$2x\sin x + x^2\cos x - \frac{1}{2}\pi = 0$$

$$x \approx .7928$$

using the calculator.

Example 2   The position (feet traveled) of a car is given by the equation $s(t) = 4t^2 + 4t$. Find the time when the car is going the same speed as its average speed over the interval 0 to 10 seconds.

Tip   Think of this graphically as a tangent line that is parallel to a secant line.

Answer   The average speed over the interval is given by

$$m = \frac{s(10) - s(0)}{10 - 0} = \frac{440}{10} = 44\,\frac{\text{ft}}{\text{sec}}.$$

The velocity at any time t is given by $v(t) = s'(t) = 8t + 4$.

This is equal to 44 when $8t + 4 = 44$, so the solution is: $t = 5$ seconds.

Multiple-Choice Exercises

1. If a function $f(x)$ gives temperature in degrees as a function of time in minutes, then the function $f'(x)$ gives:

A.   degrees as a function of time in minutes.

B.   time in minutes as a function of temperature in degrees.

C.   minutes per degree as a function of time in minutes.

D.   degrees per minute as a function of time in minutes.

E.   degrees per minute as a function of degrees.

Answer   D. To see this, note that the function $f'(x)$ represents the rate of change of $f(x)$ (which is temperature) with respect to x (which is time). So the units of $f'(x)$ will be degrees per minute. Also, $f'(x)$ is still a function of time in minutes, since the independent variable x is unchanged.

---

2.   Suppose you have a differentiable function $f(x)$, and the function $f'(x)$ gives the rate of change in an average person's height (in cm/year) as a function of age (in years). Then $f(x)$ gives:

A.   an average person's age (in years) as a function of height (in cm).

B.   an average person's height (in cm) as a function of height (in cm).

C.   an average person's height (in cm) as a function of age (in years).

D.   the rate of change of an average person's height (in cm/year) as a function of height (in cm).

E.   the rate of change of an average person's height (in cm/year) as a function of age (in years).

Answer   C. To confirm that this is true, just go back the other way: If $f(x)$ gives an average person's height (in cm) as a function of age (in years), then $f'(x)$ gives the rate of change in an average person's height (in cm/year) as a function of age (in years), which is what you want!

The next two problems refer to the following graph of $y = f(x)$.

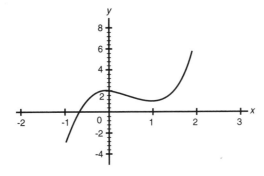

---

3.   From this graph, determine which value(s) of x give $f'(x) = 0$.

A. $x = -.65$

B. $x = 2$

C. $x = 0, x = 1$

D. $x = 0$

E. $x = -.65, x = 0, x = 1$

Answer C. At $x = 0$, the slope of the curve $f(x)$ is 0, since the tangent line to the graph of $y = f(x)$ at $x = 0$ is horizontal. But the slope of the curve IS the derivative of the function, so $f'(0) = 0$. The same holds for $x = 1$.

---

4. On which domain is $f'(x)$ a positive function?

A. $(0, 1)$

B. $[-1, 0) \cup (1, 2]$

C. $(-.65, 2]$

D. $(0, 2]$

E. $[-2, -.65] \cup (-.65, 2]$

Answer B. When x is in this set, the slope of the above curve is positive, which means that $f'(x)$ is positive.

---

5. Which of the following rules is incorrect?

A. $\dfrac{d}{dx}(f(x) + h(x)) = f'(x) + h'(x)$

B. $\dfrac{d}{dx}(ax^n) = ax^{n-1}$

C. $\dfrac{d}{dx}(a) = 0$

D. $\dfrac{d}{dx}(a \sin x) = a \cos x$

E. $\dfrac{d}{dx}(a \cos x) = -a \sin x$

The answer is: $\dfrac{d}{dx}(ax^n) = ax^{n-1}$. The correct rule is $\dfrac{d}{dx}(ax^n) = anx^{n-1}$.

---

6. Suppose $g(x) = 4x^5 - 7x^3 + 2x - 4$. Then $g'(x) =$

A. $4x^4 - 7x^2 + 2$

B. $20x^4 - 21x^2 + 2x - 4$

C. $20x^5 - 21x^3 + 2x$

D. $4x^4 - x^2 + 2x - 4$

E. $20x^4 - 21x^2 + 2$

Answer   E. Using the rule for computing derivatives of power functions, you get

$$\frac{d}{dx}(4x^5 - 7x^3 + 2x - 4)$$
$$= \frac{d}{dx}(4x^5) + \frac{d}{dx}(-7x^3) + \frac{d}{dx}(2x) + \frac{d}{dx}(-4)$$
$$= 20x^4 - 21x^2 + 2x^0 + 0 = 20x^4 - 21x^2 + 2.$$

---

7. Find $\frac{d}{dx}(7x^{-\frac{1}{3}} + \sqrt[4]{x})$

A. $\frac{d}{dx}(7x^{-\frac{1}{3}} + \sqrt[4]{x}) = -\frac{7}{3}x^{-\frac{4}{3}} + \frac{1}{4}x^{-\frac{3}{4}}$

B. $\frac{d}{dx}(7x^{-\frac{1}{3}} + \sqrt[4]{x}) = -\frac{7}{3}x^{-\frac{2}{3}} + \frac{1}{4}x^{-\frac{3}{4}}$

C. $\frac{d}{dx}(7x^{-\frac{1}{3}} + \sqrt[4]{x}) = -7x^{-\frac{4}{3}} + 4x^{-\frac{3}{4}}$

D. $\frac{d}{dx}(7x^{-\frac{1}{3}} + \sqrt[4]{x}) = -7x^{-\frac{2}{3}} + 4x^{-\frac{3}{4}}$

E. $\frac{d}{dx}(7x^{-\frac{1}{3}} + \sqrt[4]{x}) = -\frac{7}{3}x^{-\frac{4}{3}} + \frac{1}{4}x^{-\frac{1}{4}}$

Answer   A. This follows from the rules for computing derivatives of power functions:

$$\frac{d}{dx}\left(7x^{-\frac{1}{3}} + \sqrt[4]{x}\right) = \frac{d}{dx}\left(7x^{-\frac{1}{3}}\right) + \frac{d}{dx}(\sqrt[4]{x})$$
$$= -\frac{1}{3} \times 7x^{-\frac{1}{3}-1} + \frac{d}{dx}\left(x^{\frac{1}{4}}\right) = -\frac{7}{3}x^{-\frac{4}{3}} + \frac{1}{4}x^{\frac{1}{4}-1}$$
$$= -\frac{7}{3}x^{-\frac{4}{3}} + \frac{1}{4}x^{-\frac{3}{4}}.$$

---

8. Suppose $g(x) = 2\sin x - 3\cos x$. Then $g'(x) =$

A. $2\sin x + 3\cos x$

B. $2\cos x - 3\sin x$

C. $2\cos x + 3\sin x$

D. $-2\cos x + 3\sin x$

E. $-2\cos x - 3\sin x$

Answer   C. Use the rules for derivatives $\sin x$ and $\cos x$:

$$g'(x) = \frac{d}{dx}(2\sin x - 3\cos x)$$
$$= \frac{d}{dx}(2\sin x) + \frac{d}{dx}(-3\cos x)$$
$$= 2\cos x - (-3)\sin x = 2\cos x + 3\sin x.$$

9. Let $f(x) = 2\sin x$. Then $\dfrac{d}{dx}(f'(x)) =$

A. $2\cos x$

B. $2\sin x$

C. $-2\cos x$

D. $-2\sin x$

E. None of the above

Answer   D. First find $f'(x) = 2\cos x$, so that $\dfrac{d}{dx}(f'(x)) = \dfrac{d}{dx}(2\cos x) = -2\sin x$.

10. Below is the graph of $y = g(x)$.

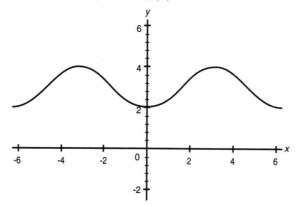

Which of the following correctly describes $g'(x)$?

A. $g'(x) = \sin x + 3$

B. $g'(x) = -\sin x + 3$

C. $g'(x) = -\cos x + 3$

D. $g'(x) = -\sin x$

E. $g'(x) = \sin x$

Answer   E. Notice that $g(x) = -\cos x + 3$. Then just take its derivative:

$$g'(x) = \frac{d}{dx}(-\cos x + 3) = \sin x + 0 = \sin x.$$

11. Below is the graph of $y = f'(x)$.

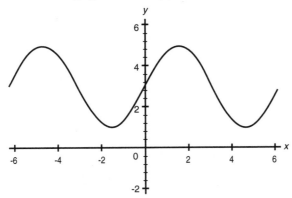

Which of the following could be the function $f(x)$?

A. $f(x) = 2\sin x + 3$

B. $f(x) = 2\cos x + 3$

C. $f(x) = -2\cos x + 3$

D. $f(x) = -2\cos x + 3x$

E. $f(x) = 2\cos x + 3x$

Answer D. One way to find this is to first determine what $f'(x)$ is. From the graph above, it's clear that $f'(x) = 2\sin x + 3$. So you need a function $f(x)$ that makes this true. It turns out that if you take $f(x) = -2\cos x + 3x$, then $f'(x) = 2\sin x + 3$.

Derivative Practice

Find the derivative of each of the following functions.

A. x^7

B. $3x^{-3}$

C. $4x^3 - 11x^{-2}$

D. $12x^{\frac{2}{3}} + 7$

E. $5x^{-\frac{3}{4}} + 3x$

F. $4\sqrt{x} - \frac{1}{5}x^{-\frac{4}{5}}$

G. $8\sqrt[3]{x} - \frac{3}{2}x^4 + 3$

H. $(2x + 3)^2$

I. $(2x + 3)^3$

J. $(x^{\frac{1}{2}} + 3x^{\frac{2}{3}})^2$

K. $(8x)^{\frac{1}{3}} + (3x)^2$

L. $4\cos x$

M. $-\frac{3}{5}\cos x - 2\sin x$

N. $\frac{7}{3}\sin x + x^{\frac{1}{3}} - 4x$

O. $(\frac{1}{2}x^2 + 15)^2 - 3\cos x$

P. $\frac{4}{3}x(12x + \frac{7}{x} + x^{-\frac{4}{3}}) - 12\sin x + 2\cos x$

Q. $x^2(4x - 2x^{-1})(x^{\frac{2}{3}}) + 4\cos x - 3$

Answers to Derivative Practice

A. $\frac{d}{dx}(x^7) = 7x^6$

B. $\frac{d}{dx}(3x^{-3}) = -9x^{-4}$

C. $\frac{d}{dx}(4x^3 - 11x^{-2}) = 12x^2 + 22x^{-3}$

D. $\frac{d}{dx}(12x^{\frac{2}{3}} + 7) = 8x^{-\frac{1}{3}}$

E. $\frac{d}{dx}(5x^{-\frac{3}{4}} + 3x) = -\frac{15}{4}x^{-\frac{7}{4}} + 3$

F. $\frac{d}{dx}(4\sqrt{x} - \frac{1}{5}x^{-\frac{4}{5}}) = 2x^{-\frac{1}{2}} + \frac{4}{25}x^{-\frac{9}{5}}$

G. $\frac{d}{dx}(8\sqrt[3]{x} - \frac{3}{2}x^4 + 3) = \frac{8}{3}x^{-\frac{2}{3}} - 6x^3$

H. $\frac{d}{dx}[(2x + 3)^2] = \frac{d}{dx}(4x^2 + 12x + 9) = 8x + 12$

I. $\frac{d}{dx}[(2x + 3)^3] = \frac{d}{dx}(8x^3 + 36x^2 + 54x + 27) = 24x^2 + 72x + 54$

J. $\frac{d}{dx}[(x^{\frac{1}{2}} + 3x^{\frac{2}{3}})^2] = \frac{d}{dx}(x + 6x^{\frac{7}{6}} + 9x^{\frac{4}{3}}) = 1 + 7x^{\frac{1}{6}} + 12x^{\frac{1}{3}}$

K. $\frac{d}{dx}[(8x)^{\frac{1}{3}} + (3x)^2] = \frac{d}{dx}(2x^{\frac{1}{3}} + 9x^2) = \frac{2}{3}x^{-\frac{2}{3}} + 18x$

L. $\frac{d}{dx}(4\cos x) = -4\sin x$

M. $\frac{d}{dx}(-\frac{3}{5}\cos x - 2\sin x) = \frac{3}{5}\sin x - 2\cos x$

N. $\frac{d}{dx}(\frac{7}{3}\sin x + x^{\frac{1}{3}} - 4x) = \frac{7}{3}\cos x + \frac{1}{3}x^{-\frac{2}{3}} - 4$

O. $\frac{d}{dx}[(\frac{1}{2}x^2 + 15)^2 - 3\cos x] = \frac{d}{dx}(\frac{1}{4}x^4 + 15x^2 + 225 - 3\cos x) = x^3 + 30x + 3\sin x$

P. $\frac{d}{dx}[\frac{4}{3}x(12x + \frac{7}{x} + x^{-\frac{4}{3}}) - 12\sin x + 2\cos x] = \frac{d}{dx}(16x^2 + \frac{28}{3} + \frac{4}{3}x^{-\frac{1}{3}} - 12\sin x + 2\cos x)$
$= 32x - \frac{4}{9}x^{-\frac{4}{3}} - 12\cos x - 2\sin x$

Q. $\frac{d}{dx}[x^2(4x - 2x^{-1})(x^{\frac{2}{3}}) + 4\cos x - 3] = \frac{d}{dx}[(4x^{\frac{11}{3}} - 2x^{\frac{5}{3}}) + 4\cos x - 3] = \frac{44}{3}x^{\frac{8}{3}} - \frac{10}{3}x^{\frac{2}{3}} - 4\sin x$

Practice Free-Response Questions

---

1. Show that the rate of change of the volume of a sphere with respect to its radius is equal to the surface area of the sphere.

Answer First, $V = \frac{4}{3}\pi r^3$.

Next, the rate of change of the volume of a sphere with respect to its radius is

$$\frac{dV}{dr} = \frac{d}{dr}(\frac{4}{3}\pi r^3) = 4\pi r^2 = \text{surface area of the sphere.}$$

---

2. Suppose that an oil spill is increasing in such a way that the surface of the spill is always perfectly circular. Find the rate at which the spill's surface area is increasing with respect to the spill's radius. Show that this rate is equal to the circumference of the spill.

Answer First, $A = \pi r^2$.

Then, the rate at which the spill's surface area is increasing with respect to the spill's radius is $\frac{dA}{dr} = \frac{d}{dr}(\pi r^2) = 2\pi r = \text{circumference of the circular spill.}$

---

3. Anna is selling kazoos. Her marketing team finds that if she charges $2.00 per kazoo, she'll sell 200 kazoos, and for each 10 cents she raises the price, she'll sell 12 fewer kazoos. Similarly, for each 10 cents she lowers the price, she'll sell 12 more kazoos.

A. Come up with an equation that represents the number of kazoos Anna will sell as a function of price x.

Answer $N = -120(x - 2) + 200 = -120x + 440$.

There is a linear relationship between kazoos sold and price. You have a point $(2, 200)$ and a slope ($m = -\frac{12}{.10} = -120$), so you can easily write the equation of the line.

B. Using your answer from part a, derive a formula that represents Anna's kazoo revenue R as a function of selling price x.

Answer Revenue is just selling price times the number of things sold. So, using part a, you get $R(x) = x(-120x + 440) = -120x^2 + 440x$.

C. Using your revenue function from part b, find the rate of change in revenue R with respect to selling price x.

What is the rate of change in revenue with respect to selling price when the price is $1.50 per kazoo? $2.00 per kazoo? $2.50 per kazoo? What do these numbers mean?

In order to make the most money, should Anna charge more than $1.50 per kazoo? Why or why not? In order to make the most money, should Anna charge more than $2.00 per kazoo? Why or why not?

Answer $\dfrac{dR}{dx} = -240x + 440.$ $R'(1.50) = \$80.$ $R'(2.00) = -\$40.$

$R'(2.50) = -\$160.$

These numbers tell you how the revenue changes relative to an adjustment of the selling price. Anna should charge more than $1.50 per kazoo, because the rate of change in revenue is positive at that point. (We assume here that Anna is trying to make the most money.) Anna should not charge more than $2.00 per kazoo, because the rate of change in revenue is negative at that point.

D. Suppose that it costs Anna $0.85 per kazoo (she buys them from a kazoo wholesaler), she only buys as many as she'll sell, and she has to pay a fixed cost of $25.00 no matter what (to pay for her kazoo booth and fancy neon "Kazoos Here" sign).

Come up with an equation that gives Anna's **profit** P as a function of kazoo selling price x. (Remember, profit is just revenue minus costs, so you need a cost function first.)

Answer The cost function should be $c(x) = 25 + .85 \,(\text{number of kazoos}) = 25 + .85(-120x + 440)$. Thus, profit is $P(x) = -120x^2 + 542x - 399.$

E. What is the rate of change in profit P with respect to selling price x?

What is the rate of change in the profit with respect to selling price when the selling price is $1.50 per kazoo? $2.00 per kazoo? $2.50 per kazoo?

Should Anna charge more than $2.00 per kazoo? Why or why not?

Should Anna charge more than $2.50 per kazoo? Why or why not?

Answer $P'(x) = -240x + 542$ $P'(1.50) = 182.$ $P'(2.00) = 62.$ $P'(2.50) = -58.$

Anna should charge more than $2.00 per kazoo, since the rate of change of profit with respect to selling price is positive there.

Anna should not charge more than $2.50 per kazoo, since the rate of change in profit with respect to selling price is negative there.

Multiple-Choice Practice on Derivatives

1. Suppose $y = x^2 - \dfrac{1}{x^2} + 3\pi$. Then $\dfrac{dy}{dx} =$

A. $2x + \dfrac{2}{x}$

B. $2x + \dfrac{2}{x^3}$

C. $2x - \dfrac{1}{x^2} + \dfrac{2}{x^3}$

D. $2x - \dfrac{1}{x^2} + \dfrac{2}{x^3} + 3\pi$

E. $2x - \dfrac{2}{x^3}$

Answer B. Using the rule for sums and differences, you can break things up into

$$\frac{dy}{dx} = \frac{d}{dx}\left(x^2 - \frac{1}{x^2} + 3\pi\right) = \frac{d}{dx}(x^2) - \frac{d}{dx}\left(\frac{1}{x^2}\right) + \frac{d}{dx}(3\pi).$$

Then notice that $\dfrac{d}{dx}(x^2) = 2x$, $-\dfrac{d}{dx}\left(\dfrac{1}{x^2}\right) = -\dfrac{d}{dx}(x^{-2}) = -(-2x^{-3}) = \dfrac{2}{x^3}$, and

$\dfrac{d}{dx}(3\pi) = 0$.

So, $\dfrac{dy}{dx} = \dfrac{d}{dx}(x^2) - \dfrac{d}{dx}\left(\dfrac{1}{x^2}\right) + \dfrac{d}{dx}(3\pi) = 2x + \dfrac{2}{x^3}$.

---

2. Suppose $g(x) = x^2 + x^{-2} + 3\sin x + \tan x + \sqrt{2}$. Then $g'(x) =$

A. $2x - 2x^{-1} + 3\cos x + \sec x$

B. $2x - 2x^{-1} + 3\cos x + \sec^2 x$

C. $2x - 2x^{-3} + 3\cos x + \sec x$

D. $2x - 2x^{-3} + 3\cos x + \sec^2 x$

E. $2x - 2x^{-3} - 3\cos x + \sec^2 x$

Answer D. You just use the fact that the derivative of a sum is the sum of the derivatives:

$$g'(x) = \frac{d}{dx}(x^2 + x^{-2} + 3\sin x + \tan x + \sqrt{2})$$
$$= \frac{d}{dx}(x^2) + \frac{d}{dx}(x^{-2}) + \frac{d}{dx}(3\sin x) + \frac{d}{dx}(\tan x) + \frac{d}{dx}(\sqrt{2})$$
$$= 2x - 2x^{-3} + 3\cos x + \sec^2 x + 0.$$

3. Suppose $v = 3t^3 \cos t$. Then $v' =$

A. $-9t^2 \sin t$

B. $9t^2 \sin t$

C. $9t^2 \cos t - 3t^3 \sin t$

D. $9t^2 \cos t + 3t^3 \sin t$

E. $9t^2 \cos t + 9t^2 \sin t$

Answer   C. Use the product rule here, which gives:

$$v' = \frac{d}{dt}(3t^3 \cos t) = \frac{d}{dt}(3t^3) \times \cos t + 3t^3 \times \frac{d}{dt}(\cos t).$$

After simplifying, we get $v' = 9t^2 \cos t + 3t^3(-\sin t) = 9t^2 \cos t - 3t^3 \sin t$.

4. Suppose $f(x) = x \tan x + x^2 \sin x$. Then $f'(x) =$

A. $x \sec^2 x + \tan x + x^2 \cos x + 2x \sin x$

B. $\tan x + x^2 \cos x + 2x \sin x$

C. $x \tan x + x^2 \cos x + 2x \sin x$

D. $x \sec^2 x + x^2 \cos x$

E. $x \sec^2 x + 2x \cos x$

Answer   A. Here you use the rules for differentiating sums and products. You get

$$\begin{aligned} f'(x) &= \frac{d}{dx}(x \tan x + x^2 \sin x) \\ &= \frac{d}{dx}(x \tan x) + \frac{d}{dx}(x^2 \sin x) \\ &= x \times \frac{d}{dx}(\tan x) + \frac{d}{dx}(x) \times \tan x + x^2 \times \frac{d}{dx}(\sin x) + \frac{d}{dx}(x^2) \times \sin x \\ &= x \sec^2 x + \tan x + x^2 \cos x + 2x \sin x. \end{aligned}$$

5. $\frac{d}{dx}\left(\frac{3-x}{x^3}\right) =$

A. $\frac{-1}{3x^2}$

B. $\frac{2}{x^3} - \frac{9}{x^4}$

C. $-\dfrac{4}{x^3} - \dfrac{9}{x^4}$

D. $3x^{-3} - x^{-2}$

E. $2x^{-3}$

Answer B. You can derive this using the quotient rule, which states $\left(\dfrac{u}{v}\right)' = \dfrac{vu' - uv'}{v^2}$:

$$\frac{d}{dx}\left(\frac{3-x}{x^3}\right) = \frac{x^3 \times \dfrac{d}{dx}(3-x) - (3-x)\dfrac{d}{dx}(x^3)}{(x^3)^2}$$

$$= \frac{x^3 \times (-1) - (3-x)(3x^2)}{x^6}$$

$$= \frac{-x^3 - 9x^2 + 3x^3}{x^6} = \frac{2x^3 - 9x^2}{x^6}$$

$$= \frac{2}{x^3} - \frac{9}{x^4}$$

---

6. Which of the following is incorrect?

A. $\dfrac{d}{dx}(\sec x) = \sec x \tan x$

B. $(\cot x)' = -\csc^2 x$

C. $\dfrac{d}{dx}(\csc x) = -\csc x \cot x$

D. $(\tan x)' = \sec x \tan x$

E. Both $\dfrac{d}{dx}(\csc x) = -\csc x \cot x$ and $(\tan x)' = \sec x \tan x$

Answer D. Notice that

$$= \left(\frac{\sin x}{\cos x}\right)' = \frac{\cos x(\sin x)' - \sin x(\cos x)'}{\cos^2 x}$$

$$= \frac{\cos^2 x + \sin^2 x}{\cos^2 x} = \frac{1}{\cos^2 x} = \sec^2 x.$$

---

7. Suppose that f, g, h, and s are differentiable functions of x. Then $\left(\dfrac{f+gh}{s}\right)' =$

A. $\dfrac{f' + g'h + gh'}{s^2}$

B. $\dfrac{sf' + sg'h' - fs' + ghs'}{s^2}$

C. $\dfrac{sf' + sgh' - fs' + ghs'}{s^2}$

D. $\dfrac{sf' - fs' + sg'h + sh'g - ghs'}{s^2}$

E. None of the above

Answer D. Using the rules for differentiating sums, products and quotients

$$\left(\frac{f+gh}{s}\right)' = \frac{s(f+gh)'-(f+gh)s'}{s^2} \quad \text{(by the quotient rule)}$$

$$= \frac{s(f'+(gh)')-(f+gh)s'}{s^2} \quad \text{(by the sum rule)}$$

$$= \frac{s(f'+g'h+h'g)-(f+gh)s'}{s^2} \quad \text{(by the product rule)}$$

$$= \frac{sf' - fs' + sg'h + sh'g - ghs'}{s^2}$$

---

8. $\dfrac{d}{dx}\left(\dfrac{x}{2+x}\right) =$

A. $\dfrac{1-x}{(2+x)^2}$

B. $\dfrac{2+x-x^2}{(2+x)^2}$

C. $\dfrac{1}{2+x}$

D. $\dfrac{2}{(2+x)^2}$

E. $\dfrac{-x}{(2+x)^2}$

Answer D. Using the quotient rule, $\dfrac{d}{dx}\left(\dfrac{x}{2+x}\right) = \dfrac{(2+x) - x(1)}{(2+x)^2} = \dfrac{2}{(2+x)^2}.$

---

9. $\dfrac{d}{dx}\left(\dfrac{x^2+3}{x^3-2}\right) =$

A. $\dfrac{-x^4 - 9x^2 - 4x}{(x^3 - 2)^2}$

B. $\dfrac{2x^4 - 4x - 9x^2}{(x^3 - 2)^2}$

C. $\dfrac{-x^4 - 4x}{(x^3 - 2)^2}$

D. $\dfrac{2x^4 - 9x^2}{(x^3 - 2)^2}$

E. $\dfrac{-9x^2}{(x^3 - 2)^2}$

Answer   A. Using the quotient rule,

$$\frac{d}{dx}\left(\frac{x^2 + 3}{x^3 - 2}\right) = \frac{(x^3 - 2)(2x) - (x^2 + 3)(3x^2)}{(x^3 - 2)^2}$$

$$= \frac{2x^4 - 4x - 3x^4 - 9x^2}{(x^3 - 2)^2} = \frac{-x^4 - 9x^2 - 4x}{(x^3 - 2)^2}$$

---

10. $\left(\dfrac{2\pi x}{x^2 - 4}\right)' =$

A. $\dfrac{2\pi x^2 - 4\pi^2 x - 8\pi}{(x^2 - 4)^2}$

B. $\dfrac{-2\pi x^2 - 8\pi}{(x^2 - 4)^2}$

C. $\dfrac{2\pi x^2}{(x^2 - 4)^2}$

D. $\dfrac{2\pi x^2 + 8\pi x}{(x^2 - 4)^2}$

E. $\dfrac{2\pi x^2 + 8\pi}{(x^2 - 4^3)}$

Answer   B. Using the quotient rule,

$$\left(\frac{2\pi x}{x^2 - 4}\right)' = \frac{(x^2 - 4)2\pi - 2\pi x(2x)}{(x^2 - 4)^2} = \frac{2\pi x^2 - 8\pi - 4\pi x^2}{(x^2 - 4)^2} = \frac{-2\pi x^2 - 8\pi}{(x^2 - 4)^2}.$$

---

11. $\dfrac{d}{dx}(3x^{\frac{1}{2}} \sin x) =$

A. $3x^{-\frac{1}{2}} \sin x + \frac{3}{2}x^{\frac{1}{2}} \cos x$

B. $\frac{3}{2}x^{-\frac{1}{2}} \sin x + 3x^{\frac{1}{2}} \cos x$

C. $3x^{-\frac{1}{2}} \sin x + 3x^{\frac{1}{2}} \cos x$

D. $3x^{\frac{1}{2}}\sin x + 3x^{-\frac{1}{2}}\cos x$

E. $\frac{3}{2}x^{\frac{1}{2}}\sin x + 3x^{-\frac{1}{2}}\cos x$

Answer   B. By the product rule, $\frac{d}{dx}(3x^{\frac{1}{2}}\sin x) = \frac{3}{2}x^{-\frac{1}{2}}\sin x + 3x^{\frac{1}{2}}\cos x$.

---

12. $\dfrac{d}{dx}(3x^{\frac{5}{2}}\tan x) =$

A. $\frac{15}{2}x^{\frac{3}{2}}\sec^2 x + 3x^{\frac{5}{2}}\tan x$

B. $\frac{15}{2}x^{\frac{3}{2}}\tan x - 3x^{\frac{5}{2}}\sec^2 x$

C. $\frac{15}{2}x^{\frac{3}{2}}\tan x + 3x^{\frac{5}{2}}\cot x$

D. $\frac{15}{2}x^{\frac{3}{2}}\tan x - 3x^{\frac{5}{2}}\cot x$

E. $\frac{15}{2}x^{\frac{3}{2}}\tan x + 3x^{\frac{5}{2}}\sec^2 x$

Answer   E. By the product rule, $\frac{d}{dx}(3x^{\frac{5}{2}}\tan x) = \frac{15}{2}x^{\frac{3}{2}}\tan x + 3x^{\frac{5}{2}}\sec^2 x$.

---

13. $\dfrac{d}{dx}(4\sec x \tan x) =$

A. $4(\sec^2 x \tan x + \sec^3 x)$

B. $4(\sec x \tan^2 x + \sec^3 x)$

C. $4(\sec^2 x \tan x - \sec^3 x)$

D. $4\cot^2 x \tan x + 4\sec^2 x \tan x$

E. $4\cot^2 x \tan x + 4\sec^2 x$

Answer   B. Using the product rule,

$$\frac{d}{dx}(4\sec x \tan x) = 4(\sec x \tan x)\tan x + 4\sec x(\sec^2 x)$$
$$= 4(\sec x \tan^2 x + \sec^3 x).$$

---

14. $\dfrac{d}{dx}(x^2 \cot x) =$

A. $-x^2 \csc^2 x$

B. $-x^2 \csc^2 x + x^2 \cot x$

C. $2x \cot x - x^2 \csc x$

D. $2x \cot x + x^2 \csc x$

E. $2x \cot x - x^2 \csc^2 x$

Answer E. We have $\dfrac{d}{dx}(x^2 \cot x) = 2x \cot x + x^2(-\csc^2 x) = 2x \cot x - x^2 \csc^2 x$.

---

15. $\dfrac{d}{dx}(x^{\frac{1}{2}} \csc x \sin x) =$

A. $\frac{1}{2}x^{-\frac{1}{2}}$

B. $\frac{1}{2}x^{-\frac{1}{2}} \csc x \sin x + x^{\frac{1}{2}}(\csc x \cot x)\sin x + x^{\frac{1}{2}} \csc x \cos x$

C. $\frac{1}{2}x^{-\frac{1}{2}} \csc x \sin x + x^{\frac{1}{2}}(\csc x \tan x)\sin x - x^{\frac{1}{2}} \csc x \cos x$

D. $\frac{1}{2}x^{-\frac{1}{2}} \csc x \sin x - x^{\frac{1}{2}}(\csc x \cot x)\cos x + x^{\frac{1}{2}} \csc x \cos x$

E. $\frac{1}{2}x^{-\frac{1}{2}} \csc x \sin x + x^{\frac{1}{2}}(\csc x \tan x)\sin x + x^{\frac{1}{2}} \csc x \cos x$

Answer A. $\dfrac{d}{dx}(x^{\frac{1}{2}} \csc x \sin x) = \dfrac{d}{dx}\left(x^{\frac{1}{2}}\dfrac{1}{\sin x}\sin x\right) = \dfrac{d}{dx}(x^{\frac{1}{2}}) = \frac{1}{2}x^{-\frac{1}{2}}$.

Or, if you didn't simplify first, you get

$$\frac{d}{dx}(x^{\frac{1}{2}} \csc x \sin x) = \frac{1}{2}x^{-\frac{1}{2}} \csc x \sin x - x^{\frac{1}{2}}(\csc x \cot x)\sin x + x^{\frac{1}{2}} \csc x \cos x$$

$$= \frac{1}{2}x^{-\frac{1}{2}}\left(\frac{1}{\sin x}\right)\sin x - x^{\frac{1}{2}}\left(\frac{1}{\sin x} \times \frac{\cos x}{\sin x}\right)\sin x + x^{\frac{1}{2}}\frac{1}{\sin x}\cos x$$

$$= \frac{1}{2}x^{-\frac{1}{2}} - x^{\frac{1}{2}}\left(\frac{\cos x}{\sin x}\right) + x^{\frac{1}{2}}\frac{\cos x}{\sin x} = \frac{1}{2}x^{-\frac{1}{2}}$$

---

16. $\dfrac{d}{dx}(\sin^3 x) = \dfrac{d}{dx}(\sin x \sin x \sin x) =$

A. $\cos x \sin^2 x$

B. $2\cos x \sin^2 x + \sin x \cos x$

C. $3\sin^2 x \cos x$

D. $2\sin x \cos x + \sin^2 x \cos x$

E. $1 + 2\sin^2 x \cos x$

Answer C.

$$\frac{d}{dx}(\sin^3 x) = \frac{d}{dx}(\sin x \sin x \sin x)$$
$$= \cos x(\sin x \sin x) + \sin x(\cos x)(\sin x) + \sin x \sin x(\cos x)$$
$$= 3\sin^2 x \cos x$$

17. $\dfrac{d}{dx}\left(\dfrac{3x^3}{\tan x}\right) =$

A. $\dfrac{9x^2 \tan x + 3x^3 \sec^2 x}{\tan^2 x}$

B. $\dfrac{9x^2}{\sec^2 x}$

C. $\dfrac{9x^2 \tan x + 3x^3 \sec^2 x}{\sec^2 x}$

D. $\dfrac{9x^2 \tan x - 3x^3 \sec^2 x}{\sec^2 x}$

E. $\dfrac{9x^2 \tan x - 3x^3 \sec^2 x}{\tan^2 x}$

Answer   E. $\dfrac{d}{dx}\left(\dfrac{3x^3}{\tan x}\right) = \dfrac{\tan x (9x^2) - 3x^3 (\sec^2 x)}{(\tan x)^2} = \dfrac{9x^2 \tan x - 3x^3 \sec^2 x}{\tan^2 x}.$

18. Below is the graph of $y = h'(x)$:

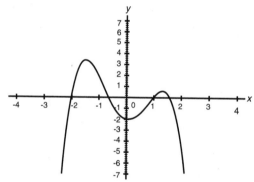

On what interval(s) is the function h positive?

A. $-2 < x < -.6$

B. $(-2, -.6) \cup (1, 1.6)$

C. $(-\infty, -2)$

D. All x values

E. Impossible to tell

Answer   E. Remember, the graph of $y = h'(x)$ can only tell the shape of the graph of $y = h(x)$; it can't tell "where" the graph is.

19. Below is the graph of $y = g(x)$:

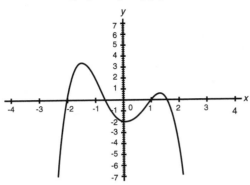

What are the zeros of $g'(x)$?

A. $x = -2, -0.6$

B. $x = -2, -0.6, 1, 1.6$

C. $x = -1.6, 1.3$

D. $x = -1.6, 0.1, 1.3$

E. Impossible to tell

Answer   D. The zeros of $g'(x)$ are at the extrema (maxima or minima) of $g(x)$, and those are at $x = -1.6, 0.1, 1.3$.

---

20. Below is the graph of $y = h(x)$ (the asymptotes are the dotted lines):

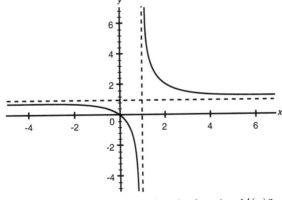

What can we expect concerning the function $h'(x)$?

A. h' has no zeros.

B. h' is undefined at $x = 1$.

C. h' is always negative on its domain.

D. h' has no smallest possible value.

E. All of the above.

Answer   E. The function h' will be undefined at $x = 1$, because h is undefined there. The function h' is always negative on its domain, because the slope of h is always negative on its domain. Finally, the function h' has no smallest possible value, since when x approaches 1 from the left, h' becomes less than any specified negative number.

---

21.   Below is the function $y = f(x)$:

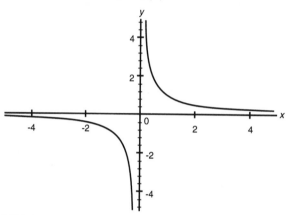

Which graph is the graph of $y = f'(x)$?

Graph I

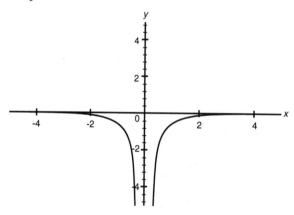

Graph II

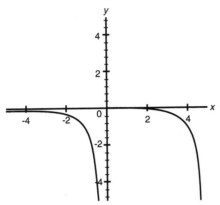

Graph III

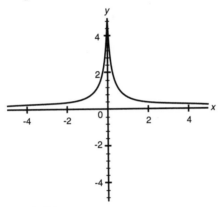

Graph IV

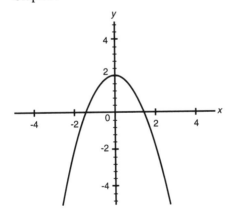

Graph V

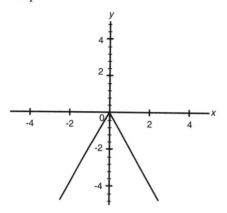

A.   Graph I

B.   Graph II

C.   Graph III

D.   Graph IV

E.   Graph V

Answer   A. Graph I.

The analysis follows: when x is very negative, the slope of f is negative but small; as x increases toward 0, the slope of f becomes very negative; at $x = 0$ the f is undefined so there is no slope; when x is positive and small, the slope of f is very negative; as x increases from 0, the slope of f becomes smaller, while remaining negative.

This is precisely what Graph I depicts.

22.   Below is the function $y = f(x)$:

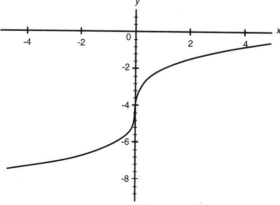

Which graph is the graph of $y = f'(x)$?

Graph I

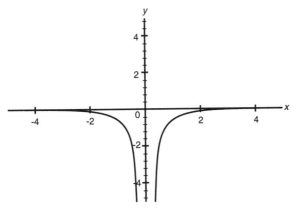

Graph II

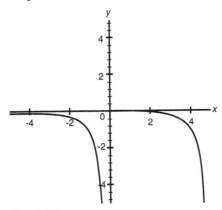

Graph III

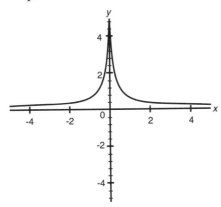

Graph IV

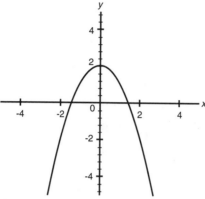

Graph V

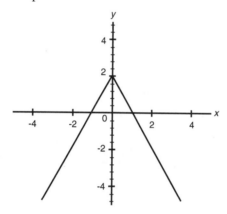

A. Graph I

B. Graph II

C. Graph III

D. Graph IV

E. Graph V

Answer   C. Graph III.

The analysis follows: When x is very negative, the slope of f is positive but small; as x increases toward 0, the slope of f gets bigger and bigger until it's infinite at $x = 0$; when x is barely greater than 0, the slope of f is VERY positive; as x becomes very positive, the slope of f is positive but gets smaller and smaller until it's near 0.

Graph III follows that pattern perfectly.

23. Below is the function $y = f(x)$:

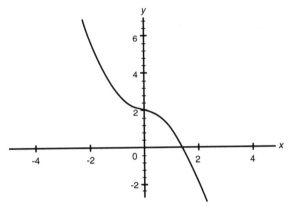

Which graph is the graph of $y = f'(x)$?

Graph I

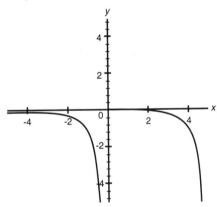

Graph II

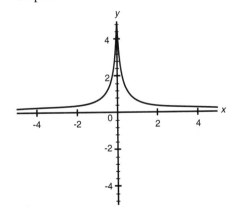

Graph III

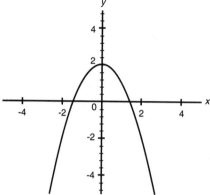

Graph IV

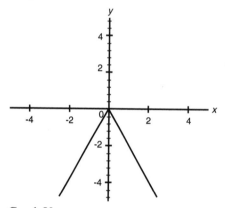

Graph V

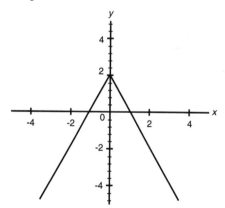

A. Graph I

B. Graph II

C. Graph III

D. Graph IV

E. Graph V

Answer   D. Graph IV.

The analysis follows: When x is very negative, the slope of f is very negative; as x increases toward 0, the slope of f increases toward 0; when x is slightly positive, the slope of f is slightly negative; and as x increases to being very positive, the slope of f gets very negative.

Graph IV follows that pattern perfectly.

---

24.   Below is the function $y = f(x)$:

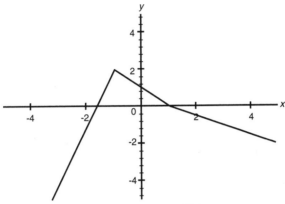

Which graph is the graph of $y = f'(x)$?

Graph I

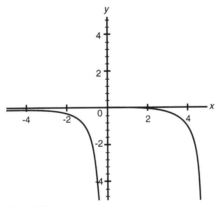

Graph II

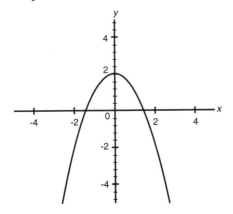

Graph III

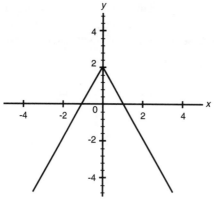

Graph IV

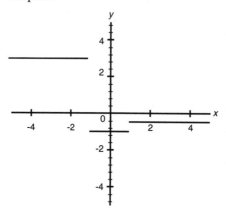

Graph V

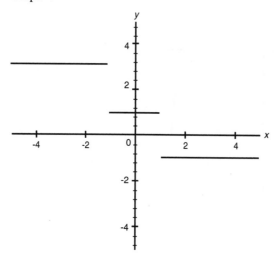

A. Graph I

B. Graph II

C. Graph III

D.   Graph IV

E.   Graph V

Answer   D. Graph IV.

The analysis follows: For x very negative, the slope of f is a constant which is about 3; for x between -1 and 1, the slope of f is a constant -1; for x greater than 1, the slope of f is a constant -0.5.

Graph IV follows this pattern precisely.

---

25.   Consider the function g(x) whose graph is below:

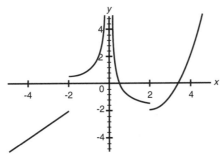

For what values of x is the function clearly NOT differentiable?

A. $x = -2$

B. $x = 0$

C. $x = -2, 0$

D. $x = -2, 0, 2$

E. $x = -2, 0, 2, 4$

Answer   D. The function is discontinuous at $x = -2$, 0, and 2. Therefore, the function is not differentiable at $x = -2$, 0, and 2.

---

26.   Suppose $g(x) = \begin{cases} x^3 + 3x^2 + 2x & \text{if } x \leq 0 \\ -x^2 & \text{if } x > 0. \end{cases}$ Then:

A. g is continuous and differentiable at $x = 0$.

B. g is continuous at $x = 0$, and g is not differentiable at $x = 0$.

C. g is not continuous at $x = 0$, but g is differentiable at $x = 0$.

D. g is not continuous at $x = 0$, and g is not differentiable at $x = 0$.

E.   Nothing can be said about the differentiability of g at $x = 0$.

Answer   B. g is continuous at $x = 0$, since $\lim\limits_{x \to 0-} g(x) = \lim\limits_{x \to 0+} g(x) = g(0) = 0$.

To see that g is not differentiable at $x = 0$, notice that $\lim\limits_{h \to 0-} \dfrac{g(0+h) - g(0)}{h} = 2$, while $\lim\limits_{h \to 0+} \dfrac{g(0+h) - g(0)}{h} = 0$. Thus, $\lim\limits_{h \to 0-} \dfrac{g(0+h) - g(0)}{h} \neq \lim\limits_{h \to 0+} \dfrac{g(0+h) - g(0)}{h}$, and therefore, the function g is not differentiable at $x = 0$.

---

27.   Suppose $g(x) = \begin{cases} x^2 + kx & \text{if } x < 0 \\ \sin x & \text{if } x \geq 0. \end{cases}$ What value for k will make g differentiable at $x = 0$?

A. -1

B. 0

C. 1

D. π

E.   Both -1 and 1

Answer   C. You need to choose k so that $\lim\limits_{h \to 0-} \dfrac{g(0+h) - g(0)}{h} = \lim\limits_{h \to 0+} \dfrac{g(0+h) - g(0)}{h}$.

(This is the definition of the derivative at $x = 0$). Setting $k = 1$ gives us

$\lim\limits_{h \to 0-} \dfrac{g(0+h) - g(0)}{h} = 2(0) + k = 1$ and $\lim\limits_{h \to 0+} \dfrac{g(0+h) - g(0)}{h} = \cos(0) = 1$.

Therefore, the function is differentiable at $x = 0$ when $k = 1$.

---

28.   The hypotheses for both Rolle's Theorem and the Mean Value Theorem involve identical statements concerning continuity and differentiability of the function f in question.

These hypotheses are:

A. f is continuous on (a, b) and differentiable on $[a, b]$.

B. f is continuous on $[a, b)$ and differentiable on $(a, b]$.

C. f is continuous on $[a, b]$ and differentiable on (a, b).

D. f is continuous on $[a, b]$ and differentiable on $[a, b]$.

E. f is continuous on (a, b) and differentiable on (a, b).

Answer   C. These are very important to remember! The open interval (a, b) means that the function need not be differentiable at the endpoints a and b. However, the closed interval $[a, b]$ means that the function MUST be continuous at the endpoints.

29. To which of the following can the Mean Value Theorem (MVT) be applied?

A. $f(x) = |x|$ on $[-3, 3]$

B. $h(x) = x^{\frac{1}{3}}$ on $[-1, 8]$

C. $g(x) = x^{-2}$ on $[2, 5]$

D. $m(x) = x^{\frac{1}{2}}$ on $[-1, 4]$

E. None of the above

Answer C. The function $g(x) = x^{-2}$ on $[2, 5]$ satisfies the hypotheses required to apply the MVT: the function is continuous on $[2, 5]$ and differentiable on $(2, 5)$.

---

30. If $f(-4) = f(3) = 0$ and $f(x)$ is continuous on $[-4, 3]$ and differentiable on $(-4, 3)$, then:

A. $f'(0) = 0$.

B. there exists a c between -4 and 3 for which $f(c) = 0$.

C. the function attains a local (relative) maximum somewhere in the interval $(-4, 3)$.

D. there exists a c between -4 and 3 for which $f'(c) = 0$.

E. there exists a c between a and b for which $f(c) = \dfrac{f(b) - f(a)}{b - a}$.

Answer D. The function satisfies the hypotheses of Rolle's Theorem: $f(-4) = f(3) = 0$, $f(x)$ is continuous on $[-4, 3]$ and differentiable on $(-4, 3)$, so there exists a c between -4 and 3 for which $f'(c) = 0$.

---

31. Suppose f is a continuous and differentiable function of x on the interval $[-5, 2]$. Suppose also that we have the following table of data for the function:

x	$f(x)$
-5	12
-3	300
0	-4
.01	300

Which of the following statements is NOT necessarily true?

A. There exists a c between -5 and -3 for which $f'(c) = 144$.

B. There exists a c between -5 and 2 for which $f'(c) = -\dfrac{304}{3}$.

C. There exists a c between -3 and $.01$ for which $f'(c) = 0$.

D. There exists a c between 0 and 2 for which $f'(c) = 30,400$.

E. There exists a c between -3 and 0 for which $f'(c) = -3.2$.

Answer   E. By the Mean Value Theorem, there exists a c between -5 and 0 for which

$$f'(c) = \frac{f(0) - t(-5)}{0 - (-5)} = \frac{-4 - 12}{0 - (-5)} = \frac{-16}{5} = -3.2.$$

However, we do not know if c is between -3 and 0. We only know that this c is between -5 and 0.

---

32.   Suppose f is a continuous and differentiable function of x on the intervals $[-4, 0)$ and $(0, 1]$. Suppose also that we have the following table of data for the function:

x	$f(x)$
-4	$.5$
-2	$.75$
$.5$	2
1	$.75$

Which of the following statements is NOT necessarily true?

A. There exists a c between -2 and 1 for which $f'(x) = -2.5$.

B. There exists a c between $.5$ and 1 for which $f'(x) = .75$.

C. There exists a c between -4 and 0 for which $f'(c) = .125$.

D. There exists a c between -4 and -2 such that $f(c) = .7$.

E. There exists a c between $.5$ and 1 such that $f(c) = 1$.

Answer   B. All we know about the interval $[.5, 1]$ is that (by the Mean Value Theorem) there exists a c between $.5$ and 1 for which

$$f'(c) = \frac{f(1) - f(.5)}{1 - .5} = \frac{.75 - 2}{1 - .5} = -2.5.$$

33. Below is the graph of $y = f(x) = x^4 - 4x^2 + x$:

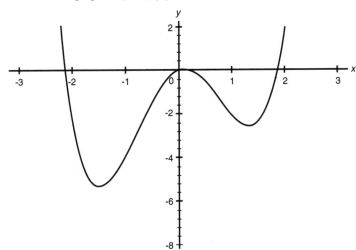

Notice that $f(-2) = -2$, and $f(2) = 2$. The Mean Value Theorem assures us that there is at least one value c between -2 and 2 for which $f'(c) = \dfrac{f(2) - f(-2)}{2 - (-2)}$.

Looking at the graph, it's clear that there is more than one such c.

How many are there, and what are they?

A. 2 values for c; $c = -1.473, .126$

B. 2 values for c; $c = -\sqrt{2}, +\sqrt{2}$

C. 3 values for c; $c = 0, -\sqrt{2}, +\sqrt{2}$

D. 3 values for c; $c = -1.473, 1.350, .126$

E. 4 values for c; $c = 0, -2.115, 1.861, .254$

Answer   C. You need to find c for which $f'(c) = \dfrac{f(2) - f(-2)}{2 - (-2)} = \dfrac{2 - (-2)}{2 - (-2)} = 1.$

The derivative of f is $f'(x) = 4x^3 - 8x + 1$. Setting $f'(x) = 1$ gives us:

$$f'(x) = 4x^3 - 8x + 1 = 1 \Rightarrow 4x^3 - 8x = 0$$
$$\Rightarrow x(x^2 - 2) = 0 \Rightarrow x = 0, -\sqrt{2}, +\sqrt{2}$$

34. Below is the graph of $y = f(x) = x^2 - 4$:

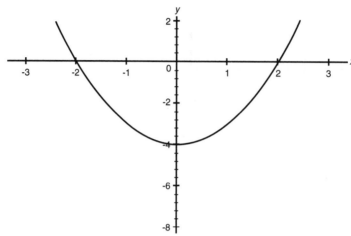

Find a constant $b > 1$ such that $f'(1) = \dfrac{f(b) - f(0)}{b - 0}$.

A. $b = 3$

B. $b = 2$

C. $b = -2$

D. $b = 4$

E. $b = 5$

Answer   B. We have $\dfrac{f(2) - f(0)}{2 - 0} = \dfrac{0 - (-4)}{2} = 2 = f'(1)$.

Free-Response Questions on the Mean Value Theorem

The Mean Value Theorem (MVT) and the Extreme Value Theorem (EVT) are two existence theorems—they guarantee that a solution exists, but they don't (usually) help you find it. Some problems will ask you to find a place where the conclusion of one theorem is satisfied; this is to make sure you understand what the conclusions are. Some problems will ask you to say why one theorem doesn't apply; this is to make sure you understand what the hypotheses are. The last couple of problems ask you to use the theorem(s) to prove something. Remember to check that all the hypotheses hold for the function you're investigating; if they don't, perhaps you should look at a different function.

1. If $h(x) = 3x^3 - 9x$, at what point on the interval $[0, \sqrt{3}]$, if any, is the tangent to the curve parallel to the secant line from $x = 0$ to $x = \sqrt{3}$?

2. Explain why the Mean Value Theorem can't be applied to the following functions on the specified intervals.

A. $f(x) = \sin x + x^2 - 4x^{\frac{3}{5}}$ on $[-2, 5]$.

B. $f(x) = \tan x + |x|$ on $[1, 3]$.

C. $f(x) = \dfrac{x}{x^2 - 4} + |x|$ on $[-1, 1]$.

---

3. While driving through New York state, a motorist passed through a Thruway toll booth and got a ticket punched for 12:15 P.M. After driving 73 miles, the motorist exited the Thruway and handed in her ticket, which was then punched at 1:09 P.M.

   The motorist was promptly handed a speeding ticket for driving in excess of 80 miles per hour on the 65 mph Thruway. Is the speeding ticket justifiable? Explain.

---

4. Answer the questions about the graphs below. (As an aid, the black points on the graph are at $x = 1$ and $x = 4.4$.)

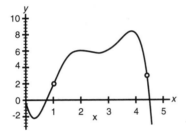

A. On what intervals does the function graphed below have a negative derivative?

   On what intervals does the function have a positive derivative?

B. At how many points on the interval $[1, 4.4]$ does the function graphed above satisfy the conclusion of the Mean Value Theorem?

Answers to Free-Response Questions on the Mean Value Theorem

---

1. Work: Slope of secant is $\dfrac{h(\sqrt{3}) - h(0)}{\sqrt{3} - 0} = \dfrac{0 - 0}{\sqrt{3} - 0} = 0$.

   The slope of the tangent line is $h'(x) = 9x^2 - 9$, and this is 0 when $x = 1$.

2.

A. The MVT can't be applied here because the function is not differentiable at $x = 0$, owing to the $4x^{\frac{3}{5}}$ term.

B. The MVT can't be applied here because the function is not continuous at $x = \frac{\pi}{2}$, owing to the $\tan x$ term.

C. The MVT can't be applied here because the function is not differentiable at $x = 0$, owing to the $|x|$ term.

3. Yes. Note that the driver had an average speed of $\dfrac{73}{(54/60)} = 81.111$ mph.

The situation of the car driving from one toll booth to the next satisfies the conditions of the MVT, since the distance function would be continuous and differentiable (no "jumps" across space, no "instantaneous jumps" in velocity). Thus, there must be some time between 12:15 and 1:09 when the motorist was going 81.111 miles per hour.

4.

A. (Answers are approximate). Negative derivative on $0 < x < .35$, $2.1 < x < 2.6$, and $3.8 < x$. Positive derivative on $.35 < x < 2.1$ and $2.6 < x < 3.8$.

B. There are three places. To find them, you could lay a clear plastic ruler on the graph so that it acts like the secant line between the two points. Move the ruler on the graph, taking care to keep it parallel to the secant line. The places where the ruler seems to just touch the graph are the places where the tangent line has that same slope. These points are at about $x = 1.9$, $x = 2.7$, and $x = 3.8$.

C. Higher-Order and Implicit Derivatives

►Objective 1 Determine higher order derivatives of functions in families that have been covered in this unit.

Example Find $\dfrac{d^4}{dx^4}g(x)$ for each of the following:

A. $g(x) = 3x^4 - 3x^2 + 4x - 5$

B. $g(x) = \sin x$

C. $g(x) = \cos x - 4x^3$

Tip Look for patterns. For example, in C, notice that the fourth derivative will involve cos again. Also, the x^3 term will drop out by the fourth derivative.

Answers

A.

$$g(x) = 3x^4 - 3x^2 + 4x - 5 \implies g'(x) = 12x^3 - 6x + 4$$
$$\implies g''(x) = 36x^2 - 6$$
$$\implies g'''(x) = 72x$$
$$\implies g^{(4)}(x) = 72$$

B. $g(x) = \sin x \implies g'(x) = \cos x \implies g''(x) = -\sin x \implies g'''(x) = -\cos x$

$$\implies g^{(4)}(x) = \sin x$$

C. $g(x) = \cos x - 4x^3 \implies g^{(4)}(x) = \cos x$

►Objective 2   Identify places where a graph is concave up or down, both from a graph and from a formula.

Example   State where the following functions are concave down. Graph the function to verify the reasonableness of your answer.

A. $h(x) = x^4 - 2x^3 - 12x^2 + x - 9$

B. $g(t) = \sin t$ on the domain $[-\pi, 2\pi]$

Tip   When the second derivative is positive, then the original graph is concave up. When the second derivative is negative, the original curve is concave down.

Answer

A.

$$h(x) = x^4 - 2x^3 - 12x^2 + x - 9$$
$$h'(x) = 4x^3 - 6x^2 - 24x + 1$$
$$h''(x) = 12x^2 - 12x - 24 = 12(x+1)(x-2)$$

Note that $12(x+1)(x-2)$ is negative when x is greater than -1 but less than 2. Thus, the graph of h is concave down when $-1 < x < 2$.

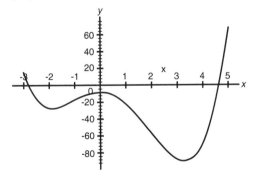

B. $g(t) = \sin t \implies g'(t) = \cos t \implies g''(t) = -\sin t$

Note that $-\sin t$ is negative when $0 < t < \pi$ on the domain $[-\pi, 2\pi]$.

The graph of $g(t) = \sin t$ is concave down when $0 < t < \pi$.

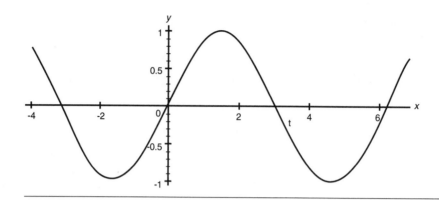

▶Objective 3 Using the second derivative, find the inflection points in a graph.

Example Find the inflection points for

A. $h(x) = x^4 - 2x^3 - 12x^2 + x - 9$

B. $g(t) = \sin t$ on the domain $(0, 2\pi)$

Tip Inflection points occur when the second derivative changes sign. It's not enough to simply find the zeros of the second derivative—you must check that the sign changes, too.

Answers

A. $h(x) = x^4 - 2x^3 - 12x^2 + x - 9 \implies h''(x) = 12x^2 - 12x - 24 = 12(x+1)(x-2)$

From the previous problem we know that the second derivative changes sign at $x = -1$ and $x = 2$.

To find the y-coordinates of these points we evaluate $h(-1) = -19$ and $h(2) = -55$.

The inflection points are at $(-1, -19)$ and $(2, -55)$.

B. Since $g(t) = \sin t$ on the domain $(0, 2\pi)$, $g''(t) = -\sin t$ changes sign at $t = \pi$. The inflection point is $(\pi, 0)$.

▶Objective 4 Use higher-order derivatives to analyze rates of change.

Example A car traveling at 60 ft per second at $t = 0$ begins braking. Its position is given by $s(t) = -8.25t^2 + 66t$. What does the second derivative tell you about this situation?

Answer The first derivative is the rate of change of the position, which is velocity:

$$v(t) = s'(t) = -16.5t + 66.$$

The second derivative tells the rate of change of the velocity.

$$a(t) = v'(t) = s''(t) = -16.5 \frac{\text{ft}}{\text{sec}^2}$$

is always negative and that tells you the car is slowing down.

(The rate of change of the rate of change of the position of the car is decreasing.)

The rate of change of the velocity is called the acceleration of the object.

The Chain Rule and Implicit Differentiation

▶Objective 5   Write the Chain Rule from memory (using $\dfrac{dy}{dx}$ notation and prime notation).

Using prime notation: If $y = f(g(x))$, then

$$y' = f'(g(x))g'(x).$$

Using Leibniz notation: If y is a differentiable function of u and u is a differentiable function of x, then

$$\frac{dy}{dx} = \frac{dy}{du} \times \frac{du}{dx}.$$

▶Objective 6   Use the Chain Rule in combination with the other derivative rules to find derivatives of functions.

Example   Find the derivative of:

A. $f(\theta) = \sin(\theta^2)$

B. $g(x) = \sqrt{x \sin x}$

Tip   Think of the Chain Rule as the derivative of the 'outside' times the derivative of the 'inside.'

Answers

A. $f(\theta) = \sin(\theta^2) \quad \Longrightarrow \quad f'(\theta) = 2\theta \cos(\theta^2)$

B. $g(x) = \sqrt{x \sin x} \quad \Longrightarrow \quad g'(x) = \dfrac{1}{2\sqrt{(x \sin x)}}(\sin x + x \cos x)$

▶Objective 7   Identify whether an equation is given implicitly or explicitly.

Example   Identify which equation defines y explicitly as a function of x.

A. $xy = x^2 - 3$

B. $y = \sin x + x^2$

Tip   Explicit equations have the dependent variable isolated on one side of the equal sign.

Answer

A. $xy = x^2 - 3$ defines y *implicitly* as a function of x.

B. $y = \sin x + x^2$ defines y *explicitly* as a function of x.

▶Objective 8   Determine the derivative for implicitly defined curves and relationships.

Example   Find y' for each of the following:

A. $xy = x^2 - 3$

B. $\cos(xy) = y^3 - x + 3$

Tip   Use the Chain Rule when doing implicit derivatives.

We used the y' notation here, and it's important to realize that $y' = \dfrac{dy}{dx}$ in this case.

Remember that y is implicitly considered to be a function of your independent variable (the one that appears in the "denominator" of the differential form of the derivative) and should be treated as such with the chain rule.

Answer

A.

$$xy = x^2 - 3$$
$$xy' + y = 2x$$
$$y' = \frac{2x - y}{x}$$

B.

$$\cos(xy) = y^3 - x + 3$$
$$-\sin(xy)(xy' + y) = 3y^2 y' - 1$$
$$-xy'(\sin(xy)) - y(\sin(xy)) = 3y^2 y' - 1$$
$$-3y^2 y' - xy' \sin(xy) = y\sin(xy) - 1$$
$$y'(-3y^2 - x\sin(xy)) = y\sin(xy) - 1$$
$$y' = \frac{y\sin(xy) - 1}{-3y^2 - x\sin(xy)}$$

▶Objective 9   Calculate the slope of the tangent line at points on an implicitly defined curve.

Example 1

A.   Find the slope of the tangent line to the curve $x^3 + y^3 - 6xy = 0$ at the point $\left(\frac{4}{3}, \frac{8}{3}\right)$.

B.   Write an equation for this tangent line.

Answers

A. To find the slope, you have to find $\dfrac{dy}{dx}$:

$$3x^2 + 3y^2y' - 6xy' - 6y = 0.$$

(Don't forget the product rule.)

$$y' = \frac{6y - 3x^2}{3y^2 - 6x} = \frac{2y - x^2}{y^2 - 2x}$$

Thus y' at the point $\left(\frac{4}{3}, \frac{8}{3}\right)$ has the value

$$y' = \frac{2y - x^2}{y^2 - 2x} = \frac{2\left(\frac{8}{3}\right) - \left(\frac{4}{3}\right)^2}{\left(\frac{8}{3}\right)^2 - 2\left(\frac{4}{3}\right)} = \frac{4}{5}.$$

B. From the point slope form, we have

$$y - \frac{8}{3} = \frac{4}{5}\left(x - \frac{4}{3}\right).$$

The illustration below shows the curve $x^3 + y^3 - 6xy = 0$ and is called the "Folium of Descartes." The straight line is the tangent line we found.

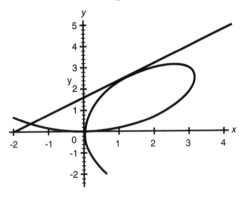

Example 2 Which points on the curve of $y^2 + xy - x^2 = 5$ have horizontal tangent lines?

Tip $y' = 0$ for horizontal tangent lines.

Answer $2yy' + xy' + y - 2x = 0 \implies y' = \dfrac{2x - y}{2y + x}.$

Thus, $y' = 0$ when $2x = y$ and next we find the x values where this occurs:

$$y^2 + xy - x^2 = 5$$
$$(2x)^2 + x(2x) - x^2 = 5$$
$$4x^2 + 2x^2 - x^2 = 5$$
$$5x^2 = 5,$$
$$x = -1, 1.$$

Since $y = 2x$ when $y' = 0$, the points are $(-1, -2)$ and $(1, 2)$. The graph of $y^2 + xy - x^2 = 5$ is shown in the illustration below:

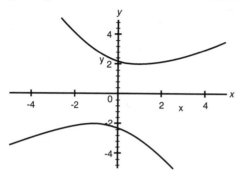

Example 3   The function $y = x^4 + ax^2 + 8x - 5$ has a horizontal tangent and a point of inflection for the same value of x. What is the value of a?

Answer $y' = 4x^3 + 2ax + 8 \quad \Longrightarrow \quad y'' = 12x^2 + 2a$.

At an inflection point $y'' = 0$, so $a = -6x^2$. If there also is a horizontal tangent line there, then $y' = 0$, meaning $4x^3 + 2ax + 8 = 0$.

Substituting $a = -6x^2$ in this equation and solving for x:

$$4x^3 + 2(-6x^2)x + 8 = 0$$
$$4x^3 - 12x^3 + 8 = 0$$
$$-8x^3 = -8$$
$$x = 1$$
$$a = -6x^2 = -6.$$

Check to see if this is reasonable:

$$y = x^4 - 6x^2 + 8x - 5$$

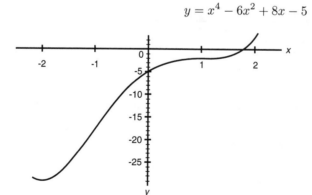

Multiple-Choice Exercises on Higher-Order and Implicit Derivatives

Recall that if $y = f(x)$ is a differentiable function of x, then the nth derivative of f with respect to x (if it exists) is denoted by $f^{(n)}$, where n is a positive integer.

When n is small ($n = 1, 2,$ or 3) you can use prime notation: f', f'', and f'''.

1. Suppose $f(x) = 3x^4 - 2x^2 + 3x$. Then $f''(x) =$

A. $12x^2 - 4$

B. $12x^3 - 4x + 3$

C. $36x^2 - 4x + 3$

D. $36x^2 - 4$

E. $12x^2 - 4x$

Answer D. First we get $f'(x) = 12x^3 - 4x + 3$. Next, we get $f''(x) = 36x^2 - 4$.

---

2. Suppose $f(x) = -x^{\frac{1}{2}}$. Then $f''(x) =$

A. $-x^{-\frac{3}{3}}$

B. $\frac{1}{2}x^{-\frac{3}{2}}$

C. $-\frac{1}{2}x^{-\frac{3}{2}}$

D. $-\frac{1}{4}x^{-\frac{3}{2}}$

E. $\frac{1}{4}x^{-\frac{3}{2}}$

Answer E. First, $f'(x) = -\frac{1}{2}x^{-\frac{1}{2}}$. Then, $f''(x) = \frac{1}{4}x^{-\frac{3}{2}}$.

---

3. Suppose $f(x) = \frac{1}{3}x^5 + x^{-2} + 3$. Then $f'''(x) =$

A. $20x^2$

B. $\frac{20}{3}x^2 - 24x^{-5}$

C. $\frac{20}{3}x^3 - 24x^{-4}$

D. $20x^2 + 24x^{-5}$

E. $20x^2 - 24x^{-5}$

Answer E. First, $f'(x) = \frac{5}{3}x^4 - 2x^{-3}$. Then $f''(x) = \frac{20}{3}x^3 + 6x^{-4}$.
Finally, $f'''(x) = 20x^2 - 24x^{-5}$.

---

4. Suppose $f(x) = 3x^{\frac{3}{4}} + \frac{1}{4}x^{\frac{7}{3}} + 3$. Then $f''(x) =$

A. $-\frac{9}{16}x^{-\frac{5}{4}} + \frac{28}{36}x^{\frac{1}{3}}$

B. $-\frac{9}{4}x^{-\frac{5}{4}} + \frac{28}{12}x^{\frac{1}{3}}$

C. $-\frac{9}{16}x^{-\frac{3}{4}} + \frac{7}{36}x^{-\frac{1}{3}}$

D. $-\frac{9}{4}x^{-\frac{1}{4}} + \frac{7}{3}x^{\frac{1}{3}}$

E. $-\frac{9}{16}x^{-\frac{5}{4}} + \frac{28}{36}x^{-\frac{1}{3}}$

Answer   A. First, $f'(x) = \frac{9}{4}x^{-\frac{1}{4}} + \frac{7}{12}x^{\frac{4}{3}}$. Then $f''(x) = -\frac{9}{16}x^{-\frac{5}{4}} + \frac{28}{36}x^{\frac{1}{3}}$.

---

5.   Suppose $f(x) = 15x^6 + \frac{1}{4}x^2 + 2\sin x$. Then $f''(x) =$

A. $450x^4 + \frac{1}{2} - 8\sin x$

B. $450x^4 + \frac{1}{2}x - 8\sin x$

C. $450x^4 + \frac{1}{2} - 2\sin x$

D. $450x^4 + \frac{1}{2} + 2\sin x$

E. $450x^4 + \frac{1}{2}x + 2\sin x$

Answer   C. First we have $f'(x) = 90x^5 + \frac{1}{2}x + 2\cos x$. Next, $f''(x) = 450x^4 + \frac{1}{2} - 2\sin x$.

---

6. $\frac{d^3}{dx^3}\left(4x^{\frac{5}{2}} - 12x^{\frac{1}{4}} + 7\cos x\right) =$

A. $15x^{\frac{1}{2}} + \frac{9}{4}x^{-\frac{7}{4}} - 7\cos x$

B. $\frac{15}{2}x^{\frac{1}{2}} + \frac{9}{4}x^{-\frac{11}{4}} - 7\sin x$

C. $15x^{-\frac{1}{2}} + \frac{63}{16}x^{-\frac{11}{4}} + 7\sin x$

D. $\frac{15}{2}x^{-\frac{1}{2}} - \frac{63}{16}x^{-\frac{11}{4}} + 7\sin x$

E. $\frac{15}{2}x^{\frac{1}{2}} - \frac{63}{4}x^{-\frac{7}{4}} - 7\sin x$

Answer   D. First we have

$$\frac{d}{dx}\left(4x^{\frac{5}{2}} - 12x^{\frac{1}{4}} + 7\cos x\right) = 10x^{\frac{3}{2}} - 3x^{-\frac{3}{4}} - 7\sin x.$$

Next we get

$$\frac{d^2}{dx^2}\left(4x^{\frac{5}{2}} - 12x^{\frac{1}{4}} + 7\cos x\right)$$
$$= \frac{d}{dx}\left(10x^{\frac{3}{2}} - 3x^{-\frac{3}{4}} - 7\sin x\right)$$
$$= 15x^{\frac{1}{2}} + \frac{9}{4}x^{-\frac{7}{4}} - 7\cos x.$$

Finally,

$$\frac{d^3}{dx^3}\left(4x^{\frac{5}{2}} - 12x^{\frac{1}{4}} + 7\cos x\right)$$
$$= \frac{d}{dx}\left(15x^{\frac{1}{2}} + \frac{9}{4}x^{-\frac{7}{4}} - 7\cos x\right)$$
$$= \frac{15}{2}x^{-\frac{1}{2}} - \frac{63}{16}x^{-\frac{11}{4}} + 7\sin x.$$

---

7. $\dfrac{d^2}{dx^2}(\sec x) =$

A. $\sec x \tan x + \sec^3 x$

B. $\sec x \tan^2 x + \sec^3 x$

C. $\sec x \tan^2 x + \sec^2 x$

D. $\sec x \tan^2 x + \sec^2 x \tan x$

E. $\tan^2 x + \sec x \tan x$

Answer B. First, $\dfrac{d}{dx}(\sec x) = \sec x \tan x$. Then

$$\frac{d^2}{dx^2}(\sec x) = \frac{d}{dx}(\sec x \tan x)$$
$$= \sec x \tan x \tan x + \sec x \sec^2 x$$
$$= \sec x \tan^2 x + \sec^3 x.$$

---

8. $\dfrac{d^2}{dx^2}\left(\dfrac{1}{x} + \cos x \sin x\right) =$

A. $\dfrac{2}{x^3}$

B. $-\dfrac{2}{x^3} - 2\cos x \sin x$

C. $\dfrac{2}{x^3} - 4\cos x \sin x$

D. $\dfrac{2}{x^3} - 2\cos x \sin x$

E. $-\dfrac{2}{x^3} + 4\cos x \sin x$

Answer   C. First we have

$$\frac{d}{dx}\left(\frac{1}{x} + \cos x \sin x\right) = -\frac{1}{x^2} - \sin x \sin x + \cos x \cos x$$

(Remember to use the product rule here.)

Next, we have $\dfrac{d^2}{dx^2}\left(\dfrac{1}{x} + \cos x \sin x\right) = \dfrac{d}{dx}\left(-\dfrac{1}{x^2} - \sin x \sin x + \cos x \cos x\right)$

$$= \frac{2}{x^3} - \cos x \sin x - \sin x \cos x + (-\sin x)\cos x + \cos x(-\sin x) = \frac{2}{x^3} - 4\cos x \sin x.$$

Note the product rule in use again here.

---

9. $\dfrac{d^2}{dx^2}(4x \sec x) =$

A. $8 \sec x \tan x + 4x \sec x \tan^2 x$

B. $8 \sec x \tan x + 4x \sec x(\tan^2 x + \sec^2 x)$

C. $4 \sec x \tan x + 4x \sec x(\tan^2 x + \sec^2 x)$

D. $4 \sec x \tan x + 8x \sec x \tan^2 x$

E. $8 \sec x \tan x + 4x \sec x \tan x + 4x \sec^2 x$

Answer   B. First we have $\dfrac{d}{dx}(4x \sec x) = 4 \sec x + 4x \sec x \tan x$.

Then, we have $\dfrac{d^2}{dx^2}(4x \sec x) = \dfrac{d}{dx}(4 \sec x + 4x \sec x \tan x)$

$= 4 \sec x \tan x + 4 \sec x \tan x + 4x(\sec x \tan x)\tan x + 4x \sec x(\sec^2 x)$

$= 8 \sec x \tan x + 4x \sec x(\tan^2 x + \sec^2 x).$

---

10. $\dfrac{d^3}{dx^3}(x^n) =$

A. $n^3 x^{n-3}$

B. $n(n-1)x^3$

C. $n(n-1)x^{n-2}$

D. $n(n-1)(n-2)x^{n-3}$

E. $n(n-1)(n-2)x^{-3}$

Answer D.

First, we have $\frac{d}{dx}(x^n) = nx^{n-1}$.

Next,

$$\frac{d^2}{dx^2}(x^n) = \frac{d}{dx}(nx^{n-1}) = (n-1)nx^{n-2}.$$

Finally,

$$\frac{d^3}{dx^3}(x^n) = \frac{d}{dx}((n-1)nx^{n-2})$$
$$= (n-2)(n-1)nx^{n-3} = n(n-1)(n-2)x^{n-3}.$$

---

11. Suppose we have a function $y = f(x)$. Which of the following does NOT represent the second derivative of this function with respect to x:

A. y''

B. $f''(x)$

C. $\dfrac{d^2y}{dx^2}$

D. $\left(\dfrac{dy}{dx}\right)^2$

E. $\dfrac{d^2f}{dx^2}$

Answer D. The expression $\left(\dfrac{dy}{dx}\right)^2$ does NOT represent the second derivative of the function with respect to x—it represents the first derivative squared.

---

12. Suppose $f(x) = x^4 - 3x^3 + 2x^{\frac{1}{3}}$. Then $f''(x) =$

A. $4x^3 - 9x^2 + \frac{2}{3}x^{-\frac{2}{3}}$

B. $12x^2 - 18x - \frac{4}{3}x^{-\frac{4}{3}}$

C. $12x^2 - 18x - \frac{4}{9}x^{-\frac{5}{3}}$

D. $4x^3 - 9x^2 + \frac{2}{3}x^{-\frac{1}{3}}$

E. $4x^2 - 18x - \frac{4}{9}x^{-\frac{4}{3}}$

Answer C.

$$f''(x) = \frac{d^2}{dx^2}(x^4 - 3x^3 + 2x^{\frac{1}{3}}) = \frac{d}{dx}(4x^3 - 9x^2 + \frac{2}{3}x^{-\frac{2}{3}})$$
$$= 12x^2 - 18x - \frac{4}{9}x^{-\frac{5}{3}}.$$

13. $\dfrac{d^2}{dx^2}(\sec x) =$

A. $\sec x(\tan^2 x + \sec^2 x)$

B. $\sec x \tan x$

C. $\sec x \tan^2 x$

D. $\sec^3 x$

E. $\sec^2 x(\tan^2 x + \sec^2 x)$

Answer   A.

$$\frac{d^2}{dx^2}(\sec x) = \frac{d}{dx}(\sec x \tan x)$$
$$= \frac{d}{dx}(\sec x) \times \tan x + \sec x \times \frac{d}{dx}(\tan x)$$
$$= \sec x \tan x \tan x + \sec x \sec^2 x$$
$$= \sec x(\tan^2 x + \sec^2 x)$$

14.   Below is the graph of $y = g(x)$:

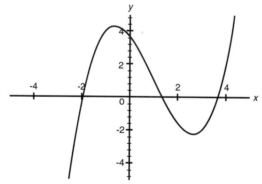

Which of the following is true concerning the concavity of the graph of $y = g(x)$?

A.   The graph is concave up for $x < 0$ and concave down for $x > 2$.

B.   The graph is concave up.

C.   The graph is concave down.

D.   The graph is concave up for $x > 2$ and concave down for $x < 0$.

E.   Cannot say anything about concavity without knowing the function

Answer   D.

Remember, a graph is concave up where it resembles a cup or bowl that could hold water, and a graph is concave down where is resembles an upside-down cup or bowl. To the left of $x = 0$, the graph is shaped like an upside-down bowl, and to the right of $x = 2$ the graph is shaped like a bowl that could hold water. So the graph is concave up for $x > 2$ and concave down for $x < 0$.

---

15.   Below is the graph of $y = h(x)$:

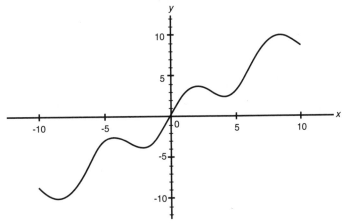

Which of the following statements is true?

A. $h''(-4)$ and $h''(2)$ are positive, while $h''(-2)$ is negative.

B. $h''(-4)$ is positive, while $h''(2)$ and $h''(-2)$ are negative.

C. $h''(-8)$ and $h''(5)$ are positive, while $h''(8)$ and $h''(-4)$ are negative.

D. $h''(-2)$ and $h''(8)$ are both negative.

E. $h''(-8)$ is positive, while $h''(2)$ and $h''(5)$ are negative.

Answer   C. Near $x = -8$ and $x = 5$, the graph is concave up, so the second derivatives should be positive at those points. Near $x = 8$ and $x = -4$ the graph is concave down, so the second derivative should be negative there.

16. Below is the graph of $y = f(x)$:

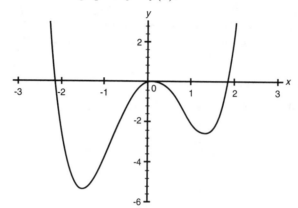

The graph of $y = f(x)$ appears to have how many inflection points?

A. 0

B. 1

C. 2

D. 3

E. 4

Answer   C. Inflection points occur where concavity changes. The graph changes concavity twice; from concave up to concave down (at around $x \approx -0.8$) and then from concave down to concave up (at around $x \approx 0.8$).

17. Suppose $f(x) = x^5 - \dfrac{10}{3}x^3 + 2$. The inflection points for the function $f(x)$ are at:

A. $x = 0$

B. $x = +\sqrt{2}, -\sqrt{2}$

C. $x = 1, -1$

D. $x = 0, +\sqrt{2}, -\sqrt{2}$

E. $x = 0, -1, 1$

Answer   E. To find the inflection points, first you need the second derivative: $f''(x) = 20x^3 - 20x$. Then you must set $f''(x) = 0$ and solve for x:

$$20x^3 - 20x = 0 \quad \Rightarrow \quad 20x(x^2 - 1) = 0 \quad \Rightarrow \quad x = 0, -1, 1.$$

Finally, you must check to see whether any of these are inflection points, by seeing if the sign of the second derivative changes.

	$x < -1$	$-1 < x < 0$	$0 < x < 1$	$1 < x$
$f''(x)$	neg	pos	neg	pos
concavity	down	up	down	up

You see here that you have inflection points at $x = 0, -1, 1$.

---

18. Suppose $f(x) = x^6 - 5x^4 + 15x^2 - \pi x$. Then f has inflection points at:

A. $x = 0$

B. $x = -1, 1$

C. $x = 0, -1, 1$

D. $x = 0, \sqrt{2}, \sqrt{2}$

E. Nowhere; f has no inflection points.

Answer   E. To find inflection points you need to set the second derivative equal to zero and solve for x:

$$f''(x) = 30x^4 - 60x^2 + 30 = 0 \Rightarrow 30(x^4 - 2x^2 + 1) = 0$$
$$\Rightarrow 30(x^2 - 1)^2 = 0 \Rightarrow x = -1, 1$$

as possible inflection points. But notice that the second derivative $f''(x) = 30(x^2 - 1)^2$ is never negative, so the concavity never changes. So there are no inflection points.

---

19. Suppose $y = (3x + 7)^3$. Then $\dfrac{dy}{dx} =$

A. $3(3x + 7)^2$

B. $9(3x + 7)^2$

C. $3(3x + 7)^3$

D. $3(3x + 7)^2 + 3$

E. $9x(3x + 7)^2$

Answer   B. By the chain rule:

$$\frac{dy}{dx} = 3(3x + 7)^2 \times \frac{d}{dx}(3x + 7) = 3(3x + 7)^2 \times 3 = 9(3x + 7)^2.$$

If it helps, try setting $u(x) = 3x + 7$. Then $y = u^3$, and

$$\frac{dy}{dx} = \frac{dy}{du} \times \frac{du}{dx} = \frac{d}{du}(u^3) \times \frac{d}{dx}(3x + 7)$$
$$= (3u^2) \times (3) = 9u^2 = 9(3x + 7)^2.$$

20. Suppose $y = \cos^3 x$. Then $y' =$

A. $-3\cos^2 x \sin^2 x$

B. $3\sin^2 x \cos x$

C. $-3\cos^3 x$

D. $3\cos^2 x$

E. $-3\cos^2 x \sin x$

Answer E. You have

$$y' = \frac{dy}{dx} = \frac{d}{dx}(\cos^3 x) = 3(\cos^2 x) \times \frac{d}{dx}(\cos x) \text{ (by the Chain Rule)}$$
$$= 3\cos^2 x \times (-\sin x) = -3\cos^2 x \sin x.$$

21. Suppose $y = (x^3 + 2)^2$. Then $\dfrac{dy}{dx} =$

A. $(3x^2 + 1)(x^3 + 2)$

B. $6x^2(x^2 + 2)$

C. $6x^2(x^3 + 2)$

D. $(3x^2 + 2)(x^2 + 2)$

E. $2(3x^2 + 2)$

Answer C. You have

$$\frac{dy}{dx} = \frac{d}{dx}((x^3 + 2)^2) = 2(x^3 + 2) \times \frac{d}{dx}(x^3 + 2) \text{ (by the Chain Rule)}$$
$$= 2(x^3 + 2) \times (3x^2) = 6x^2(x^3 + 2).$$

22. Suppose $y = \sin(\cos(x))$. Then $y' =$

A. $-\cos(\cos(x))\sin x$

B. $-\sin(\cos(x))\sin x$

C. $\sin(-\sin(x))$

D. $\sin(\cos(x))\cos(x)$

E. $\cos(-\sin x)$

Answer   A. Since this one can look confusing, set $u(x) = \cos x$, so that $y = \sin u$. Then

$$y' = \frac{dy}{dx} = \frac{dy}{du} \times \frac{du}{dx} = \frac{d}{du}(\sin u) \times \frac{d}{dx}(\cos x)$$
$$= \cos u \times (-\sin x) = -\cos u \times \sin x = -(\cos(\cos x)) \sin x.$$

---

23.   Suppose $y = \left(\frac{x^2+1}{-x}\right)^3$. Then $\frac{dy}{dx} =$

A. $3\left(\frac{x^2+1}{-x}\right)^2 \left(\frac{2x}{-1}\right)$

B. $3\dfrac{(x^2+1)^2(-3x^2-1)}{x^2}$

C. $3\left(\frac{x^2+1}{-x}\right)^2$

D. $3\dfrac{(x^2+1)^2(1-x^2)}{x^4}$

E. $3\left(\frac{x^2+1}{-x}\right)^2 \left(\frac{2x-1}{x^2}\right)$

Answer   D. This is a complicated one. Set $u(x) = \dfrac{x^2+1}{-x}$, so that $y = u^3$. Then you have

$$\frac{dy}{dx} = \frac{d}{du}(u^3) \times \frac{d}{dx}\left(\frac{x^2+1}{-x}\right) \quad \text{(by the Chain Rule)}$$
$$= 3u^2 \times \left(\frac{-x(2x) - (x^2+1)(-1)}{(-x)^2}\right)$$

(note that you used the quotient rule here)

$$= 3u^2 \times \left(\frac{-2x^2 + x^2 + 1}{x^2}\right) = 3u^2 \times \left(\frac{1-x^2}{x^2}\right)$$
$$= 3\left(\frac{x^2+1}{-x}\right)^2 \times \left(\frac{1-x^2}{x^2}\right) = 3\frac{(x^2+1)^2}{x^2} \times \left(\frac{1-x^2}{x^2}\right)$$
$$= 3\frac{(x^2+1)^2(1-x^2)}{x^4}.$$

---

24.   Suppose $y = \sin^2(x^2+2)$. Then $y' =$

A. $4x \cos(x^2+2)$

B. $4x \sin(x^2+2) \cos(x^2+2)$

C. $4x \sin(x^2+2)$

D. $\sin(x^2 + 2)\cos(2x)$

E. $\sin(x^2 + 2)\cos(2x)2x$

Answer   B. Suppose that $y = f(v)$, $v = g(u)$, and $u = h(x)$.

By the Chain Rule you know that $\dfrac{dy}{dx} = \dfrac{dy}{dv} \times \dfrac{dv}{dx}$.

Then, by the Chain Rule, you also know $\dfrac{dv}{dx} = \dfrac{dv}{du} \times \dfrac{du}{dx}$.   So

$$\frac{dy}{dx} = \frac{dy}{dv} \times \frac{dv}{dx} = \frac{dy}{dv}\left(\frac{dv}{du} \times \frac{du}{dx}\right)$$

(you've just substituted from above)

$$= \frac{dy}{dv} \times \frac{dv}{du} \times \frac{du}{dx}.$$

Here, with $y = f(v) = v^2$, $v = g(u) = \sin u$, and $u = h(x) = x^2 + 2$,

$$\frac{dy}{dx} = \frac{dy}{dv} \times \frac{dv}{du} \times \frac{du}{dx} = (2v)(\cos u)2x = [2\sin(x^2 + 2)]\cos(x^2 + 2)(2x)$$

$$= 4x\sin(x^2 + 2)\cos(x^2 + 2) = 4x\sin(x^2 + 2)\cos(x^2 + 2)$$

If you are comfortable using the Chain Rule twice without writing down the variables v and u, you can say that

$$\frac{dy}{dx} = 2\sin(x^2 + 2)\frac{d}{dx}\sin(x^2 + 2) =$$

$$2\sin(x^2 + 2)\cos(x^2 + 2)2x =$$

$$4x\sin(x^2 + 2)\cos(x^2 + 2).$$

---

25.   Suppose $y = w^3 + \sin w$, and $w = 3x^2 - x$. Then $\dfrac{dy}{dw} =$

A. $3w^2 + \cos w$

B. $3(3x^2 - x)^2 + 6x\cos(3x^2 - x) - 1$

C. $3(3x^2 - x)^2 + (6x - 1)\cos(3x^2 - x)$

D. $3(3x^2 - x)^2(6x - 1) + (6x\cos -1)(3x^2 - x)$

E. $w^2 + \cos w$

Answer   A. This is a bit of a trick question, since you're asked only to find the derivative of y with respect to w, and thus the function w doesn't come into play, nor do you need the Chain Rule! $\dfrac{dy}{dw} = \dfrac{d}{dw}(w^3 + \sin w) = 3w^2 + \sin w$. This is just a reminder that you must always pay close attention to what variables are at play and what is asked for.

26. Of the following equations, which defines y explicitly as a function of x, and which defines y implicitly in terms of x?

I. $y = 3x^2 + 7x - 1$

II. $y = x^y - \sin^2 x$

III. $3x + \ln x + \dfrac{1}{x} = y$

IV. $x^5 + 13x^4 - 7y^2 + y^4 = 11$

A. Explicitly: I only. Implicitly: II, III and IV

B. Explicitly: I and II. Implicitly: III and IV

C. Explicitly: II and IV. Implicitly: I and III

D. Explicitly: I and III. Implicitly: II and IV

E. Explicitly: I, II and III. Implicitly: IV only

Answer   D. Equations I and III are explicit because they're in the form $y = f(x)$ where $f(x)$ is a function involving x only. Equations II and IV are NOT in this form, so they're implicit.

27. Here is the curve generated by the equation $x^3 + y^3 = 6xy$:

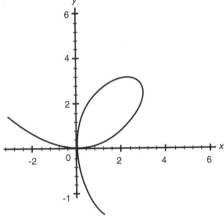

Find the slope of the tangent line to this curve at the point $(3, 3)$.

A. Slope $= -3$

B. Slope $= -\frac{1}{3}$

C. Slope $= -1$

D. Slope $= 3$

E.   Not enough information

Answer   C. Differentiate the equation implicitly with respect to x, treating y as a function of x:

$$\frac{d}{dx}(x^3 + y^3) = 3x^2 + 3y^2\frac{dy}{dx} = 6\left(y + x\frac{dy}{dx}\right) = \frac{d}{dx}(6xy)$$

.

Now solve this equation for $\frac{dy}{dx}$. After dividing both sides by 3, you get

$$\frac{dy}{dx}(y^2 - 2x) = 2y - x^2 \Rightarrow \frac{dy}{dx} = \frac{2y - x^2}{y^2 - 2x}.$$

Now plug in the given values $(x, y) = (3, 3)$ to get

$$\frac{dy}{dx} = \frac{6 - 9}{9 - 6} = -\frac{3}{3} = -1.$$

---

28.   Below is the curve generated by the equation $x^{\frac{2}{3}} + y^{\frac{2}{3}} = 4$:

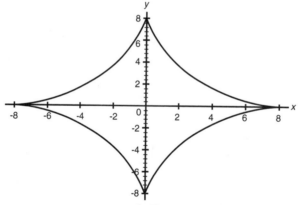

Find the slope of the tangent to the curve at the point $(\sqrt{27}, -1)$.

A.   Slope $= \frac{\sqrt{3}}{3}$

B.   Slope $= -\frac{\sqrt{3}}{3}$

C.   Slope $= \frac{1}{3}$

D.   Slope $= -\frac{1}{3}$

E.   Not enough information

Answer A. First check to see if the point $(\sqrt{27}, -1)$ is on the curve. (It is.)

Next, differentiate with respect to x:

$$\frac{d}{dx}\left(x^{\frac{2}{3}}\right) + \frac{d}{dx}\left(y^{\frac{2}{3}}\right) = \frac{d}{dx}(4)$$

$$\Rightarrow \frac{2}{3}x^{-\frac{1}{3}} + \frac{2}{3}y^{-\frac{1}{3}}\frac{dy}{dx} = 0$$

$$\Rightarrow x^{-\frac{1}{3}} + y^{-\frac{1}{3}}\frac{dy}{dx} = 0.$$

Now plug in the specific point $x = \sqrt{27}$, $y = -1$, and get

$$(\sqrt{27})^{-\frac{1}{3}} + (-1)^{-\frac{1}{3}}\frac{dy}{dx} = 0 \Rightarrow \frac{1}{(\sqrt{27})^{\frac{1}{3}}} - 1\frac{dy}{dx} = 0$$

$$\Rightarrow \frac{1}{\sqrt{27^{\frac{1}{3}}}} = \frac{dy}{dx} \Rightarrow \frac{1}{\sqrt{3}} = \frac{dy}{dx} \Rightarrow \frac{\sqrt{3}}{3} = \frac{dy}{dx}.$$

So the slope of the tangent line to the curve at $(\sqrt{27}, -1)$ is $\frac{\sqrt{3}}{3}$.

---

29. Suppose $xy = 1$. Find the value of $\frac{dy}{dx}$ implicitly at $x = -2$.

A. $\frac{3}{4}$

B. $\frac{1}{2}$

C. 0

D. $-\frac{1}{4}$

E. Not enough information

Answer D. First, find the appropriate y value when $x = -2$.

That value is $y = -\frac{1}{2}$, since $-2 \times \left(-\frac{1}{2}\right) = 1$. Next, differentiate implicitly:

$$\frac{d}{dx}(xy) = \frac{d}{dx}(1) \Rightarrow y + x\frac{dy}{dx} = 0.$$

Then plug in $x = -2$, $y = -\frac{1}{2}$:

$$-\frac{1}{2} + (-2)\frac{dy}{dx} = 0 \Rightarrow \frac{dy}{dx} = -\frac{1}{4}.$$

---

30. Suppose $x^2 - 2y^2 = 8$. Find the value of $\frac{dy}{dx}$ at $x = 4$.

A. 1

B. 2

C. 0

D. -1

E. Not enough information

Answer E. You can certainly differentiate implicitly here, to get

$$\frac{d}{dx}(x^2) - \frac{d}{dx}(2y^2) = \frac{d}{dx}(8) \Rightarrow 2x - 4y\frac{dy}{dx} = 0.$$

But then you have to plug in values for x and y. You know $x = 4$ but you can't determine what y is, since both $y = 2$ and $y = -2$ satisfy the equation. You haven't been given enough information here.

---

31. Find $\frac{dy}{dx}$ when $x^2 - xy + y^4 = 17$.

A. $\frac{dy}{dx} = \frac{-2x}{4y^3 - x}$

B. $\frac{dy}{dx} = \frac{y - 2x}{4y^3 - x}$

C. $\frac{dy}{dx} = \frac{y - 2x}{4y^3}$

D. $\frac{dy}{dx} = \frac{17 - 2x}{4y^3 - x}$

E. Not enough information

Answer B. Differentiating implicitly and solving for $\frac{dy}{dx}$, you get

$$\frac{d}{dx}(x^2) - \frac{d}{dx}(xy) + \frac{d}{dx}(y^4) = \frac{d}{dx}(17)$$
$$\Rightarrow 2x - \left(y + x\frac{dy}{dx}\right) + 4y^3\frac{dy}{dx} = 0$$
$$\Rightarrow (4y^3 - x)\frac{dy}{dx} = y - 2x \Rightarrow \frac{dy}{dx} = \frac{y - 2x}{4y^3 - x}.$$

---

32. Find $\frac{dy}{dx}$ when $\cos(x + 3y) = x\tan(y)$.

A. $\frac{dy}{dx} = \frac{-\tan(y)}{x\sec^2(y) + 3\sin(x + 3y)}$

B. $\frac{dy}{dx} = -\frac{\sin(x + 3y)}{x\sec^2(y) + 3\sin(x + 3y)}$

C. $\dfrac{dy}{dx} = -\dfrac{\sin(x+3y) + \tan(y)}{x \sec^2(y)}$

D. $\dfrac{dy}{dx} = -\dfrac{\sin(x+3y) + \tan(y)}{\sec^2(y) + \cos(x+3y)}$

E. $\dfrac{dy}{dx} = -\dfrac{\sin(x+3y) + \tan(y)}{x \sec^2(y) + 3\sin(x+3y)}$

Answer E.

$$\frac{d}{dx}(\cos(x+3y)) = \frac{d}{dx}(x\tan(y))$$

$$\Rightarrow -\sin(x+3y)\left(1 + 3\frac{dy}{dx}\right) = \tan(y) + x\sec^2(y)\frac{dy}{dx}$$

$$\Rightarrow -\sin(x+3y) - \tan(y) = x\sec^2(y)\frac{dy}{dx} + 3\sin(x+3y)\frac{dy}{dx}$$

$$\Rightarrow -(\sin(x+3y) + \tan(y)) = (x\sec^2(y) + 3\sin(x+3y))\frac{dy}{dx}$$

$$\Rightarrow \frac{dy}{dx} = -\frac{\sin(x+3y) + \tan(y)}{x\sec^2(y) + 3\sin(x+3y)}.$$

Free-Response Questions on the Chain Rule and Implicit Differentiation

---

1. Air is being pumped into a spherical weather balloon. The radius (in feet) of the balloon at any time $t \geq 0$ is given by the (differentiable) function $r(t)$, where t is in minutes. The volume (in cubic feet) of the balloon is a function of the radius, denoted by $V(r)$.

A. Explain what each of the derivatives $\dfrac{dr}{dt}$, $\dfrac{dV}{dr}$, and $\dfrac{dV}{dt}$ represent. Be sure to include the units for each of the derivatives.

B. Express $\dfrac{dV}{dt}$ in terms of $\dfrac{dr}{dt}$ and r.

---

2. The length-mass relationship for Pacific halibut is well represented by the equation

$$M = 10.375L^3,$$

where M is mass in kilograms and L is the length in meters. In addition, the formula $.18(2 - L)$ represents the rate of growth in length, $\dfrac{dL}{dt}$, where t is time in years.

A. Describe what each of the derivatives $\dfrac{dM}{dL}$ and $\dfrac{dM}{dt}$ represent. Be sure to include the units for each of the derivatives.

B. Find $\dfrac{dM}{dt}$ in terms of L.

C.   Find the rate of growth (with respect to time) of the mass of a Pacific halibut at the instant that it has a mass of 15 kilograms.

---

3.   For each of the following, express both y and $\dfrac{dy}{dx}$ in terms of x.

For example, given $y = u^2$, $u = 2x - 5$, you'd write

$$y = u^2 = (2x - 5)^2, \text{ and } \frac{dy}{dx} = \frac{dy}{du}\frac{du}{dx} = (2u)(2) = 4u = 4(2x - 5).$$

A. $y = u^3; u = 3x^2 + 2x$

B. $y = u^{\frac{1}{2}}; u = \sin x$

C. $y = \sin u; u = -2\cos x$

D. $y = u \sec u; u = 3x$

---

4.   For each of the following, find $\dfrac{dy}{dx}$.

A. $y = (x^2 + 3)^6$

B. $y = \sqrt{2x^{-3} - 7x}$

C. $y = 2\tan(-5x^{-2})$

---

5.   Suppose f is a differentiable function of x. Use the chain rule to prove each of the following.

A.   If f is an odd function, $f'(x)$ is an even function.

B.   If f is an even function, then $f'(x)$ is an odd function.

---

6.   Suppose f and g are differentiable functions of x on the domain of all real numbers. The table below gives some values for the functions and their derivatives at a few values of x.

x	$f(x)$	$f'(x)$	$g(x)$	$g'(x)$
-2	-2	0	5	-2
0	1	-2	3	-3
2	5	2	4	-2
4	-2	-4	-5	2

A.   If $h = f \circ g$, what is the value of $h'(2)$?

B.   If $h = f \circ f$, what is the value of $h'(4)$?

C.   If $h(x) = f(x^3 + 1)$, then what is the value of $h'(1)$?

D. If $h(x) = \sqrt{g(x)}$, what is the value of $h'(-2)$?

___

7. For each of the following, find $\dfrac{dy}{dx}$.

A. $y = \sin^2(\cos(3x^{-2}))$

B. $y = \sqrt{1 + \sqrt{1 + \sqrt{x}}}$

C. $y = \tan\left(\dfrac{(x+4)^7}{x}\right)$

___

8. Answer the questions below.

A. What is the 100th derivative of $\sin(3x)$. (Hint: Find the pattern here.)

B. If $y = 4\sec(3x^{-3} - 1) + \pi$, what is the derivative of y with respect to $(3x^{-3} - 1)$?

___

9. Find $\dfrac{dy}{dx}$ for each of the following equations.

A. $4x^2 + y^2 = 20$

B. $x^2 + 5x^2y + y^3 = 12$

C. $x^2 + \sqrt{xy} = 13$

D. $\cos(xy) = \sqrt{\sin(y)}$

E. $xy = \sec(y)$

___

10. Find the slope of the tangent line to the curve at the given point.

A. $\dfrac{x^2}{25} + \dfrac{y^2}{36} = 1$, at the point $\left(-3, \dfrac{24}{5}\right)$.

B. $5x^2 + 4y^2 = 56$, at the point $(2, -3)$.

C. $xy + 16 = 0$, at the point $(-8, 2)$

11. Consider the curve generated by the equation $y^2 = x^3(2 - x)$, depicted below.

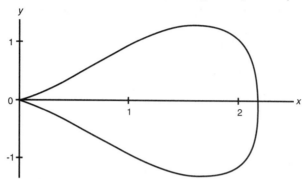

This classic curve is called the piriform. Find the points on the piriform other than $(0, 0)$ where the tangent is horizontal.

12. Two curves are called *orthogonal* if at each point of intersection their tangent lines are perpendicular. Show that the two curves $xy = c$ and $x^2 - y^2 = k$ are orthogonal, for any constants $c \neq 0$ and $k \neq 0$.

13. Consider the curve defined by $x^2 + xy + y^2 = 27$.

A. Write an expression for the slope $\dfrac{dy}{dx}$ of the curve at any point (x, y).

B. Determine whether the lines tangent to the curve at the x-intercepts of the curve are parallel. Show the analysis that leads to your conclusion.

C. Find the points on the curve where the lines tangent to the curve are vertical.

Free-Response Questions: Answers

1.

A. $\dfrac{dr}{dt}$ represents the rate of change in the radius with respect to time. It has units "feet per minute."

$\dfrac{dV}{dr}$ represents the rate of change in the volume with respect to the radius. It has units "cubic feet per foot."

$\dfrac{dV}{dt}$ represents the rate of change in the volume with respect to the time. It has units "cubic feet per minute."

B.

$$V = \frac{4}{3}\pi r^3.$$

Thus

$$\frac{dV}{dt} = \frac{dV}{dr}\frac{dr}{dt} = 4\pi r^2 \frac{dr}{dt}.$$

2.

A. $\frac{dM}{dL}$ represents the rate of change in mass with respect to the length of the fish. The units here are "kilograms per meter."

$\frac{dM}{dt}$ represents the rate of change in mass with respect to time (age of fish). The units here are "kilograms per year."

B.

$$\frac{dM}{dt} = \frac{dM}{dL}\frac{dL}{dt}$$
$$= 3(10.375)L^2(.18(2 - L))$$
$$= 31.125L^2(.36 - .18L).$$

C. Since $M = 10.375L^3$, $L = \sqrt[3]{\frac{M}{10.375}}$, so when $M = 15$, $L = \sqrt[3]{\frac{15}{10.375}} \approx \sqrt[3]{1.44578} \approx$ 1.1308. Then, by the formula from part B, $\frac{dM}{dt} = 31.125L^2(.36-.18L) \approx 6.227$ kilograms per year.

3.

A.

$$y = (3x^2 + 2x)^3, \quad \text{while } \frac{dy}{dx} = \frac{dy}{du}\frac{du}{dx} = (3u^2)(6x + 2) = 3(3x^2 + 2x)^2(6x + 2).$$

B.

$$y = \sin^{\frac{1}{2}} x, \quad \text{while } \frac{dy}{dx} = \frac{dy}{du}\frac{du}{dx} = \left(\frac{1}{2}u^{-\frac{1}{2}}\right)(\cos x) = \frac{1}{2}\left(\sin^{-\frac{1}{2}} x\right)(\cos x).$$

C.

$$y = \sin(-2\cos x), \quad \text{while } \frac{dy}{dx} = \frac{dy}{du}\frac{du}{dx} = (\cos u)(2\sin x) = 2\sin(x)\cos(-2\cos x).$$

D.

$$y = (3x)\sec(3x), \quad \text{while } \frac{dy}{dx} = \frac{dy}{du}\frac{du}{dx} = (\sec u + u\sec u\tan u)(3)$$
$$= 3(\sec(3x) + (3x)\sec(3x)\tan(3x)).$$

4.

A.

$$\frac{dy}{dx} = 6(x^2 + 3)^5(2x)$$

B.

$$y = (2x^{-3} - 7x)^{\frac{1}{2}}, \ \text{ so } \frac{dy}{dx} = \frac{1}{2}(2x^{-3} - 7x)^{-\frac{1}{2}}(-6x^{-4} - 7).$$

C.

$$\frac{dy}{dx} = 2\sec^2(-5x^{-2})(10x^{-3})$$

5.

A. Since f is odd, $f(-x) = -f(x)$. Differentiating both sides of the equation you get $f'(-x)(-1) = -f'(x)$ (by chain rule). So, $f'(-x) = f'(x)$, and $f'(x)$ is an even function.

B. Since f is even, $f(-x) = f(x)$. Differentiating both sides of the equation, you get $f'(-x)(-1) = f'(x)$ (by the chain rule). So, $f'(-x) = -f'(x)$, and $f'(x)$ is an odd function.

(You can see evidence that these are true by looking at even and odd degrees of x^n and the trig functions, and their graphs.)

6.

A.

$$h'(2) = f'(g(2)) \times g'(2) = (f'(4))(-2) = (-4)(-2) = 8.$$

B.

$$h'(4) = f'(f(4)) \times f'(4) = (f'(-2))(-4) = 0(-4) = 0.$$

C.

$$h'(x) = f'(x^3 + 1)(3x^2).$$
$$h'(1) = (f'(2))(3) = 2(3) = 6.$$

D.

$$h'(x) = \frac{1}{2}(g(x))^{-\frac{1}{2}}g'(x).$$
$$h'(-2) = \frac{1}{2}(g(-2))^{-\frac{1}{2}}g'(-2) = \frac{1}{2}(5)^{-\frac{1}{2}}(-2) = -\frac{1}{\sqrt{5}} \approx -0.447.$$

7.

A.

$$\frac{dy}{dx} = 2\sin(\cos(3x^{-2}))\cos(\cos(3x^{-2}))(-\sin(3x^{-2}))(-6x^{-3})$$
$$= 12x^{-3}\cos(\cos(3x^{-2}))\sin(\cos(3x^{-2}))\sin(3x^{-2})$$

B.

$$y = \left(1 + \left(1 + x^{\frac{1}{2}}\right)^{\frac{1}{2}}\right)^{\frac{1}{2}}.$$

So, $\frac{dy}{dx} = \frac{1}{2}\left(1 + \left(1 + x^{\frac{1}{2}}\right)^{\frac{1}{2}}\right)^{-\frac{1}{2}}\left(\frac{1}{2}\left(1 + x^{\frac{1}{2}}\right)^{-\frac{1}{2}}\left(\frac{1}{2}x^{-\frac{1}{2}}\right)\right)$

C.

$$\frac{dy}{dx} = \sec^2\left(\frac{(x+4)^7}{x}\right)\left(\frac{7x(x+4)^6 - (x+4)^7}{x^2}\right)$$

8.

A.

$$3^{100}\sin(3x).$$

B.

$4\sec(3x^{-3} - 1)\tan(3x^{-3} - 1).$

This follows by setting $u = 3x^{-3} - 1$. Then $y = 4\sec(u) + \pi$, and

you're just looking for $\frac{dy}{du} = 4\sec u \tan u.$

9.

A.

$$8x + 2y\frac{dy}{dx} = 0 \Rightarrow \frac{dy}{dx} = -\frac{4x}{y}.$$

B.

$$2x + 10xy + 5x^2\frac{dy}{dx} + 3y^2\frac{dy}{dx} = 0 \Rightarrow \frac{dy}{dx} = -\frac{2x + 10xy}{5x^2 + 3y^2}.$$

C.

$$2x + \frac{1}{2}(xy)^{-\frac{1}{2}}\left(y + x\frac{dy}{dx}\right) = 0 \Rightarrow \frac{dy}{dx} = \frac{-4x(xy)^{\frac{1}{2}} - y}{x}.$$

D.

$$-\sin(xy)\left(y + x\frac{dy}{dx}\right) = \frac{1}{2}(\sin(y))^{-\frac{1}{2}}\cos(y)\frac{dy}{dx}$$

$$\Rightarrow \frac{dy}{dx} = \frac{-y\sin(xy)}{\frac{1}{2}(\sin(y))^{-\frac{1}{2}}\cos(y) + x\sin(xy)}.$$

E.

$$y + x\frac{dy}{dx} = \sec(y)\tan(y)\frac{dy}{dx}$$

$$\Rightarrow \frac{dy}{dx} = \frac{y}{\sec(y)\,\tan(y) - x}.$$

10.

A. $\frac{9}{10}$. Note that $\frac{2x}{25} + \frac{2y}{36}\frac{dy}{dx} = 0 \Rightarrow \frac{dy}{dx} = \frac{-36x}{25y}$.

Thus, the slope of the tangent line to the curve at the point $(-3, \frac{24}{5})$ is $\frac{-36(-3)}{25(\frac{24}{5})} = \frac{9}{10}$.

B. $\frac{5}{6}$. Note that $10x + 8y\frac{dy}{dx} = 0 \Rightarrow \frac{dy}{dx} = \frac{-5x}{4y}$.

Thus, the slope of the tangent line to the curve at the point $(2, -3)$ is $\frac{-5(2)}{4(-3)} = \frac{5}{6}$.

C. $\frac{1}{4}$. Note that $y + x\frac{dy}{dx} = 0 \Rightarrow \frac{dy}{dx} = \frac{-y}{x}$.

Thus, the slope of the tangent line to the curve at the point $(-8, 2)$ is $\frac{-(2)}{-8} = 0.25$.

11. The points are $(\frac{3}{2}, \frac{3\sqrt{3}}{4})$ and $(\frac{3}{2}, -\frac{3\sqrt{3}}{4})$.

Differentiate the equation $y^2 = x^3(2 - x)$ implicitly to find y':

$$2yy' = 6x^2 - 4x^3 \Rightarrow y' = \frac{6x^2 - 4x^3}{2y}.$$

The tangent line is horizontal when the numerator is 0 and the denominator is not.

Set $6x^2 - 4x^3 = 0 \Rightarrow x^2(6 - 4x) = 0 \Rightarrow x = 0, \frac{3}{2}$.

When $x = 0$, the equation of the curve gives you $y = 0$, but we want points other than $(0, 0)$ where the tangent is horizontal.

When $x = \frac{3}{2}$, you have $y^2 = (\frac{3}{2})^3(2 - \frac{3}{2}) = \frac{27}{8} \times \frac{1}{2} = \frac{27}{16} \Rightarrow y = \pm\frac{3\sqrt{3}}{4}$.

(And here y' really is zero.) So the points are $(\frac{3}{2}, \frac{3\sqrt{3}}{4})$ and $(\frac{3}{2}, -\frac{3\sqrt{3}}{4})$.

(And remember, we asked for points, so you need to have found both the x- and the y-values.)

12. For the first curve, $x\frac{dy}{dx} + y = 0 \Rightarrow \frac{dy}{dx} = \frac{-y}{x}$.

For the second curve, $2x - 2y\frac{dy}{dx} = 0 \Rightarrow \frac{dy}{dx} = \frac{x}{y}$.

The derivatives are negative reciprocals of each other, therefore the tangent lines to the curves at identical points are perpendicular.

13.

A.

$$2x + y + x\frac{dy}{dx} + 2y\frac{dy}{dx} = 0 \Rightarrow \frac{dy}{dx} = \frac{-2x - y}{x + 2y}.$$

B. The slope at the x-intercepts is given by $\frac{-2x - 0}{x + 2(0)} = \frac{-2x}{x} = -2$, as long as x is not zero.

The x-intercepts are $x = \pm 3\sqrt{3}$.

(You need to check this to be sure that an x-intercept doesn't occur at $x = 0$, where the derivative is not well defined. You can also simply observe that the equation of the curve clearly doesn't have solution $x = 0, y = 0$.)

Thus the lines tangent to the curve at the x-intercepts of the curve are parallel, since all the slopes are the same.

C. The points on the curve where the tangent lines are vertical are $(6, -3)$ and $(-6, 3)$.

You need to look for places where the derivative $\frac{dy}{dx}$ is undefined. That is, you want to look for where the denominator $x + 2y$ is 0 and the numerator is not.

Set $x = -2y$ and plug into the equation for the curve. You get

$$(-2y)^2 + (-2y)y + y^2 = 27 \Rightarrow 4y^2 - 2y^2 + y^2 = 27 \Rightarrow 3y^2 = 27 \Rightarrow y = \pm 3,$$

when $y = -3$, $x = 6$ and when $y = 3$, $x = -6$.

Chapter 5   *Differential Calculus*

A. Extrema and Optimization

▶Objective 1   Know the meanings of relative and absolute extrema.

Example   Refer to the graph below. The domain for this function is $[-4, 4]$.

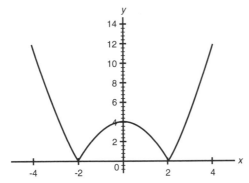

True or False: There's an absolute maximum at $x = 0$.

Tip   To find absolute extrema, compare the y-values at the critical points and at the endpoints (if any).

Answer   False. This graph has a relative max at $x = 0$, but the y-values are higher at both $x = -4$ and $x = 4$.

►Objective 2   Identify when a relative maximum or minimum will occur in a function (find the critical points, which are the points at which $y' = 0$ or y' is undefined).

Example   Find all the critical points of $y = \sec^3 x$ for x between -3.5 and 3.5.

Tip   A critical point can be either where $y' = 0$ or where y' is undefined (but the function is defined).

Answer   The three critical points are $(-\pi, -1)$, $(0, 1)$, and $(\pi, -1)$.

You need to find all the places where $y' = 0$ or y' is undefined on this interval. Here, the derivative of $y = \sec^3 x$ is $y' = 3\sec^3 x \tan x$, which is zero whenever $\sec x = 0$ (never) or $\tan x = 0$ (at $x = -\pi, 0,$ and π) and is undefined (at $\pm\frac{\pi}{2}$). But the original function is also undefined at $\pm\frac{\pi}{2}$, so these don't give critical points. The three critical points are $(-\pi, -1)$, $(0, 1)$, and $(\pi, -1)$. The graph of the function confirms this.

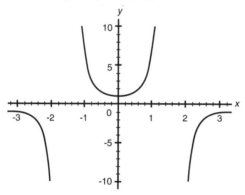

►Objective 3   Use a first derivative number-line test to identify relative extrema (analyze critical points).

Example   Analyze the three critical points found in the problem above ($y = \sec^3 x$ for x between -3.5 and 3.5) using the first derivative test.

Tips   It isn't enough to just find a critical point—you have to analyze it to be sure you've found a max or min. Think of the graph of $y = x^3$—it has a critical point at $x = 0$, but that's neither a max nor a min.

In explaining your analysis, be sure to label your number-line tests with y' or f'. Also, be sure to find the actual minimum or maximum y-values when asked for extreme values—just the x-values alone won't be enough.

Divide the number-line into intervals indicating the critical points, and points of discontinuity. Check the sign of the derivative on each interval (label with $+$ or $-$), and make sure to indicate that these correspond to values of y'.

Answer   Your function is $y = \sec^3 x$, and your critical points are $(-\pi, -1)$, $(0, 1)$, and $(\pi, -1)$.

The three critical points, together with the places where y has asymptotes, break the x-axis into several intervals:

x	$x < -\pi$	$x = -\pi$	$-\pi < x < -\frac{\pi}{2}$	$x = -\frac{\pi}{2}$	$-\frac{\pi}{2} < x < 0$
y'	pos	0	neg	undef	neg
y	inc	MAX	dec	undef	dec

x	$x = 0$	$0 < x < \frac{\pi}{2}$	$\frac{\pi}{2} < x < \pi$	$x = \pi$	$\pi < x$
y'	0	pos	pos	0	neg
y	MIN	inc	inc	MAX	dec

The sign chart, above, tells you that y has a relative max at $x = -\pi$, a relative min at $x = 0$, and a relative max at $x = \pi$. This matches what you see in the graph of y.

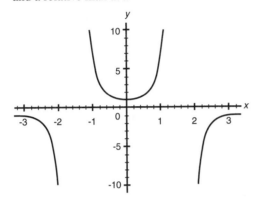

►Objective 4   Use the second derivative test to identify relative extrema (analyze critical points).

Example   Find all relative extrema of $y = x^5 - 5x^4$.

Tips   Remember that the second derivative test won't always work! If $y'' = 0$, it means the test failed, and you have to use the first derivative test instead.

$y'' = 0$ does not mean that there is no max or min. It simply means the test has failed, and you have to try something else.

You can use the second derivative test only on critical points where $y' = 0$. At critical points where the derivative is undefined, you should use the first derivative test.

Answer   First, you find the critical points $y' = 5x^4 - 20x^3 = 0$ at $x = 0$ and at $x = 4$. This y' is never undefined. Now, use the second derivative test to analyze these critical points: $y'' = 20x^3 - 60x^2$. At $x = 0$, you get $y'' = 0$, so the second derivative test fails. Using the first derivative test now, at $x = 0$, you find that the derivative goes from positive to negative, so there's a relative maximum at $x = 0$. At $x = 4$, $y'' > 0$, so there's a relative minimum at $x = 4$. (The second derivative test works here.) The relative minimum is -256, at $x = 4$, and the relative maximum is 0, at $x = 0$. This matches what the graph looks like:

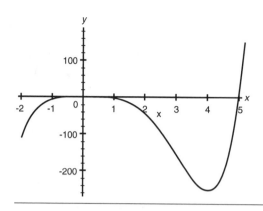

▶Objective 5   Use a second derivative number-line test to find inflection points.

Example   Find all inflection points of $y = x^5 - 5x^4$.

Tips   An inflection point is where y'' changes sign. It isn't enough to look for the zeros of y''—you have to check that the sign changes. Remember the graph of $y = x^4$. There $y'' = 0$ at $x = 0$, but it's a minimum point, not an inflection point.

Answer   First, find where $y'' = 0$. $y'' = 20x^3 - 60x^2 = 0$ at $x = 0$ and at $x = 3$.

Now, check the signs of y'' on each side:

x	$x < 0$	$x = 0$	$0 < x < 3$	$x = 3$	$x > 3$
y''	neg	0	neg	0	pos
y	con. down	NO pt of inflection	con. down	point of inflection	con. up

Note that y'' does not change sign at $x = 0$, so there is no inflection point there. But y'' does change sign at $x = 3$, so y changes concavity there, and $(3, -162)$ is an inflection point.

▶Objective 6   Analyze curves using a combination of the first and second derivative number-line tests.

Example   Consider this graph:

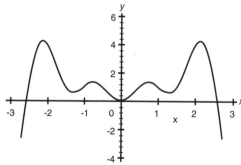

This graph is the graph of h', the derivative of a function h. The domain of the function h is the set of all real numbers x such that $-2.7 \le x \le 2.7$. (So the entire graph of the derivative is shown here.)

A. For what values of x does h have a relative maximum? Justify your answer.

B. On what intervals is the graph of h concave upward? Justify your answer.

Tips   Pay attention to whether a graph is of a function, its derivative, or its second derivative.

Do your best to read a graph correctly. But if the graph is hard to read points from, as this one is, don't worry about your points being slightly different from the answer key.

Answers

A. h will have a relative maximum when h' (shown here) goes from positive to negative (this is the first derivative test). In this case, h will have a relative maximum at about $x = 2.6$.

B. h will be concave upward when h'' is positive, which is when h' is increasing. So h will be concave upward on the intervals $(-2.7, -2.1)$, $(-1.3, -0.8)$, $(0, 0.8)$, and $(1.3, 2.1)$.

(All the numbers here are approximations from the graph—your approximations might be slightly different.)

▶**Objective 7**   Solve optimization problems (including those dealing with volume, area, time, and distance).

Example   A square piece of cardboard, 8 inches on each side, is used to make an open top box by cutting out a small square from each corner and bending up the sides. What size square should be cut from each corner for the box to have the maximum volume?

Tip   Remember the steps to solve such a problem:

1. Identify the quantity to be optimized (look for *-est* words) and the independent variables that will change that quantity.

2. Write the equation that relates the variables to what you want to optimize (drawing a picture will probably help).

3. Consolidate your information until you have a function of just one variable (usually).

4. Use calculus to find its max and min.

5. Remember to check that you have found a max or min, using the first or second derivative test.

6. Be sure to answer the question that's asked (are you asked for dimensions or volume?). Include the correct units.

Answer   You're trying to maximize volume.

If x is the length of the cut-out square, then the equation for volume is

$$V = (8 - 2x)(8 - 2x)x.$$

Solve $V' = 12x^2 - 64x + 64 = 0$ to find $x = 4$ and $x = \frac{4}{3}$.

The domain here is $(0, 4)$, so you only need to check $x = \frac{4}{3}$.

Test $x = \frac{4}{3}$ (with first or second derivative test) and find that it gives a maximum.

The answer: If you cut a square of length $\frac{4}{3}$ inches from each corner and fold up the sides, you get the box with the largest volume.

Exercises on Extrema and Optimization

The first four questions refer to the following graph of a function.

___

1. Given that the domain for this function is $[-1.5, 2.5]$, how many critical points are shown on this graph?

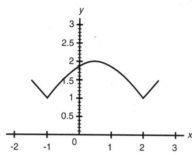

A. 0

B. 1

C. 2

D. 3

E. 5

Answer   E. The critical points are wherever the derivative is 0 or is undefined, including the endpoints of the domain. Here, there are 5. They are at $x = -1$, $x = \dfrac{1}{2}$, and $x = 2$, and at the endpoints of the domain, $x = -1.5$ and $x = 2.5$.

True or False?

___

2. Given that the domain for this function is $[-1.5, 2.5]$, there is a local maximum at $x = 2.5$.

Answer   True. The function is higher at $x = 2.5$ than at all the nearby points, so it is a local maximum.

___

3. Given that the domain for this function is $[-1.5, 2.5]$, there is a global maximum at $x = 2.5$.

Answer   False. A global maximum must be the biggest of all the values of the function over the whole domain. The function is higher at $x = \dfrac{1}{2}$ than at $x = 2.5$, so this is not a global max.

(However, there is a global maximum at $x = \dfrac{1}{2}$.)

___

4. Given that the domain for this function is $[-1.5, 2.5]$, this function has no global minimum.

Answer   False. It has two global minimums: one at $x = -1$ and one at $x = 2$. The global minimum doesn't have to be lower than every other point, it just has to be the lowest possible function value. It's possible for a function to have infinitely many global extrema (think of $y = \sin x$, for example).

5. Suppose you know that f is a differentiable function and the domain is all real numbers, and that $f' < 0$ for x in $[-a, 0)$, $f' = 0$ at $x = 0$, and $f' > 0$ for x in $(0, a]$ for some positive constant a. Then you know that f has a _____ at $x = 0$.

A. local minimum

B. local maximum

C. global minimum

D. global maximum

E. None of these

Answer   A. The function is decreasing on the left of $x = 0$ and increasing on the right, so there's a local minimum.

6. Suppose you know that f is a differentiable function on the interval $(-h, h)$ where $h > 0$ that f'' exists on the interval $(-h, h)$ and that $f'' < 0$ at $x = 0$. Then you know that f has a _____ at $x = 0$.

A. local minimum

B. local maximum

C. global minimum

D. global maximum

E. None of these

Answer   B. The function has a horizontal tangent and is concave down, so there's a local maximum there.

The next two problems refer to the graph of the function illustrated below.

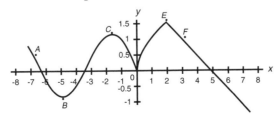

7. At which points is the first derivative equal to 0?

A. B, C, D, E, F

B. B, C

C. D, E

D. B, C, D, E

E. A, F

Answer   B. B and C are the only points where the derivative exists, and the slope of the tangent line is 0.

8. At point B, the second derivative is:

A. 0

B. negative

C. does not exist

D. positive

E. decreasing

Answer   D. The first derivative is getting bigger as you move from left to right in the neighborhood of point B, so the second derivative is positive.

9. Consider the function $y = 3x^5 - 25x^3 + 60x + 1$. Find the critical points of this function.

A. None

B. $-2, -1, 0, 1, 2$

C. $-2, -1, 1, 2$

D. $1, 2$

E. $-2, 2$

Answer   C. You find the critical points by computing the derivative of this function,

$$y' = 15x^4 - 75x^2 + 60 = 15(x^4 - 5x^2 + 4)$$
$$= 15(x^2 - 4)(x^2 - 1)$$
$$= 15(x + 2)(x - 2)(x + 1)(x - 1),$$

and then finding where y' is zero or undefined. Here, it is defined everywhere, so you need to find only the zeros: at $x = -2$, $x = -1$, $x = 1$, and $x = 2$.

10. Consider the function $y = 3x^5 - 25x^3 + 60x + 1$. If the first derivative test is used to decide whether this function has a maximum at $x = 1$, which of the following statements describes the result?

A. The derivative is positive to the left of $x = 1$ and negative to the right of $x = 1$, so the function has a relative minimum at $x = 1$.

B. The derivative is positive to the left of $x = 1$ and negative to the right of $x = 1$, so the function has a relative maximum at $x = 1$.

C. The derivative is positive to the left of $x = 1$ and positive to the right of $x = 1$, so the function has neither a relative maximum nor a minimum at $x = 1$.

D. The derivative is negative to the left of $x = 1$ and positive to the right of $x = 1$, so the function has a relative minimum at $x = 1$.

E. None of these apply.

Answer   B. If the function is increasing on the left of a critical point and then decreasing on the right, it has a relative maximum at that critical point.

The Second Derivative Test

If $y'' > 0$ at a critical point where $y' = 0$, then y has a relative minimum at that critical point.

If $y'' < 0$ at a critical point where $y' = 0$, then y has a relative maximum at that critical point.

If $y'' = 0$ at a critical point, then the test fails. (So you have to use the first derivative test, after all.)

The Second Derivative Test can often be used more quickly than the First Derivative Test to determine whether a critical point where $y' = 0$ is a relative maximum or minimum.

---

11. Consider the function $y = 3x^5 - 25x^3 + 60x + 1$. If the second derivative test is used to determine whether this function has a relative maximum or minimum at $x = -1$, which of the following statements describes the result?

A. $x = -1$ isn't a critical point, so there's neither a max nor a min there.

B. $y'' > 0$ there, so there is a relative minimum at $x = -1$.

C. $y'' < 0$ there, so there is a relative maximum at $x = -1$.

D. $y'' = 0$ there, so there is neither a max nor a min there.

E. $y' = 0$ there, so the second derivative test fails.

Answer   B. $y'' = 60x^3 - 150x$. At $x = -1$, $y'' = 90 > 0$, so the second derivative test says there is a relative minimum at $x = -1$.

12. Consider the function $y = 3x^5 - 25x^3 + 60x + 1$. How many relative maxima does it have?

A. 0

B. 1

C. 2

D. 3

E. 4

Answer   C. This function has a two relative maxima, one at $x = 1$ and one at $x = -2$.

13. Consider the function $y = 3x^5 - 25x^3 + 60x + 1$.
    Find the absolute maximum value of y on the interval $[0, 3]$.

A. 39

B. 3

C. 235

D. -322

E. -34

Answer   C. The absolute maximum value 235 is attained at the endpoint $x = 3$.

14. The function f given by $f(x) = x^3 + 12x - 24$ is:

A. increasing for $x < -2$, decreasing for $-2 < x < 2$, increasing for $x > 2$.

B. decreasing for $x < 0$, increasing for $x > 0$.

C. increasing for all x.

D. decreasing for all x.

E. decreasing for $x < -2$, increasing for $-2 < x < 2$, decreasing for $x > 2$.

Answer   C. The derivative here, $f'(x) = 3x^2 + 12$, is always positive, so this function is always increasing.

15. True or False: A differentiable function must have a relative minimum between any two relative maxima.

Answer True. Think about the first derivative test. Suppose your two maxima are at $x = a$ and $x = b$, where $a < b$. Then the derivative must go from positive to negative at $x = a$, and again from positive to negative at $x = b$. So in between a and b, the derivative must have gone from negative to positive somewhere—that place is a relative minimum. (This is actually the Intermediate Value Theorem again!)

---

16. What is the greatest number of relative maxima that a polynomial of degree 15 can have?

A. 4

B. 6

C. 7

D. 9

E. 14

Answer C. The derivative of this polynomial will have degree 14, so it can have at most 14 zeros, so our polynomial can have at most 14 critical points.

Because critical points have to alternate between maxima and minima, no more than half of them, or 7, can be relative maxima. (Of course, there don't have to be as many as 7 relative maxima—the derivative doesn't have to have 14 distinct zeros, so you may not have 14 critical points. And some of the critical points may turn out to be neither a maximum nor a minimum.)

Free-Response Practice on Extrema

Note: It's most important to get the connections here between the derivative and second derivative and the original function. You need to have the first and second derivative test down cold. Many students forget to use any kind of test when they find extrema—they find the critical points and go straight to the answer. But you have to be able to justify that you've really found a max or min (especially when you get to the exams). In fact, it's a good idea to write down your number line test, properly labeled, whenever you're doing any kind of max/min problem.

The mechanics, like correctly finding a derivative and correctly solving an equation, are important, but not as important as understanding what the process is here.

---

1. Let f be the function given by $f(x) = 12x^5 - 15x^4 - 20x^3 + 30x^2 + 8$.

A. Find f'.

B. Find all critical points of f.

C. Use the second derivative test to determine whether these points are relative maxima or minima. Show the test for each critical point. If the test fails, say, "The second derivative test fails."

D. Use the first derivative test to nail down any point for which the second derivative test didn't work.

E. Find the absolute maximum and minimum of f on $[-1.5, 1.5]$. Show how you know (using calculus, not your calculator).

---

2. Let g be the function given by $g(x) = x^3 - 5x^2 + 3x + k$, where k is a constant.

A. On what intervals is g increasing? Justify your answer.

B. On what intervals is g concave downward? Justify your answer.

C. Find the value of k for which g has 11 as its relative minimum.

---

3. Consider the graph below:

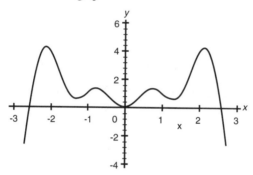

This graph is of h', the derivative of a function h. The domain of the function h is the set of all real numbers x, such that $-2.7 \le x \le 2.7$ (so, the entire graph of the derivative is shown here).

Warning for this entire problem: Read the question carefully. You're looking at a graph of the derivative, while answering questions about the unseen function.

A. Find the critical points of h.

B. For what values of x does h have a relative maximum? Justify your answer.

C. For what values of x does h have a relative minimum? Justify your answer.

D. On what intervals is the graph of h concave upward? Justify your answer.

Free-Response Practice on Extrema—Answers

1.

A. $y' = 60x^4 - 60x^3 - 60x^2 + 60x$.

B. Solve $y' = 0$. This y' actually factors as $y' = 60x(x-1)^2(x+1)$, so the critical values are $x = 0$, $x = 1$, and $x = -1$. Of course, it would be OK to solve this with your calculator.

C. $y'' = 240x^3 - 180x^2 - 120x + 60$. Second derivative test works to show that there is a minimum at $x = 0$, and a maximum at $x = -1$. Second derivative test fails at $x = 1$. (At this point, it is *incorrect* to say "so there is no max or min at $x = 1$." You have to use some other test.)

D. First derivative test shows y' is positive on both sides of $x = 0$, so there is neither a max nor a min at $x = 0$.

E. All the critical points are inside this interval, so compare the values of y at $x = -1.5$, $x = -1$, $x = 0$, and $x = 1.5$. Absolute max is at $x = -1$, where $y = 31$, and absolute min is at $x = -1.5$, where $y = -24$, or so. It would be OK to use the calculator graph to pick which of your critical points and endpoints were absolute max and min, but you have to use calculus to know what those points of interest are. For example, it wouldn't be correct to find them by tracing your calculator graph.

2.

A. $y' = 3x^2 - 10x + 3$. Solve for critical points ($x = 3$ and $x = \frac{1}{3}$), and check signs. Choose intervals where y' is positive: $(-\infty, \frac{1}{3}) \cup (3, \infty)$. Open or closed intervals are fine.

B. $y'' = 6x - 10$. Solve for where $y'' = 0$ (at $x = \frac{5}{3}$), and check signs. Choose interval where y'' is negative: $(-\infty, \frac{5}{3})$. Open or closed intervals are fine.

C. From part A, you know that the derivative changes from negative to positive at $x = 3$. So the relative minimum is at $x = 3$. Plugging it in, you get $g(3) = -9 + k = 11$, or $k = 20$.

3.

A. The critical points of h are where h' is zero or undefined. That's at about $x = -2.6$, $x = 0$, and $x = 2.6$, as well as the endpoints, $x = -2.7$ and $x = 2.7$. Because the graph is hard to read clearly, there is some leeway on the -2.6 and 2.6—but the zero should be easy to read, and the endpoints of the domain are listed in the problem.

B. The maximum will be where h' changes sign from positive to negative. This occurs at about $x = 2.6$.

C. The minimum will be where h' changes sign from negative to positive. This occurs at about $x = -2.6$.

D. The graph of h will be concave upward whenever the graph is increasing: for
    $$-2.7 < x < -2, \quad -1.4 < x < -0.8, \quad 0 < x < 0.8, \quad 1.4 < x < 2.$$
    Either open or closed intervals are fine since the actual endpoints of these intervals are estimates from the graph—some variation from these numbers is fine.

Multiple-Choice Exercises on Optimization

Problem: Suppose you want to find the dimensions of the lightest cylindrical can containing 0.25 liters (250 cubic centimeters) if the top and bottom are made of a material that is three times as heavy (per unit area) as the material used for the side.

1. What is the quantity to be maximized or minimized?

A. Weight

B. Volume

C. Height

D. Radius

E. Density

Answer   A. You want to find the dimensions of the can with the smallest weight.

2. What one word in the problem is your key to whether you're maximizing or minimizing?

A. Lightest

B. Dimensions

C. Containing

D. Heavy

E. Find

Answer   A. The key words are often *-est* words. In this problem, you want to find the *lightest* can, so you're minimizing weight.

3. What variable(s) will affect the weight?

I. Height

II. Radius

III. Surface Area

IV. Volume

V. Choice of materials

VI. Density

A. IV

B. II

C. I and II

D. II and III

E.   All of the above

Answer   C. Height and radius are the two variables in this optimization problem.

___

4.   Choose the equation that relates the variables r (radius), h (height), and k (density of the sides) to the quantity you're trying to minimize, w (weight).

A. $w = 2k\pi r^2 + 2k\pi rh$

B. $w = 2k\pi r^2 + 6k\pi rh$

C. $w = 6k\pi r^2 + 2k\pi rh$

D. $w = 3k\pi r^2 + 2k\pi rh$

E.   None of the above

Answer   C. The top and bottom of the can are circles, and their weight is three times that of the wall. The area of each lid is πr^2, you have two of them, and they each have a density of $3k$ for the heavier weight, so they contribute $6k\pi r^2$ to the total weight. The wall of the can has area $2\pi rh$ with a density of k, so it contributes $2k\pi rh$ to the weight. So $w = 6k\pi r^2 + 2k\pi rh$.

___

5.   Choose the correct equation representing the weight of the can as a function involving only k and one other variable. (Use r as the variable.)

A. $w = 6k\pi r^2 + 500k\pi r$

B. $w = 6k\pi r^2 + 2k\pi rh$

C. $w = 6k\pi r^2 + kr\sqrt{\frac{250}{\pi}}r$

D. $w = 6k\pi r^2 + \frac{500k}{r}$

E. $w = 6k\pi r^2 + \frac{500k}{\sqrt{r}}$

Answer   D. The fixed volume $\pi r^2 h = 250$ lets you solve for $h = \frac{250}{\pi r^2}$. Then you can substitute this expression into the expression for weight to get a function of just the one variable r.

___

6.   Find the dimensions r and h of the can, carried out to three places past the decimal point.

A. $r = 2.367 \quad h = 7.101$

B. $r = 2.367 \quad h = 14.202$

C. $r = 2.3 \quad h = 14.2$

D. $r = 0 \quad h = 28$

E. $r = 1.19 \quad h = 14.202$

Answer   B. Find a minimum, using the first or second derivative test.

B. Tangent and Normal Lines

▶Objective 1   Write the equation of the tangent line to a curve at a point (using implicit differentiation when necessary).

Example   Find the line tangent to the curve $x^2 + 16y^2 + 96y + 76 = 0$, at the point $(2, -1)$.

Tips   What do you need to write the equation of a line? A point and a slope. The slope of the tangent line is the value of the derivative at that point.

You don't have to put your equation into any particular form—any correct answer is fine (unless you're given specific instructions for some reason).

Answer   Find $\dfrac{dy}{dx}$ (implicitly) to find the slope: $\dfrac{dy}{dx} = -0.0625$. That slope, combined with the given point, lets you write the equation of the line: $y + 1 = -0.0625(x - 2)$.

▶Objective 2   Write the equation of the normal line to a curve at a point (using implicit differentiation when necessary).

Example   Find the line normal to the curve $x^2 + 16y^2 + 96y + 76 = 0$ at the point $(2, -1)$.

Tip   The slope of the normal line is just the negative reciprocal of the derivative (the slope of the tangent line).

Answer $y + 1 = 16(x - 2)$.

Find $\dfrac{dy}{dx}$ (implicitly) to find the slope of the tangent line: $\dfrac{dy}{dx} = -0.0625$. The slope you want is the negative reciprocal of that, or $m = 16$. That slope, combined with the given point, lets you write the equation of the line: $y + 1 = 16(x - 2)$.

▶Objective 3   Write the equation of the normal or tangent line, given the curve and a point not on the curve.

Example   Find an equation for a normal line to the curve $y = 4x^3 + 12x - 3$ that passes through the point $(1, 1)$.

Tip   Use the slope (found by looking at the derivative) to set up the equations to solve.

Answer   To find this line, first take the derivative ($y' = 12x^2 + 12$), then use its negative reciprocal in the point-slope form of the line.

This gives you an equation that you solve for x (with your calculator):

$$4x^3 + 12x - 4 = -\frac{1}{12x^2+12}(x - 1).$$

Here's one solution, at about $x = 0.32601$. At that value of x, the slope of the line here is -0.07533, so the equation for the line is

$$y - 1 = -0.07533(x-1).$$

▶Objective 4 Identify areas of local linearity (and absences of local linearity) on a graph; explain the concept of local linearity.

Example Here's the graph of a continuous function $g(x)$. Find a point where $g(x)$ is not locally linear and explain why $g(x)$ is not locally linear.

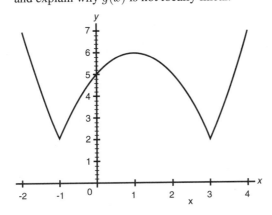

Tips A differentiable function is locally linear because, up close, the curve and its tangent line are very close to each other.

This concept will often be tested as part of some other question—for example, a question involving the tangent line approximation.

Answer There are two points here, $(-1, 2)$ and $(3, 2)$.

Local linearity talks about a curve and its tangent line being close. So this can occur only where the function has a tangent line, where the function is differentiable.

▶Objective 5 Use the tangent line approximation to find approximate values of functions.

Example A rectangular field has one side along a straight river and the other three sides fenced. There are exactly 100 feet of fencing available. The river side is about 40 feet and the area of the field is 1,200 square feet. But the measurement isn't exactly accurate. Use the tangent line approximation to find a linear relationship between small variations for the length of the river side (from 40) and the resulting (approximate) variations in area. About how big is the field if the river is actually 40.3 feet long?

Tips This is just using the equation of the tangent line to find an approximation for the y value instead of evaluating the function itself.

You'd most likely be asked to use the tangent line approximation in a case where evaluating the function directly or with your calculator would be difficult or impossible.

If you have the information you need to write the equation of the tangent line (a point and a slope, the derivative), you have everything you need to use the tangent line approximation.

Answer   1,203 square feet.

In this problem, the side you're measuring is the river side—that will be your x.

The area of the field is what you'll be approximating here. Because there is exactly 100 feet of fence (no possible error here), the area function is

$$A = f(x) = x \left(\frac{100 - x}{2} \right), \quad \text{and therefore } f'(x) = 50 - x.$$

The value where you know both f and f' is $a = 40$.

The tangent line approximation will give a linear relationship:

A is approximately equal to $f(40) + f'(40)(x - 40) = 1200 + 10(x - 40)$.

So $f(40.3)$ is approximately equal to $1200 + 10(.3) = 1203$ square feet.

▶Objective 6   Use local linearity and tangent line approximation to solve problems.

Example   About how accurately should you measure the radius of a sphere for your estimate of the surface area $S = 4\pi r^2$ to be within 1% of the true value?

Answer   Here, the variable (the thing you're measuring) is r; the function you're trying to approximate is $s = 4\pi r^2$ and its derivative with respect to r is $8\pi r$.

The error in the estimate of the surface area, then, will be approximately $(8\pi r)\Delta r$.

You want $\left| \frac{\text{error}}{\text{surface area}} \right|$ to be less than .01, and you're looking for $\left| \frac{\text{error}}{\text{radius}} \right|$.

So that gives you the equation to solve for Δr:

$$\left| \frac{8\pi r \Delta r}{4\pi r^2} \right| = \left| \frac{2\Delta r}{r} \right| = \frac{2|\Delta r|}{r} \le .01, \quad \text{or} \quad \left| \frac{\Delta r}{r} \right| \le .005.$$

If you measure the radius to within $\frac{1}{2}\%$, your estimate of the surface area should be within about 1%.

Multiple-Choice Exercises on Tangent and Normal Lines

---

1.   Which of these is the equation of the tangent line to the curve $y = \sin^2 x + \cos^2 x$ at the point $(0, 1)$?

A. $y = x$

B. $y = 0$

C. $y = 1$

D. $y = 2\pi$

E. $x = 0$

Answer C. The function $y = \sin^2 x + \cos^2 x$ is just a fancy way of saying $y = 1$, so its graph is a (horizontal) straight line. It is therefore identical to its tangent line.

---

2. The equation of the tangent line to the curve $y = \frac{5}{2}x^2 - 1,080$ at the point $(0, -1,080)$ is

A. $y = -1,080$

B. $y = 0$

C. $y = 5x$

D. $x = 0$

E. $y = -x$

Answer A. When you plug the point into the derivative, you find that the slope of the tangent line is 0, so the equation is just $y = 0(x - 0) - 1,080 = -1,080$.

> The tangent line and the normal line to a curve at a point are perpendicular.

---

3. Which of the following lines passes through the point $(2, 8)$ and is perpendicular to the line $y = 2x + 4$ at this point?

A. $y = -2x + 4$

B. $y = -\frac{1}{2}x + 4$

C. $y = -2x + 12$

D. $y = -\frac{1}{2}x + 9$

E. There's not enough information to tell.

Answer D. The slope of the perpendicular line will be the negative reciprocal of the slope given here (which is 2), and it will go through the point (2, 8). So the line will be $y - 8 = -\frac{1}{2}(x - 2)$, or $y = -\frac{1}{2}x + 9$.

---

4. Find the slope of the line normal to the curve $y = 5x^2 - \sin(\pi x)$ at the point (10, 500). Write your answer as a decimal, carried out to four places.

A. -0.0097

B. -0.0103

C. 96.8584

D. -96.8584

E. 0

Answer B. The slope of the tangent line at $x = 10$ is the derivative $(y' = 10x - \pi \cos \pi x)$ evaluated at $x = 10$. The slope of the normal line is the negative reciprocal of this, or $-\frac{1}{100-\pi} = -0.0103$.

---

5. Choose the correct slope of the tangent line to the curve $xy^2 + \frac{y}{2x} = \frac{3}{2}$ at the point $(1, 1)$.

A. $\dfrac{1}{2}$

B. 5

C. $-\dfrac{1}{5}$

D. -2

E. $\dfrac{1}{3}$

Answer C. You must find the derivative implicitly:

$$2yxy' + y^2 + \frac{2xy' - 2y}{(2x)^2} = 0$$

Plug in $x = 1$ and $y = 1$ and solve for y':

$$2y' + 1 + \frac{2y' - 2}{4} = 0, \text{ or } y' = -\frac{1}{5}$$

.

---

6. Choose the correct slope of the tangent line to the curve $x^2 y^2 + \dfrac{1}{xy} = 2$ at the point $(1, 1)$.

A. -1

B. -2

C. 1

D. 0

E. $\dfrac{1}{2}$

Answer A. You must find the derivative implicitly:

$$2xy^2 + 2x^2 yy' - \frac{1}{x^2 y} - \frac{1}{xy^2}y' = 0$$

Plug in $x = 1$ and $y = 1$ and solve for y':

$$2 + 2y' - 1 - y' = 0, \quad y' = -1$$

7. What is the slope of the tangent line to the curve $xy^2 + 5\tan(xy) - 2y + x = 4$ at the point $(4, 0)$?

A. Undefined

B. -0.0556

C. $.5$

D. 0.8667

E. -0.8667

Answer B. You have to differentiate implicitly, then plug in $x = 4$ and $y = 0$. You should then solve for $\dfrac{dy}{dx}$.

$$y^2 + 2xyy' + 5\sec^2(xy)(y + xy') - 2y' + 1 = 0$$

$$0 + 0 + 0 + 5(4y') - 2y' + 1 = 0, \text{ so } y' = \frac{-1}{18} = -.0556$$

8. An equation of the line tangent to the graph of $f(x) = x(1 - 2x)^3$ at the point $(1, -1)$ is:

A. $y = -7x + 6$

B. $y = -6x + 5$

C. $y = -2x + 1$

D. $y = 2x - 3$

E. $y = 7x - 8$

Answer A. The slope of the tangent line is the derivative of f at $x = 1$.

$$f'(x) = (1 - 2x)^3 + 3x(1 - 2x)^2(-2).$$
$$f'(1) = (-1)^3 + 3(-1)^2(-2) = -7$$

Of these choices, only $y = -7x + 6$ has the right slope. A quick check shows that $(1, -1)$ does lie on this line.

10. The slope of the line *normal* to the graph of $4\sin x + 9\cos y = 9$ at the point $(\pi, 0)$ is:

A. undefined

B. -2.25

C. 0.4444

D.   −0.4444

E.   0

Answer   E. When you differentiate this implicitly to find $\dfrac{dy}{dx}$, you get

$$4\cos x - 9\sin y \frac{dy}{dx} = 0, \text{ or } \frac{dy}{dx} = \frac{4\cos x}{9\sin y}.$$

But when you plug in $(x, y) = (\pi, 0)$, you get a zero in the denominator—the tangent line is vertical. So that means the normal line is horizontal, and its slope is zero.

---

11.   Let f be the function defined by $f(x) = 3x^5 - 5x^3 + 2$. The horizontal tangent lines to the graph of $y = f(x)$ are:

A. $y = 0; y = 3$

B. $y = 2; y = 3$

C. $y = -1; y = 1$

D. $y = 0; y = 2; y = 4$

E. $y = 0; y = -1; y = 1$

Answer   D. There are three horizontal lines. To find them, find the critical points:

$f'(x) = 15x^4 - 15x^2 = 0$ when $x = 0$, $x = 1$, or $x = -1$.

The critical points here are $(-1, 4)$, $(0, 2)$, and $(1, 0)$ and the horizontal lines, then, are $y = 0$, $y = 2$, and $y = 4$.

---

12.   You want to use the tangent line approximation to approximate $\sin 95°$. A good point to use for a (the point where you know the values of the function and its derivative) would be:

A. $0°$

B. $90°$

C. $\frac{\pi}{2}$

D. $94°$

E. $\frac{47\pi}{90}$

Answer   C. Whenever you work with the trigonometric functions, you always want to use radians. $\frac{\pi}{2} = 90°$ is close to the $95°$ you're interested in, and you know the exact answers for $\sin \frac{\pi}{2}$ and $\cos \frac{\pi}{2}$, so this will be perfect.

13. The point $(-1, 1)$ lies on the curve $y^3 - \dfrac{x}{y} + \dfrac{y}{x} = 1$.

Use the tangent line to approximate the y-coordinate of the point where $x = -1.01$.

A. 1.02

B. 0.978

C. 0.965

D. 1.101

E. 0.98

Answer E. The first part is to find y' at $(x, y) = (-1, 1)$:

$$3y^2 y' - \frac{y - xy'}{y^2} + \frac{y'x - y}{x^2} = 0$$
$$3y' - (1 + y') + (-y' - 1) = 0$$
$$y' = 2$$

Use the tangent line to find y approximately equal to $1 + 2(-1.01 - (-1)) = 0.98$.

As a check, you can plug $(x, y) = (-1.01, .98)$ into the equation to make sure it holds (at least approximately):

$0.98^3 - \dfrac{-1.01}{0.98} + \dfrac{0.98}{-1.01} = 1.0015$; that's pretty close, so this checks.

14. You know the following information about a differentiable function f:
$$f(0) = 2 \quad \text{and} \quad -1 < f'(x) < 1 \text{ for all } x.$$
Which of the following statements must be true about $f(3)$?

A. $f(3) = 2$

B. $f(3) > 0$

C. $f(3) = -1$ or $f(3) = 5$

D. $-1 < f(3) < 5$

E. None of these statements has to be true.

Answer D. Since the derivative is trapped between -1 and 1, the curve itself is trapped between the lines $y = 2 - x$ and $y = 2 + x$. So the value of f at $x = 3$ has to be between -1 and 3. Here's a graph illustrating one example of this:

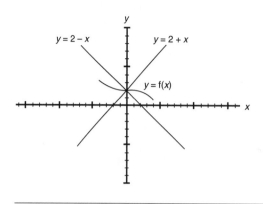

15. Suppose you have the following data about the function $f(x)$:

x	$f(x)$
0	10
1	27.18
1.1	30.04
1.2	33.20
2	73.89

Use the points $(1.1, 30.04)$ and $(1.2, 33.20)$ to estimate $f'(1.2)$.

Use that estimate and a tangent line approximation to estimate $f(1.25)$:

A. 31.62

B. 29.88

C. 34.78

D. 35.24

E. 33.90

Answer C. First, you approximate $f'(1.2)$:

$$f'(1.2) \approx \frac{30.04 - 33.20}{1.1 - 1.2} = 31.6.$$

Next, you find a tangent line approximation for $f(1.25)$:

$$f(1.25) \approx f(1.2) + f'(1.2)(1.25 - 1.2) = 33.20 + 31.6(.05) = 34.78.$$

Free-Response Questions on Tangent and Normal Lines

Note: When you're practicing with these problems, be sure to write down all your work.

When you use a derivative, write it down. When you solve an equation, write it down as well as any intermediate steps you use to solve it. Explain what you're doing, and why, in complete sentences.

1. The line $3x + b$ is tangent to the curve $y = \frac{1}{5}x^5 - \frac{3}{4}x^4 - \frac{17}{3}x^3 + \frac{39}{2}x^2 - 17x$.
   Find the point(s) of tangency and the b for each point.

2. Find the lines tangent and normal to the curve $x^2 + 16y^2 + 96y + 80 = 0$ at the point $(4\sqrt{3}, -2)$.

3. How many tangent lines to the curve $y = 4x^3 + 12x - 5$ pass through the point $(1, 1)$?

4. This line is tangent to the graph of $y = x - \dfrac{x^2}{500}$ at the point Q, as shown in the figure below. (The drawing isn't to scale).

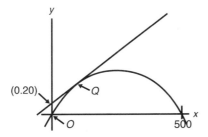

A. Find the x-coordinate of point Q.

B. Write an equation for the line.

C. Suppose the graph of $y = x - \dfrac{x^2}{500}$, shown in the figure, where x and y are measured in feet, represents a hill.

   There is a 50-foot tree growing vertically at the top of the hill.

   Does a spotlight at point $P(0, 20)$ directed along the line shown shine on any part of the tree?

5. The tangent to the curve $y = x^n$ at the point (a, b) intersects the x-axis at the point $(t, 0)$.
   Express t in terms of a and n.

Free-Response Answers

---

1. There are three points:

    $(-4, 345.867),$ $b = 357.866667$

    $(1, -3.717),$ $b = -6.716667$

    $(5, -149.583),$ $b = -164.583333$

    The first step is to find the derivative: $x^4 - 3x^3 - 17x^2 + 39x - 17.$

    Then, solve $x^4 - 3x^3 - 17x^2 + 39x - 17 = 3$ (with the calculator) to get three values of x.

    Remember that the question asks for points, so you also need to find y-values (you can use your calculator).

    Finally, use algebra to solve for b in each case.

---

2. Equation of the tangent line: $y + 2 = -\frac{1}{4}\sqrt{3}(x - 4\sqrt{3})$, or $y = -\frac{\sqrt{3}}{4}x + 1.$

    The equation of the normal line: $y + 2 = \frac{4}{\sqrt{3}}(x - 4\sqrt{3})$, or $y = \frac{4}{\sqrt{3}}x - 18.$

    First, take the derivative: $2x + 32y\dfrac{dy}{dx} + 96\dfrac{dy}{dx} = 0.$

    Then find the slope of the tangent line at $(4\sqrt{3}, -2)$. The slope equals $-\frac{1}{4}\sqrt{3}.$

    That gives you enough information to write the equation of the tangent line:

    $$y + 2 = -\frac{1}{4}\sqrt{3}(x - 4\sqrt{3}) \quad \text{or} \quad y = -\frac{\sqrt{3}}{4}x + 1.$$

    Then use the tangent slope here to find the slope of the normal line: $\frac{4}{\sqrt{3}}.$

    Write the equation of the normal line: $y + 2 = \frac{4}{\sqrt{3}}(x - 4\sqrt{3}) \quad \text{or} \quad y = \frac{4}{\sqrt{3}}x - 18.$

---

3. There is one.

    First, notice that $(1, 1)$ is not on this curve, so you'll need to find a value x that works.

    To find it, take the derivative $y' = 12x^2 + 12$ and use it as the slope in the point-slope form of the line. This gives an equation that you can solve for x:

    $$y - 1 = y'(x - 1) \quad \text{or} \quad (4x^3 + 12x - 5) - 1 = (12x^2 + 12)(x - 1).$$

    Solve this with your calculator–you'll see one solution, at about $x = 1.746.$

    As a check, you can find the equation of the tangent line to the curve at $x = 1.746$ and check that it passes through $(1, 1)$. The equation for the tangent line is $y - 37.243 = (48.582)(x - 1.746)$, and $(1, 1)$ really is on that line.

4.

A. The slope of the line is the derivative of the curve, $y' = 1 - \dfrac{x}{250}$.

Use some algebra to solve for x; one way is noticing that the slopes are equal:

$$1 - \frac{x}{250} = \frac{x - \frac{x^2}{500} - 20}{x}.$$

Solve this to find $x = 100$.

B. You can find the slope from your derivative in part a, and you also have the point $(0, 20)$. The equation of the line is $y = .6x + 20$.

C. The x-coordinate of the top of the hill, either using calculus or the vertex formula for a parabola, is $x = 250$.

The y-coordinate of the top of the hill is 125, so the top of the tree is at 175.

The y-coordinate on the line is 170, so yes, the light does hit the tree.

---

5. The derivative is the slope of the tangent line at each point and equals $\dfrac{dy}{dx} = nx^{n-1}$.

The given point on the curve is (a, a^n), so the slope of the line between (a, a^n) and $(t, 0)$ is:

$$\frac{a^n - 0}{a - t} = \frac{a^n}{a - t}.$$

Set this equal to the slope of the tangent line at $x = a$ to get $na^{n-1} = \dfrac{a^n}{a - t}$.

Solve this for t to get $t = \left(\frac{n-1}{n}\right)a$.

C. Rates of Change and Related Rates

▶Objective 1   Translate verbal descriptions involving rates of change into statements written in mathematical symbols.

Example   Identify the rate(s) in the following applied problem. Write equation(s), using derivatives, to translate the verbal descriptions of rates into math symbols.

A tank of water is in the shape of a cone, point down, with radius 4 feet and height 10 feet. The water is draining out of the tank at a rate of .2 cubic feet a minute. How fast is the level of the water falling when the water is 8 feet deep?

Tips   The units can help you decide what your symbols should be.

Be sure to use the correct sign—if the quantity is decreasing, the rate must be negative.

Answer   There are two rates mentioned in this problem. The second sentence talks about how the volume (cubic feet) is changing with respect to time (minutes); it translates into $\dfrac{dV}{dt} = -.2$ cubic feet per minute, where V is volume in cubic feet and t is time in minutes.

The last sentence asks about how the depth of the water is changing with respect to time.

If h is the depth of the water, then the last sentence translates to: Find $\dfrac{dh}{dt}$ when $h = 8$.

►Objective 2   Translate mathematical symbols involving rates of change into verbal descriptions.

Example   Let the distance traveled by a car since noon be s, in miles. Let the time since noon be t, measured in hours. Let the fuel that the car uses since noon be g, measured in gallons.

Suppose there is a bacteria colony in a Petri dish in the trunk of the car. Let the number of bacteria at time t be P. Suppose there is a cup of coffee in the cupholder inside the car. Let T be the temperature of the coffee, measured in degrees Fahrenheit.

Translate the following rates into words: $\dfrac{dg}{dt}, \dfrac{ds}{dt}, \dfrac{dT}{ds}$. Include the appropriate units.

Answer $\dfrac{dg}{dt}$ is the rate of change of fuel per time.

This is a rate of fuel consumption for the car, measured in gallons per hour.

$\dfrac{ds}{dt}$ is the rate of change of distance per time.

This is the velocity of the car, measured in miles per hour.

$\dfrac{dT}{ds}$ is the rate of change of temperature per distance.

This is the rate the temperature of the coffee is changing, in degrees Fahrenheit per mile.

►Objective 3   Solve related-rates problems.

Example   A boat is pulled to a wharf by a rope with one end attached to the front of the boat and the other end passing through a ring attached to the wharf at a point 6 feet higher than the front of the boat. The rope is being pulled through the ring at the rate of 1.2 feet per second. How fast is the boat approaching the wharf when 10 feet of rope are out?

Tips   Remember the steps to solve such a problem:

1. Translate the verbal statement of the problem into a picture (if possible) and math.

2. Identify the rates of change (derivatives) you have and the one you're looking for.

3. Find the relationship between the variables.

4. Consolidate your information until you have one expression, usually involving no more than two variables.

5. Differentiate your expression, implicitly if necessary, and solve for your target rate.

6. Make sure you answered the question. Include the correct units.

Geometric relationships are common. You should know how to use similar triangles, triangle trigonometry, the Pythagorean theorem, and also formulas for the area and circumference of a circle, area and perimeter of a triangle or a rectangle, and volume and surface area of a rectangular prism (that is, box). You won't be asked to remember any other geometric formulas.

Answer   The rope is the hypotenuse of a right triangle, with the distance between the boat and the wharf at the base, and the 6-foot height for the other leg.

Call the distance between the boat and the wharf x and the length of rope y.

You know $\dfrac{dy}{dt} = -1.2$ feet per second, and you want $\dfrac{dx}{dt}$.

The Pythagorean theorem gives you $x^2 + 36 = y^2$.

Differentiate to find $2x\dfrac{dx}{dt} = 2y\dfrac{dy}{dt}$.

When $y = 10$, $x = 8$, so you can substitute the values you know:

$$2 \times 8 \times \frac{dx}{dt} = 2 \times 10(-1.2), \quad \text{or} \quad \frac{dx}{dt} = -\frac{3}{2} = -1.5 \text{ feet per second.}$$

So the boat is approaching the wharf at 1.5 feet per second.

▶**Objective 4**   Explain the distinction between speed and velocity.

Example   The graph shown here is of the *velocity* of an object. The horizontal axis is t, in seconds, and the vertical axis is v, in feet per second. Find the time(s) where the speed of this object is greatest.

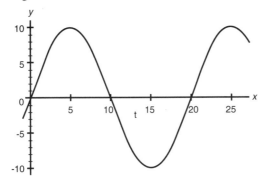

Tip   Velocity has a sign, indicating the direction of travel. Speed is what your speedometer would show.

Answer   Speed is the absolute value of velocity. So speed will be greatest when $|v|$ is greatest, so you need to look at both maxima and minima of v. In this graph, the maxima and minima of v have the same absolute value, 10 feet per second. The speed of this object is greatest at $t = 5$, $t = 15$, and $t = 25$.

▶**Objective 5**   Calculate speed, velocity, and acceleration functions from position functions (including algebraic functions, trigonometric functions, and combinations of the two).

Example An object's position is given by the equation $x(t) = 25 - \sin(\pi t)$.

Find its velocity and acceleration.

Answer The object's velocity is just the derivative, $x'(t) = -\pi \cos(\pi t)$, and the object's acceleration is just the derivative of velocity, $x''(t) = \pi^2 \sin(\pi t)$.

▶**Objective 6** Analyze rectilinear motion situations using position, speed, velocity, and acceleration functions that you've determined. For example, find distances traveled, find maximum and minimum speeds reached, and graph velocities vs. speeds for the whole function.

Example An object is thrown from the top of a tower at $t = 0$. Its vertical distance in meters from the ground after t seconds is given by the equation $s(t) = 250 + .5t - 4.9t^2$. Find its maximum velocity.

Answer Its velocity is its derivative, $v(t) = .5 - 9.8t$. This is the function you want to find the maximum value for, so take its derivative, $v'(t) = -9.8$ and find critical values.

This last expression is constant and therefore has no zeros. Thus, there are no critical values!

The velocity is always decreasing (because its derivative is negative), so the maximum value will occur at $t = 0$, the beginning of the domain of these functions. The maximum velocity is $v(0) = .5$ meters per second. Notice that this makes sense with the intuitive idea of what must be happening. The object will keep going faster and faster (down) until it hits the ground—the velocity does decrease (get more negative) the entire way.

Multiple-Choice Exercises on Rates of Change and Related Rates

In questions 1–4, identify the particular word, or words, in each passage that most directly refers to a derivative. (If no words refer to a derivative, select "None of these".)

---

1. A hot air balloon has a velocity of 50 ft/min and is flying at a constant height of 500 feet.

   A. 50

   B. ft/min

   C. constant height

   D. 500 feet

   E. None of these.

Answer B. The velocity is always a derivative of a distance or position function. And you can tell that the rate of 50 ft/min is a derivative because the units are a fraction, feet *per* minute.

---

2. Treatment of a spherical tumor is causing the radius of the tumor to decrease at a rate of 1 millimeter per month.

A. 1

B. radius

C. a rate of

D. spherical

E. causing

Answer   C. The rate of change here is identified by the word *rate*! Again, you can tell that the 1 millimeter per month is a derivative because of the fractional units.

---

3. Water is being purified by being pumped at 300 liters per hour through a conical filter.

A. 300

B. conical filter

C. liters per hour

D. being pumped

E. None of these

Answer   C. The derivative here is signaled by the units put together in a fraction—the word *per* tells you you're looking at a fractional unit, and thus a derivative. The derivative here is "300 liters per hour."

---

4. The bus went 1.4 miles before stopping.

A. 1.4

B. miles

C. stopping

D. before

E. None of these

Answer   E. This particular sentence doesn't mention a derivative. (The distance was changing, so there might have been a derivative in some other part of the story. But this sentence doesn't mention any **rate** of change.)

---

5. A lantern hangs above the center of a round table. You can raise or lower the lantern by means of a rope. Use the following letters to represent the quantities in the problem: the radius of the table is r, the length of the rope is R, the height of the lantern above the table is h, the intensity of the light cast on the table is L, the angle from the light to the point on the table is θ, and time is t. Which of the following derivatives represents the answer to the following question? "How does the intensity of the light cast on the table change as you move the lantern up or down?"

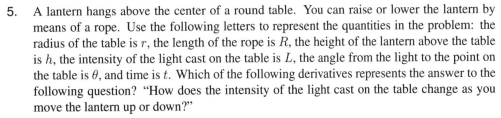

A. $\dfrac{dL}{dt}$

B. $\dfrac{dh}{dR}$

C. $\dfrac{dL}{d\theta}$

D. $\dfrac{dh}{dt}$

E. $\dfrac{dL}{dh}$

Answer E. You want to investigate the change in L with respect to h, or $\dfrac{dL}{dh}$.

---

6. Sand is falling into a pile in the shape of a cone. Which of the units shown here could go with the rate at which the sand is accumulating?

A. miles per hour

B. cubic centimeters per second

C. liters per centimeter

D. All of the above

E. None of the above

Answer D. All three choices could describe the rate mentioned here. The question is not specific about what the rate is. It could be "the rate at which the height of the cone is increasing with respect to time" (miles per hour). It could be "the rate at which the volume of the cone is increasing with respect to time" (cubic centimeters per second). It could be "the rate at which the volume of the cone is increasing with respect to the height of the cone" (liters per centimeter). Without more information, you can't eliminate any of these.

---

7. Alonah is going to be making a product, and $C(x)$ represents the cost to Alonah of x number (units) of that product. The marginal cost at a given value of x is $C'(x)$. At a production level of 100 units, Alonah knows that the marginal cost is $20. Suppose that she suddenly finds out that her fixed costs are $250 more than she previously thought. Then what is the new marginal cost at a production level of 100 units?

A. $20

B. $22.50

C. $270

D. $45

E. None of the above

Answer A. The marginal cost is the same as it was before. Mathematically, her new cost function would be given by $C(x) + 250$. Then, the marginal cost at a production level of 100 units would be $(C(x) + 250)'$, evaluated at $x = 100$, which is just $C'(100) = 20$. Practically speaking, raising the fixed costs doesn't change the "cost of producing the next unit," which is what marginal cost represents.

---

8. The cost function and revenue function for producing and selling x units of a certain product are given by $C(x)$ and $R(x)$ respectively, where both C and R are in dollars. Suppose that $C(x)$ and $R(x)$ are both differentiable functions of x, and consider the profit function $P(x) = R(x) - C(x)$, which gives you profit P in dollars for producing and selling x units. (Assume that each unit you make gets sold). Assuming that the maximum profit is attained at $x = 124$ (and this is not at an endpoint of the effective domain), what must be the relationship between marginal cost and marginal revenue at that point?

A. Marginal cost is less than marginal revenue, and both are positive.

B. Marginal cost is greater than marginal revenue, and both are positive.

C. Marginal cost is equal to marginal revenue.

D. Marginal cost is less than marginal revenue, and both are negative.

E. Marginal cost is greater than marginal revenue, and both are negative.

Answer C. Notice that $P'(x) = R'(x) - C'(x)$. Next, since profit is maximized at $x = 124$, and this is not an endpoint of the domain (and the derivative exists there), then $P'(124) = 0$. So you have $C'(124) = R'(124)$, or in other words, the marginal cost and the marginal revenue are equal at $x = 124$.

---

9. How fast does the radius of a spherical soap bubble change when you blow air into it at the rate of 15 cubic centimeters per second? So the answer to this question is:

A. $4\pi r^2$ centimeters per second.

B. $60\pi r^2$ centimeters per second.

C. $\dfrac{15}{4\pi r^2}$ centimeters per second.

D. $20\pi r^3$ centimeters per second.

E. 60π centimeters per second.

Answer   C. Our known rate is $\dfrac{dV}{dt}$, the change in volume with respect to time, which is 15 cubic

centimeters per second. The rate we want to find is $\dfrac{dr}{dt}$, the change in the radius with respect

to time. Remember that the volume of a sphere is $V = \frac{4}{3}\pi r^3$.

If we differentiate the volume relationship, we get $\dfrac{dV}{dt} = 4\pi r^2 \dfrac{dr}{dt}$.

Plug in the value that we know, $\dfrac{dV}{dt} = 15$, and solve for the rate we're after: $\dfrac{dr}{dt} = \dfrac{15}{4\pi r^2}$.

Notice that this answer depends on r; the radius is increasing faster when r is small than when r is big. And that makes sense with our intuitive understanding of ballons; the radius increases quickly when r is small, slowly when r is big.

The units make sense, too: cubic centimeters per second divided by centimeters squared is

$\dfrac{\text{cm}^3/\text{sec}}{\text{cm}^2} = \dfrac{\text{cm}}{\text{sec}}$. That gives us centimeters per second, just what we want.

---

10.   A baseball diamond is a square 90 feet on a side. A player runs from first base to second base at a rate of 15 feet per second. At what rate is the player's distance from third base decreasing when the player is halfway between first and second base?

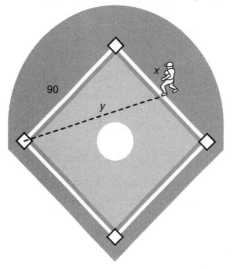

A. -6.71 feet per second

B. 6.71 feet per second

C. 8.66 feet per second

D. -8.66 feet per second

E. 0.149 feet per second

Answer   B. We've already set part of this problem up. If we let x be the distance between the player and second base, and y be the distance between the player and third base, then $\dfrac{dx}{dt} = -15$ feet per second, and $\dfrac{dy}{dt}$ will tell us what we want to know.

Use the picture to find a relationship that will help you answer the question.

The relationship here comes from the Pythagorean theorem, $x^2 + 90^2 = y^2$. If we differentiate that with respect to t, we get

$$2x\frac{dx}{dt} = 2y\frac{dy}{dt}.$$

Plugging in $\frac{dx}{dt} = -15$ and solving for $\frac{dy}{dt}$ gives us $\frac{dy}{dt} = -6.71$ feet per second.

This rate is negative, because the distance is decreasing. The question, however asked, "How fast is the distance decreasing," so the answer is, "The distance is decreasing by 6.71 feet per second." (Watch out if you didn't get a negative rate here!)

---

11. A man 2 meters tall walks at the rate of 2 meters per second toward a streetlight that's 5 meters above the ground. At what rate is the tip of his shadow moving?

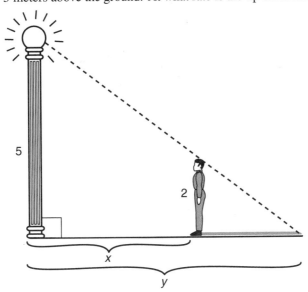

A. $-\frac{10}{3}$ meters per second

B. $-\frac{4}{5}$ meters per second

C. 2 meters per second

D. $\frac{8}{5}$ meters per second

E. $-\frac{8}{5}$ meters per second

Answer A. We've already set this up part of the way. We know that $\frac{dx}{dt} = -2$ meters per second, and we're looking for $\frac{dy}{dt}$. Use the picture to help you find the relationship between x and y, and use it to answer the question asked here.

Similar triangles give the relationship here: $\frac{y}{5} = \frac{y-x}{2}$, or $y = \frac{5}{3}x$.

Differentiate this to find $\dfrac{dy}{dt} = \dfrac{5}{3}\dfrac{dx}{dt} = -\dfrac{10}{3}$ meters per second.

So the tip of the shadow is traveling $\frac{10}{3}$ meters per second toward the lamppost. Notice that this answer doesn't depend on how far away the man is—as long as he's traveling at a constant speed, so is his shadow.

---

12. A circular oil slick of uniform thickness is caused by a spill of 1 cubic meter of oil. The thickness of the oil is decreasing at the rate of 0.1 cm/hr as the slick spreads. (Note: 1 cm = 0.01 m.) At what rate is the radius of the slick increasing when it is 8 meters?

A. 0.804 centimeters per hour

B. 1.234 meters per hour

C. 0.804 meters per hour

D. 325.201 centimeters per hour

E. 0.556 meters per hour

Answer   C. You can think of this oil slick as a very very flat cylinder; its volume is given by $V = \pi r^2 h$, where r is the radius and h is the height of this cylinder.

We know that the rate $\dfrac{dh}{dt} = -0.001$ meters per hour, and we want to know $\dfrac{dr}{dt}$.

The relationship between the variables comes from the fixed volume of the oil slick: $V = \pi r^2 h = 1$. You can solve this for either r or h, if you want, or you can differentiate this implicitly as it stands (remembering the Product Rule and the Chain Rule):

$$2\pi r h \frac{dr}{dt} + \pi r^2 \frac{dh}{dt} = 0$$

When $r = 8$, h is $\frac{1}{64\pi}$. Plugging in the known values (don't round those numbers off until the end!), we get

$$\frac{dr}{dt} = \frac{\pi r^2 \frac{dh}{dt}}{2\pi r h} = -\frac{r}{2h}\frac{dh}{dt} = -\frac{8}{2\frac{1}{64}\pi}(-0.001) = 0.804 \text{ meters per hour.}$$

The motion of any object falling to the ground can be modeled by $y(t) = -\frac{1}{2}gt^2 + v_0 t + y_0$, where $y(t)$ is the height above the ground, g is the constant acceleration due to gravity (usually 32 ft/sec^2 or 9.8 m/sec^2, depending on what units you're using), v_0 is the initial velocity, and y_0 is the initial height above the ground.

---

13. If a baseball is thrown upward at 2 meters per second from a point 5 meters off the ground, its height would be given by $y(t) =$

A. $-\frac{1}{2}(9.8) + 2t + 5 = -4.9 + 2t + 5$

B. $-\frac{1}{2}(32)t^2 + 2t + 5 = -16t^2 + 2t + 5$

C. $-\frac{1}{2}(9.8)t^2 + 2t + 5 = -4.9t^2 + 2t + 5$

D. $\frac{1}{2}(9.8)t^2 + 2t - 5 = 4.9t^2 + 2t - 5$

E. $\frac{1}{2}(9.8)t^2 + 2t + 5 = 4.9t^2 + 2t + 5$

Answer   C. The initial velocity is 2 meters per second, so $v_0 = 2$ (positive, because the velocity is upward). The initial height above the ground is $y_0 = 5$ (again, positive because the baseball starts from above the ground). And, because you're using meters, the constant g is $9.8 \frac{m}{sec^2}$.

---

14.   An object's position is given by the equation $x(t) = 3\sin(2\pi t)$. Its acceleration is given by:

A. $a(t) = -12\pi^2 \sin(2\pi t)$

B. $a(t) = -3\sin(2\pi t)$

C. $a(t) = 12\sin(2\pi t)$

D. $a(t) = -12\pi^2 \cos(2\pi t)$

E. $a(t) = 3\sin(t)$

Answer   A. Acceleration is the second derivative of the position function.

The next two problems refer to the following situation:

A girl is riding on a Ferris wheel. The sun is casting a shadow on a wall behind the Ferris wheel, and the shadow is going straight up and straight down. The *velocity* of the girl's shadow on the wall is graphed here (the horizontal axis is t, in seconds, and the vertical axis is v, in feet per second). When the shadow is going upward, the velocity is positive.

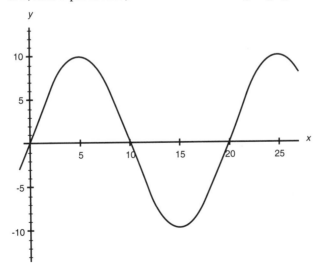

---

15.   When is the speed of the shadow at a maximum?

A. $t = 5$ and $t = 25$

B. $0 < t < 10$ and $20 < t$

C. $t = 0, t = 10$ and $t = 20$

D. $10 < t < 20$

E. $t = 5, t = 15$, and $t = 25$

Answer   E. Speed is the absolute value of velocity.

---

16.   Where is the shadow turning around?

A. $t = 5$ and $t = 25$

B. $0 < t < 10$ and $20 < t$

C. $t = 0, t = 10$ and $t = 20$

D. $10 < t < 20$

E. $t = 5, t = 15$, and $t = 25$

Answer   C. The shadow is turning around when the velocity is changing sign, which is also where the y value is zero and the graph crosses the x-axis.

---

17.   An object is moving along the x-axis. Its position at any time is given by $x(t) = 3\cos(\pi t) + 2$. Find its acceleration.

A. $a(t) = 3\sin(\pi t)$

B. $a(t) = -3\pi \sin(\pi t)$

C. $a(t) = 3\sin(\pi t) + 2$

D. $a(t) = -3\pi^2 \cos(\pi t)$

E. $a(t) = -3\cos(\pi t)$

Answer   D. The second derivative of the position function is the acceleration.

$$v(t) = x'(t) = -3\pi \sin(\pi t),$$

and

$$a(t) = x''(t) = v'(t) = -3\pi^2 \cos(\pi t).$$

  Apex Learning

18. A ball is thrown straight out at 80 feet per second from an upstairs window that's 15 feet off the ground. It turns out that its motion can be broken down into its motion in the x-direction (horizontally) and its motion in the y-direction (vertically), and that those two components can be thought of completely independently. In other words, its horizontal motion has no influence on its vertical motion, and vice versa.

$x(t)$ is its position in the x-direction, and $y(t)$ is its position in the y-direction.

Which of the following choices is correct? (Ignore air resistance.)

A. $x(t) = 80t, \quad y(t) = 15 - 16t^2$

B. $x(t) = 15 - 16t^2, \quad y(t) = 80t$

C. $x(t) = 80t - 16t^2, \quad y(t) = 15$

D. $x(t) = 80\sin t, \quad y(t) = 15 - 16t^2 + 80\cos t$

E. $x(t) = 80t, \quad y(t) = -16t^2$

Answer   A. The two components of this ball's motion can be thought of independently.

The horizontal motion is affected only by the initial velocity here, 80 feet per second.

So the ball's position in the x-direction is just that constant velocity times t, or $x(t) = 80t$.

In the y-direction, the ball is simply a falling body. The units are feet here, so $g = 32$ feet per second squared. There is no initial velocity in the y-direction, and the initial position comes from the window being 15 feet off the ground.

So $y(t) = 15 - 16t^2$.

---

19. A ball is thrown straight out at 80 feet per second from an upstairs window that's 15 feet off the ground. Find the ball's horizontal distance from the window at the moment it strikes the ground.

A. Can't be found

B. .96825 feet

C. 6.33 feet

D. 212.23 feet

E. 77.46 feet

Answer   E. The ball will land when the y-component is 0, which is at about $t = .96825$.

The x-component at that time is 77.46 feet, and this tells us how far away the ball is.

The next two problems refer to the velocity $v(t)$ of an object with the sinusoidal graph pictured below.

Apex Learning   227

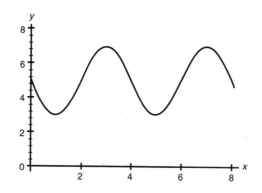

20. The object has zero acceleration at what instants?

A. $t = 2, t = 4, t = 6$, etc., when the graph crosses the line $v = 5$.

B. $t = 1, t = 3, t = 5$, etc.

C. $t = 1, t = 5, t = 9$, etc.

D. $t = 1, t = 2, t = 3$, etc.

E. Can't tell from this graph

Answer   B. Zero acceleration will correspond to the maxima and minima of this velocity function.

21. Which of these statements accurately describes the motion of the object?

A. The object is moving in a negative direction.

B. The object is moving back and forth.

C. The object is moving in a positive direction.

D. We can't tell anything from the velocity alone.

E. None of these

Answer   C. The velocity here is always positive. This means that the distance function (which we can't see) is increasing, so the object is moving in a positive direction.

Free-Response Questions on Rectilinear Motion

1. A particle moves along a straight line according to the distance formula $s(t) = t^3 - 12t^2 + 45t - 17$.

A. Find all times when the particle is at rest.

B.   Find all times at which the particle's velocity is a relative maximum or minimum.

C.   Does the particle have a minimum speed on the interval $t \geq 0$, and if so what is it and at what time(s) is it attained?

---

2.   Below is the graph of the velocity of an object moving along a straight line, for $0 \leq t \leq 5$.

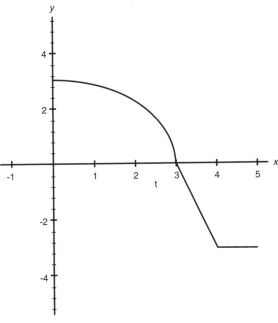

A.   What is the object's maximum acceleration, and for what value(s) of t is it attained? Explain how you found this.

B.   Does the object have a minimum acceleration, and if so, what is it and when is it attained? Explain.

C.   At what time t does the object attain its maximum positive distance, and how do you know this?

---

3.   A particle moves along a straight line with position $s(t) = t^2 \sin(\pi t)$, for $0 \leq t \leq 10$.

A.   When is the first time (AFTER $t = 0$, of course) that the particle changes directions?

B.   At what time does the particle achieves its maximum speed on this interval?

---

4.   A skydiver is about to jump. From time $t = 0$ seconds until just before time $t = 10$ seconds, the skydiver is flying in a plane at a level $5,000$ feet altitude. At time $t = 10$ seconds, the skydiver jumps out of the plane, immediately experiencing constant downward (negative) acceleration due to gravity. The skydiver continues to accelerate at this constant rate until time $t = 20$ seconds, when the parachute opens. For the next two seconds, the skydiver decelerates at a constant rate, until reaching a constant (slow) downward velocity at time $t = 22$ seconds.

Sketch three graphs.

A. Graph I should represent the altitude of the skydiver as a function of time, for time between 0 and about 30 seconds.

B. Graph II should represent the vertical component of the velocity of the skydiver as a function of time, for time between 0 and about 30 seconds.

C. Graph III should represent the acceleration of the skydiver as a function of time, for time between 0 and about 30 seconds.

You do **not** need to figure out the underlying functions for these graphs. You need only to sketch qualitatively what they look like. You must make certain that your three graphs make sense **together**—in other words, make sure your graph of altitude versus time and your graph of velocity versus time don't have contradicting qualities!

---

5. The graph below shows the velocity of an object moving along a line, for $0 \le t \le 10$, t in seconds.

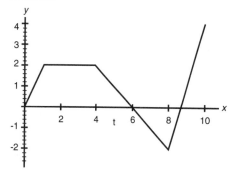

A. When does the object attain its maximum acceleration? (Remember to give an explanation.)

B. When does the object first change direction? (Give explanation.)

C. What is the object's average velocity over the interval $0 \le t \le 10$?

(Hint: Find the total displacement of the object over the interval.)

---

6. Suppose a quantity of gas (with a fixed mass) is confined in a container of volume V. Boyle's Law states that, so long as the temperature of a gas is constant, the product of the pressure P and the volume V is a constant. Suppose that, for a certain gas, $PV = 850$, where P is measured in pounds per square inch and V is measured in cubic inches.

A. Find the average rate of change of P (with respect to V) as V increases from 200 cubic inches to 300 cubic inches.

B. Suppose that the volume is increasing at 100 cubic inches per hour. Find the instantaneous rate of change of P with respect to time when $V = 200$ cubic inches.

C. Suppose that the temperature T is allowed to change as well. The expression that relates P, T, and V is $\frac{PV}{T} = $ constant.

How does P change (increase, decrease, or stay the same) if V is fixed and T increases?

D.   A rigid sealed container (so the volume is constant) contains a quantity of gas under pressure. At $t = 0$, the pressure is 4 pounds per square inch and the temperature is $15°C$. The container is heated by $5°C$ per minute. What is the rate of change of the pressure inside the container at time $t = 0$?

Free-Response Questions on Rectilinear Motion—Answers

1.

A. $t = 3, 5$. The particle is at rest when the velocity is zero.

Set $v(t) = 0$. $v(t) = s'(t) = 3t^2 - 24t + 45 = 3(t - 3)(t - 5)$.

B. $t = 4$. To find a max or min of velocity, you need to find its critical points.

Set $v'(t) = a(t) = 0$, or $a(t) = v'(t) = 6t - 24 = 0$.

There's just one critical point, at $t = 4$, and v' goes from negative to positive there—this is a relative minimum.

C.   Yes, minimum speed is 0 attained at $t = 3$ and $t = 5$, when the velocity is zero. Remember that speed is the absolute value of velocity. That means if the velocity is zero anywhere, that will be its minimum speed. The minimum velocity you found in part B will correspond to a relative maximum speed!

2.

A.   The maximum acceleration of 0 is achieved on the interval $4 \leq t \leq 5$. This is also attained at $t = 0$, but it isn't obvious from the graph that the tangent line at $t = 0$ is horizontal, so don't worry if you weren't sure. This velocity is always decreasing, so its slope is always ≤ 0, so the maximum acceleration will be zero.

B.   The object doesn't have a minimum acceleration, since the slope of the velocity curve goes to negative infinity as t goes to 3 from the left.

C.   At $t = 3$. After $t = 3$, the velocity is negative, so the distance decreases.

3.

A. $t \approx .72859$. You must set $v(t) = s'(t) = 0$ and solve to find the critical point(s).

$v(t) = 2t \sin(\pi t) + \pi t^2 \cos(\pi t)$. Solve this with your calculator for the first $t > 0$ such that $v(t) = 0$. You should get $t \approx .72859$. You must also check that the sign changes here (it does).

B. $t = 10$. Set $a(t)$ equal to 0 and solve on calculator:

$$v'(t) = a(t) = 2\sin(\pi t) + 2\pi t \cos(\pi t) + 2\pi t \cos(\pi t) - \pi^2 t^2 \sin(\pi t)$$
$$= (2 - \pi^2 t^2)\sin(\pi t) + 4\pi t \cos(\pi t).$$

There are many values t on this interval that satisfy $a(t) = 0$. We also need to check the endpoints (in fact, for this one, the maximum speed is at an endpoint). You can check them individually or observe from the graph of v that the amplitude is increasing as $t \to 10$ to narrow down your search. You do need to check at least $t = 9.044$, where the speed is about 257, and $t = 10$, the endpoint, where the speed is about 314. (Remember that you're looking at speed, not velocity.)

4.

A. Sample Graph I (altitude vs. time)

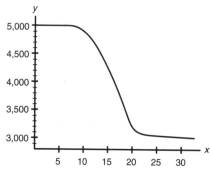

The things to look for:

*   The constant $5,000$ from time 0 to 10.

*   For $10 < t \leq 20$ (the time of downward acceleration), the graph **cannot** have a constant slope (line). It must have negative concavity (that is, be concave down).

*   For $20 < t \leq 22$ (the time of deceleration), the graph must show positive curvature (that is, be concave up).

*   For $22 < t$, the slope should be constant and negative.

B. Sample Graph II (velocity versus time)

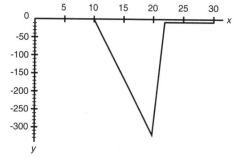

Things to look for:

*   For $0 \leq t < 10$, velocity should be 0.

*   For $10 < t \leq 20$, velocity should be a negative slope **line** (since acceleration is **constant**).

* For $20 < t \leq 22$, velocity should be a positive slope (very steep) **line** (since deceleration is constant).

* For $22 < t$, velocity is constant and negative (not very large).

C. Sample Graph III (acceleration versus time)

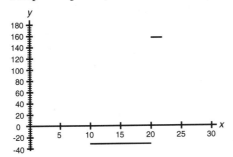

Things to look for:

* For $0 \leq t < 10$, acceleration is 0.

* For $10 \leq t < 20$, acceleration is constant negative.

* For $20 \leq t < 22$, acceleration is constant positive.

* For $22 \leq t$, acceleration is 0.

There's quite a bit of lee-way in what these graphs can look like, but they do need to be consistent with each other. Pay attention as you go to whether there are major contradictions between your graphs. For example, if you have a sharp corner in one graph, you need to show that the derivative is undefined there. If your derivative contains a line with a nonzero slope, your original function must have some curvature to it. Be sure that the sign of your acceleration graph matches the concavity of your altitude graph.

5.

A. The maximum acceleration will be from $8 < t \leq 10$, where the slope is greatest.

B. At $t = 6$. This is when the velocity goes from positive to negative.

C. Average velocity is $\dfrac{9}{10}$ units per second.

Steps:

Average velocity is the total displacement (position change) over the total time elapsed.

First find out how much displacement there is over the entire interval:

In the first second, the object moves 1 unit in a positive direction, since the average of the velocities is the average velocity of 1 on this interval.

In the interval $1 \leq t \leq 4$, the object moves 6 units in a positive direction, since velocity is 2

In the interval $4 \leq t \leq 8$, the object has net movement of 0 units, since its average velocity on that interval is 0.

On the interval $8 \leq t \leq 10$, the object moves 2 units in a positive direction, since the average velocity on this interval is 1.

Total displacement is 9 units in a positive direction. Average velocity is $\dfrac{9}{10}$ units per second.

6.

A. The average rate of change here is $\dfrac{\text{difference in } P}{\text{difference in } V}$.

When $V = 200$, $P = \frac{850}{200} = 4.25$, and when $V = 300$, $P = \frac{850}{300} \approx 2.833$.

So the average rate of change is $\dfrac{2.833 - 4.25}{300 - 200} \approx -.0142 \dfrac{\text{pounds per square inch}}{\text{cubic inches}}$.

B. Differentiate implicitly $PV = 850$ with respect to t: $\quad P\dfrac{dV}{dt} + V\dfrac{dP}{dt} = 0$.

You know $P = 4.25$, $V = 200$, and $\dfrac{dV}{dt} = 100$.

Substitute the known values and solve for

$$\frac{dP}{dt} = -\frac{(4.25)100}{200} = -2.125 \frac{\text{pounds per square inch}}{\text{hour}}.$$

C. If V is fixed and T increases, then P must also increase in order for this ratio $\dfrac{PV}{T}$ to remain constant.

D. This situation falls under the formula in part c. V is constant, so that means $\dfrac{P}{T}$ is constant.

You're given $\dfrac{dT}{dt} = 5$ degrees per minute, and you want $\dfrac{dP}{dt}$.

Just as in part b, differentiate implicitly with respect to t and plug in known values.

$$\frac{T\frac{dP}{dt} - P\frac{dT}{dt}}{T^2} = \frac{15(\frac{dP}{dt}) - 4(5)}{15^2} = 0,$$

so $\dfrac{dP}{dt} \approx 1.333 \dfrac{\text{pounds per square inch}}{\text{minute}}$.

Chapter 6   *The Definite Integral*

The derivative helps us measure *rate* of change, while the *definite integral* helps us measure *accumulated* change. Just as the derivative of a function has a nice graphical interpretation in terms of the *slope* of its graph, the definite integral has a graphical interpretation in terms of *area*. In this chapter we'll review the definition of a definite integral and the objectives for understanding its mathematical meaning and applications.

A. Area Under a Curve

▶Objective 1   Use summation notation to describe the sum of a series.

Example   Write each of these using summation notation.

A. $\frac{1}{3} + \frac{1}{3}(4) + \frac{1}{3}(9) + \frac{1}{3}(16) + \frac{1}{3}(25) + \frac{1}{3}(36) + \frac{1}{3}(49)$

B. $\frac{1}{2}g(1.5) + \frac{1}{2}g(2) + \frac{1}{2}g(2.5) + \frac{1}{2}g(3) + \frac{1}{2}g(3.5) + \frac{1}{2}g(4) + \frac{1}{2}g(4.5)$

C. $\frac{1}{3}g(4) + \frac{2}{3}g(9) + g(16) + \frac{4}{3}g(25) + \frac{5}{3}g(36) + 2g(49)$

Tips   The index k usually starts with a value of 1.

Always check **at least** three terms to see if you've found the right pattern (first, last, and one in the middle somewhere).

Answer

A. $\sum_{k=1}^{7} \frac{1}{3}k^2$. Notice that the factor $\frac{1}{3}$ can be pulled out of the sum if you want: $\frac{1}{3} \sum_{k=1}^{7} k^2$.

B. This one is a little tricky: $\dfrac{1}{2}\displaystyle\sum_{k=1}^{7} g(1 + .5k)$.

C. $\displaystyle\sum_{k=1}^{6} \dfrac{k}{3} g((k+1)^2)$. This one only had six terms. Notice that we used $k+1$ to get $g(4)$ in the first term.

▶**Objective 2**   Calculate exact areas under curves geometrically for circles, trapezoids, triangles, and for rectangles.

Example   Find the area under the curve $f(x) = \sqrt{9 - (x+2)^2}$ from $-2 \le x \le 1$.

Tip   You'll need to recognize when a function describes a shape with a well-known area.

Answer $A = \frac{1}{4}\pi(3)^2 = \frac{9}{4}\pi$, since this is one-fourth of a circle with a radius of 3.

▶**Objective 3**   Approximate area under a curve using midpoint, left-endpoint, and right-endpoint approximations.

Example   Using midpoint, left-endpoint, and right-endpoint approximations approximate the area under the curve $f(x) = e^{\sqrt{x}}$ from $0 \le x \le 2$ with 4 intervals of equal width.

(Round your calculator approximation to three digits past the decimal point.)

Tips   Make sure you write out the steps you use. If you put down only your calculator answer, you won't receive full credit.

Recognize that these are all Riemann sums, since rectangles are used to approximate the area.

Answer   Left endpoint Riemann sum L:

$L = \Delta x(f(0) + f(\frac{1}{2}) + f(1) + f(\frac{3}{2})) = \frac{1}{2}(e^{\sqrt{0}} + e^{\sqrt{\frac{1}{2}}} + e^{\sqrt{1}} + e^{\sqrt{\frac{3}{2}}}) \approx 4.575$.

Midpoint Riemann sum M:

$M = \Delta x(f(\frac{1}{4}) + f(\frac{3}{4}) + f(\frac{5}{4}) + f(\frac{7}{4})) = \frac{1}{2}(e^{\sqrt{\frac{1}{4}}} + e^{\sqrt{\frac{3}{4}}} + e^{\sqrt{\frac{5}{4}}} + e^{\sqrt{\frac{7}{4}}}) \approx 5.420$.

Right endpoint Riemann sum R:

$R = \Delta x(f(\frac{1}{2}) + f(1) + f(\frac{3}{2}) + f(2)) = \frac{1}{2}(e^{\sqrt{\frac{1}{2}}} + e^{\sqrt{1}} + e^{\sqrt{\frac{3}{2}}} + e^{\sqrt{2}}) \approx 6.131$.

▶**Objective 4**   Write Riemann sums to represent the approximations using summation notation.

Example   Write summation notation to represent the sums from the previous example.

Tips   Let x_k stand for the x-value chosen from the kth interval. Think of finding x_k as a separate part of the problem. Notice how the form of all three of these Riemann sums are identical except for the choice of x_k.

Your form for x_k may be slightly different than ours.

For example, on the midpoint answer below, $M = \frac{1}{2} \sum_{k=1}^{4} e^{\sqrt{\frac{2k-1}{4}}}$ would also be fine.

Answer   Left endpoint Riemann sum approximation:

$$A \approx L = \frac{1}{2}(e^{\sqrt{0}} + e^{\sqrt{\frac{1}{2}}} + e^{\sqrt{1}} + e^{\sqrt{\frac{3}{2}}}) = \frac{1}{2} \sum_{k=1}^{4} e^{\sqrt{x_k}} = \frac{1}{2} \sum_{k=1}^{4} e^{\sqrt{\frac{k-1}{2}}}.$$

Midpoint Riemann sum approximation:

$$A \approx M = \frac{1}{2}(e^{\sqrt{\frac{1}{4}}} + e^{\sqrt{\frac{3}{4}}} + e^{\sqrt{\frac{5}{4}}} + e^{\sqrt{\frac{7}{4}}}) = \frac{1}{2} \sum_{k=1}^{4} e^{\sqrt{x_k}}$$

$$= \frac{1}{2} \sum_{k=1}^{4} e^{\sqrt{\frac{k-1}{2}+\frac{1}{4}}} = \frac{1}{2} \sum_{k=1}^{4} e^{\sqrt{\frac{1}{2}k-\frac{1}{4}}}.$$

Right endpoint Riemann sum approximation:

$$A \approx R = \frac{1}{2}(e^{\sqrt{\frac{1}{2}}} + e^{\sqrt{1}} + e^{\sqrt{\frac{3}{2}}} + e^{\sqrt{2}}) = \frac{1}{2} \sum_{k=1}^{4} e^{\sqrt{x_k}} = \frac{1}{2} \sum_{k=1}^{4} e^{\sqrt{\frac{k}{2}}}.$$

▶Objective 5   Identify the distinction between approximations with inscribed and circumscribed rectangles.

Example 1   Label each of the three approximations from the previous problem as using inscribed, circumscribed, or some other kind of rectangles.

Tips   Inscribed rectangles have a height given by the smallest value of the function on the interval.

Circumscribed rectangles have a height given by the largest value of the function on the interval.

Answer   Since $f(x) = e^{\sqrt{x}}$ is monotonic increasing positive function we know:

Left-endpoint rectangles will be inscribed.

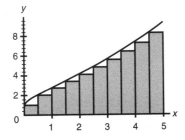

Midpoint rectangles will be neither inscribed. nor circumscribed

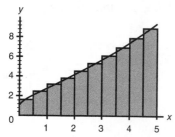

Right-endpoint rectangles will be circumscribed.

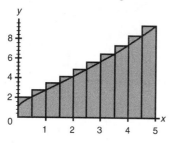

Example 2   Use four circumscribed rectangles of equal width to approximate the area under the curve $f(x) = 20 + x - x^2$ from $0 \le x \le 4$.

Answer   Since the width of each rectangle is one, we simply end up with:

$$A \approx 1 \left(f\left(\frac{1}{2}\right) + f(1) + f(2) + f(3) \right) = 20.25 + 20 + 18 + 14 = 72.25.$$

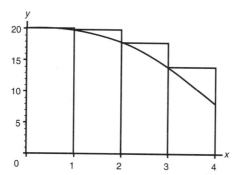

Why do we use $f(\frac{1}{2})$? Because it's the largest value of the function on the interval $0 \le x \le 1$. From our first derivative test:

x	$0 \le x \le \frac{1}{2}$	$\frac{1}{2}$	$\frac{1}{2} < x \le 1$
$f'(x) = 1 - 2x$	*Positive*	0	*Negative*
$f(x) = 20 + x - x^2$	Increasing	Relative *Max*	Decreasing

►Objective 6   Identify ways in which Riemann sum approximations can be improved.

Varying the width of the intervals. By taking more samples when the curve is changing dramatically, you can improve your estimate:

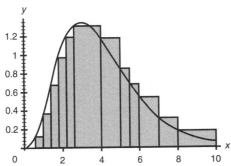

Increasing the number of intervals. No matter what method you use, if you increase the number of samples, you'll generally get a better estimate of the area:

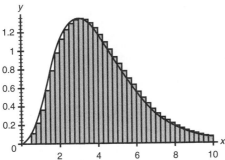

Picking sample points wisely for each interval. If you pick heights that tend to make overestimates cancel underestimates your approximation will improve:

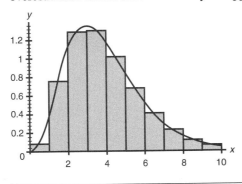

►Objective 7   Use the trapezoid rule to approximate the area under a curve.

Example   Use the trapezoidal rule to approximate the area under the curve $f(x) = e^{\sqrt{x}}$ from $0 \leq x \leq 2$ with 4 intervals of equal width.

The trapezoidal rule for approximating

$$\int_a^b f(x)\,dx$$

using n subintervals of equal length with

$$a = x_0,\ x_1,\ x_2,\ \ldots,\ x_{n-1},\ x_n = b$$

as partition points is

$$T = \frac{b-a}{2n}(f(a) + 2f(x_1) + 2f(x_2) + \times s + 2f(x_{n-1}) + f(b))$$

Tip Notice that you could get the same answer by simply averaging the left- and right-endpoint answers from above:

$$T = \frac{L+R}{2}.$$

Answer Using the trapezoidal rule,

$$T = \frac{b-a}{2n}(f(a) + 2f(x_1) + 2f(x_2) + \times s + 2f(x_{n-1}) + f(b))$$

we have

$$A \approx T = \frac{1}{4}\left(e^{\sqrt{0}} + 2e^{\sqrt{\frac{1}{2}}} + 2e^{\sqrt{1}} + 2e^{\sqrt{\frac{3}{2}}} + e^{\sqrt{2}}\right) \approx 5.353.$$

▶Objective 8 Use numerical methods to approximate the area under a curve, no matter whether the data is given in a formula, graph, or table.

Example Given that $g(x)$ is a continuous function that has values given as follows:

x	1	2	3	6	8
$g(x)$	4	12	8	6	10

Estimate the area under the curve $y = g(x)$ from $1 \le x \le 8$ using the trapezoidal approximation.

Tip While it may be useful to memorize the trapezoidal rule, you must also understand the underlying idea: using trapezoids to estimate area.

Answer Because the widths of the intervals vary, you should just calculate the area of each trapezoid:

$$A = 1\left(\frac{4+12}{2}\right) + 1\left(\frac{12+8}{2}\right) + 3\left(\frac{8+6}{2}\right) + 2\left(\frac{6+10}{2}\right) = 8 + 10 + 21 + 16 = 55$$

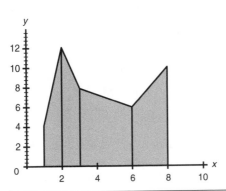

▶Objective 9   Identify the typical error present for all the approximation methods.

Assume that a curve is **non-negative**, **increasing**, and **concave up** over an interval $[a, b]$.

Let's consider the error obtained when estimating the area under such a curve using left-hand endpoints, right-hand endpoints, trapezoids, and midpoints.

Example   Using the various approximations for the definite integral $\int_0^2 3x^2 dx = 8$, describe the kind of error you get with each one.

Answer   Left-hand endpoints:

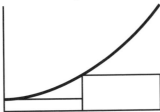

Since the graph is **increasing**, left-hand endpoints will give an **underestimate** of the area because the rectangles are inscribed.

Right-hand endpoints:

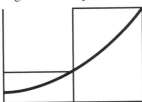

Since the graph is **increasing**, right-hand endpoints will give an **overestimate** of the area because the rectangles are circumscribed.

Trapezoids:

Since the curve is **concave up**, the trapezoidal method will give an **overestimate**.

Midpoints:

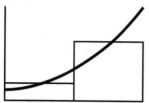

Since the curve is **concave up** the midpoint method will give a slight **underestimate**. The overestimate in the first half of the rectangle is smaller than the underestimate for the second half of the rectangle, since the slope of the curve is always increasing (concave up). Typically, the midpoint approximation is about twice as good as the trapezoidal method.

Multiple-Choice Exercises: Areas and Riemann Sums

___

1. The area under $f(x) = \sqrt{16 - x^2}$ on the domain $-4 \le x \le 4$ is:

A. 128π

B. 16π

C. 8π

D. 4π

E. You'd have to approximate this one.

Answer C. The area is $\dfrac{1}{2}$ of a circle so:

$$A = \frac{1}{2}\pi r^2 = \frac{1}{2}\pi 4^2 = 8\pi.$$

___

2. The area under $f(x) = 2x - 3$ on the domain $3 \le x \le 7$ is:

A. 28

B. 20

C. 16

D. 12

E. You would have to approximate this one.

Answer A. You have a trapezoid here with base of 4; heights of 3 and 11; $A = 4(\frac{11+3}{2}) = 28$.

3. Suppose you approximate the area under $q(x) = \sin x + 2$ on the domain $\frac{\pi}{2} \leq x \leq \frac{3\pi}{2}$ with $n = 4$ subintervals, using right endpoints. The width of each interval will be:

A. $\frac{3\pi}{8}$

B. $\frac{\pi}{8}$

C. $\frac{\pi}{4}$

D. $\frac{\pi}{2}$

E. Not enough information was given.

Answer   C.

$$\Delta x = \frac{b-a}{n} = \frac{\frac{3\pi}{2} - \frac{\pi}{2}}{4} = \frac{\pi}{4}.$$

4. Suppose you approximate the area under $q(x) = \sin x + 2$ on the domain $\frac{\pi}{2} \leq x \leq \frac{3\pi}{2}$ with $n = 4$ subintervals using right endpoints. The set of x values that you need to use are:

A. $\{0, \frac{\pi}{4}, \frac{2\pi}{4}, \frac{3\pi}{4}\}$

B. $\{\frac{\pi}{4}, \frac{2\pi}{4}, \frac{3\pi}{4}, \frac{4\pi}{4}\}$

C. $\{\frac{2\pi}{4}, \frac{3\pi}{4}, \frac{4\pi}{4}, \frac{5\pi}{4}\}$

D. $\{\frac{3\pi}{4}, \frac{4\pi}{4}, \frac{5\pi}{4}, \frac{6\pi}{4}\}$

E. Not enough information was given.

Answer   D. The first x value is Δx over from $\frac{\pi}{2}$, and the last one is equal to $\frac{3\pi}{2}$.

5. Suppose you approximate the area under $q(x) = \sin x + 2$ on the domain $\frac{\pi}{2} \leq x \leq \frac{3\pi}{2}$ with $n = 4$ subintervals using right endpoints. The approximated area is about:

A. 5.498

B. 7

C. 11.103

D. 14.137

E. Not enough information was given.

Answer   A.

$$\left(\frac{\pi}{4}\right)\left(q\left(\frac{3\pi}{4}\right) + q\left(\frac{4\pi}{4}\right) + q\left(\frac{5\pi}{4}\right) + q\left(\frac{6\pi}{4}\right)\right) = \frac{7}{4}\pi \approx 5.497787.$$

6. The rectangles used to estimate the area under the curve of $y = x^3$ on the domain $3 \leq x \leq 5$ using $n = 5$ subintervals will be:

A. inscribed

B. circumscribed

C. neither inscribed nor circumscribed

D. some inscribed and some circumscribed

E. Not enough information is given

Answer   E. You need to know where to calculate the heights of these rectangles.

7. The rectangles used to estimate the area under the curve of $y = \sqrt[3]{x}$ on the domain $3 \leq x \leq 5$ using $n = 5$ subintervals with right-hand endpoints will be:

A. inscribed

B. circumscribed

C. neither inscribed nor circumscribed

D. some inscribed and some circumscribed

E. Not enough information is given

Answer   B. Every rectangle has a height given by the largest y value over each interval.

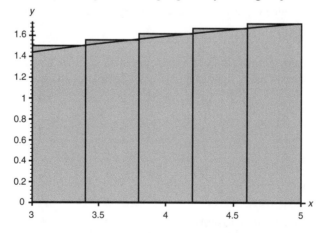

8. The rectangles used to estimate the area under the curve of $y = x^2$ on the domain $-5 \leq x \leq 5$ using $n = 5$ subintervals with right-hand endpoints will be:

A. all inscribed

B. all circumscribed

C. neither inscribed nor circumscribed

D. some inscribed and some circumscribed

E. Not enough information is given

Answer D. This has some of both inscribed and circumscribed rectangles.

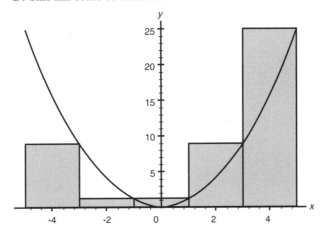

---

9. Using midpoints to approximate the area under the curve $y = t^3 e^{-t}$ between $t = 0$ and $t = 10$, with $n = 10$ subintervals, the t values that you'd use are:

A. $\{\frac{1}{2}, \frac{3}{2}, \frac{5}{2}, \frac{7}{2}, \frac{9}{2}, \frac{11}{2}, \frac{13}{2}, \frac{15}{2}, \frac{17}{2}, \frac{19}{2}\}$

B. $\{\frac{3}{2}, \frac{7}{2}, \frac{11}{2}, \frac{15}{2}, \frac{19}{2}\}$

C. $\{1, 3, 5, 7, 9\}$

D. $\{1, 2, 3, 4, 5, 6, 7, 8, 9, 10\}$

E. $\{\frac{1}{2}, 2, \frac{5}{2}, 4, \frac{11}{2}, 6, 7, 8, 9, \frac{19}{2}\}$

Answer A. You have to pick t values right in the middle of each interval.

---

10. Using midpoints to approximate the area under the curve $y = t^3 e^{-t}$ between $t = 0$ and $t = 10$, with $n = 10$ subintervals, the approximated area is about:

A. 50

B. 25

C. 5.934

D. 2.967

E. 3.967

Answer C. The width of each interval is one. So all you have to do is add up $t^3 e^{-t}$ evaluated at each member of the set $\{\frac{1}{2}, \frac{3}{2}, \frac{5}{2}, \frac{7}{2}, \frac{9}{2}, \frac{11}{2}, \frac{13}{2}, \frac{15}{2}, \frac{17}{2}, \frac{19}{2}\}$ or $\sum_{k=1}^{10} (k - \frac{1}{2})^3 e^{-(k-\frac{1}{2})} = 5.934$.

---

11. Given the Riemann sum of $S = (\frac{\pi}{4})(p(\frac{3\pi}{4}) + p(\frac{4\pi}{4}) + p(\frac{5\pi}{4}) + p(\frac{6\pi}{4}))$. Using summation notation this expression for S can be written as:

A. $\frac{\pi}{4} \sum_{k=1}^{4} p(\frac{3\pi}{4} + \frac{k\pi}{4})$

B. $\sum_{k=1}^{4} p(\frac{3\pi}{4} + \frac{k\pi}{4})$

C. $\sum_{k=1}^{4} p(\frac{\pi}{2} + \frac{k\pi}{4})$

D. $\frac{\pi}{4} \sum_{k=1}^{4} p(\frac{\pi}{2} + \frac{k\pi}{4})$

E. $\frac{\pi}{2} \sum_{k=1}^{4} p(\frac{\pi}{2} + \frac{k\pi}{4})$

D. $\Delta x = \frac{\pi}{4}, n = 4$, and $x_k = \frac{\pi}{2} + \frac{k\pi}{4}$ in the general form $\sum_{k=1}^{n} f(x_k)\Delta x$.

---

12. Suppose you approximate the area under the curve of $f(x)$ on the domain $2 \leq x \leq 5$ with $n = 6$ subintervals. If $\sum_{k=1}^{n} f(x_k)\Delta x$ is the Riemann sum, match each of the following expressions for x_k with **left-hand endpoints**, **right-hand endpoints**, or **midpoints**.

I. $x_k = 2 + \frac{k}{2}$

II. $x_k = 2 + \frac{k-1}{2}$

III. $x_k = 2 + \frac{2k-1}{4}$

A. I is left-hand endpoints, II is midpoints, and III is right-hand endpoints.

B. II is left-hand endpoints, III is midpoints, and I is right-hand endpoints.

C. II is left-hand endpoints, I is midpoints, and III is right-hand endpoints.

D. III is left-hand endpoints, II is midpoints, and I is right-hand endpoints.

E.   III is left-hand endpoints, I is midpoints, and II is right-hand endpoints.

Answer   B.

I. $x_k = 2 + \frac{k}{2}$ gives values of $\{\frac{5}{2}, \frac{6}{2}, \frac{7}{2}, \frac{8}{2}, \frac{9}{2}, \frac{10}{2}\}$ for $k = 1$ to 6, which is right-hand endpoints.

II. $x_k = 2 + \frac{k-1}{2}$ gives values of $\{\frac{4}{2}, \frac{5}{2}, \frac{6}{2}, \frac{7}{2}, \frac{8}{2}, \frac{9}{2}\}$ for $k = 1$ to 6, which is left-hand endpoints.

III. $x_k = 2 + \frac{2k-1}{4}$ gives values of $\{\frac{9}{4}, \frac{11}{4}, \frac{13}{4}, \frac{15}{4}, \frac{17}{4}, \frac{19}{4}\}$ for $k = 1$ to 6, which is midpoints.

---

13.   If you use midpoints to approximate the area A under the curve

$$y = f(x) = 5\sin(\pi x) + 2.5\cos(4\pi x)$$

on the interval $[0, 1]$ using 10 equal subdivisions, then $A \approx$

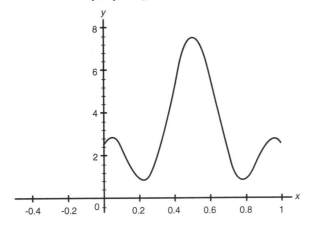

A.   3.157

B.   3.196

C.   3.407

D.   2.078

E.   2.780

Answer   B. The Riemann sum asked for is $R = \sum\limits_{k=1}^{10} f(x_k)(0.1)$, where $x_1 = .05$, $x_2 = .15$, $\ldots$, $x_{10} = .95$.

Summing these all up, you get $R = 3.196$, which is the approximation for the area.

14. If you use right-hand endpoints and 6 equal subdivisions to approximate the area A beneath the curve $y = f(x) = \frac{1}{1+x^2}$ on the interval $[0, 6]$ then $A \approx$

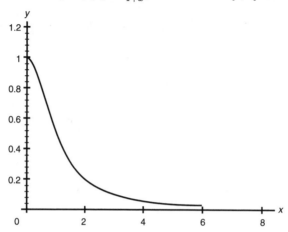

A. 0.9243

B. 1.405

C. 1.897

D. 1.021

E. 1.682

Answer A. The Riemann sum asked for is $R = \sum\limits_{k=1}^{6} f(x_k)(0.1)$, where $x_1 = 1, x_2 = 2, \ldots, x_6 = 6$.

So you have $R = 1(\frac{1}{1+1^2} + \frac{1}{1+2^2} + \times s + \frac{1}{1+6^2}) \approx 0.9243$, which is the approximation for the area.

15. The table below gives data points for the continuous function $y = f(x)$.

x	0	.2	.4	.6	.8	1.0	1.2	1.4	1.6	1.8	2.0
$f(x)$	30	44	55	68	66	54	39	37	26	25	40

Approximate the area under the curve $y = f(x)$ on the interval $[0, 2]$ using left-hand endpoints and 10 equal subdivisions. You get *Area* $\approx$

A. 96.8

B. 454

C. 88.8

D. 90.8

E. 444

Answer C. The Riemann sum you need here is $R = .2(30 + 44 + 55 + 68 + 66 + 54 + 39 + 37 + 26 + 25) = 88.8$.

---

16. Consider the curve $y = f(x) = \frac{1}{x}$, and the region under $f(x)$ between $x = 1$ and $x = 3$ which is graphed below.

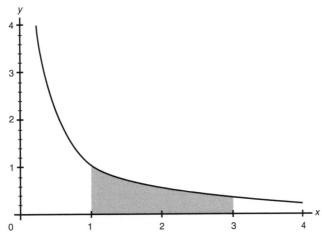

Suppose L is the left-hand endpoint Riemann sum with 15 subdivisions, R is the right-hand endpoint Riemann sum with 15 subdivisions, and A is the true area of this region. Which of the following is correct?

A. $R < L < A$

B. $L < A < R$

C. $L = A = R$

D. $R < A < L$

E. $A < R < L$

Answer D. Since the function is decreasing on this interval, the Riemann sum using left-hand endpoints will be greater than the true area, which will in turn be greater than the Riemann sum using right-hand endpoints. That is, $R < A < L$.

17.   The function $y = f(x)$ is graphed below:

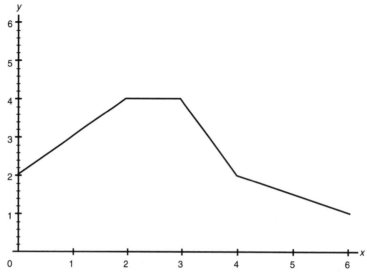

Which of the following Riemann sums yields the exact area under the curve on the interval $[0, 6]$?

A. $R = \sum_{k=1}^{6} f(x_k)\Delta x_k$, where 6 equal subdivisions and right-hand endpoints are used.

B. $R = \sum_{k=1}^{4} f(x_k)\Delta x_k$, where the subdivisions are taken to be at $\{0, 1, 3, 4, 6\}$ and right-hand endpoints are used.

C. $R = \sum_{k=1}^{6} f(x_k)\Delta x_k$, where 6 equal subdivisions and left-hand endpoints are used.

D. $R = \sum_{k=1}^{6} f(x_k)\Delta x_k$, where the subdivisions are taken to be at $\{0, 2, 3, 4, 5, 6\}$ and left-hand endpoints are used.

E. $R = \sum_{k=1}^{6} f(x_k)\Delta x_k$, where the subdivisions are taken to be at $\{0, 2, 3, 4, 6\}$ and midpoints are used.

Answer   E. The exact area is 16, which you can find by dividing the region into rectangles and triangles or trapezoids.

$R = \sum_{k=1}^{6} f(x_k)\Delta x_k$, with the same subdivisions at $\{0, 2, 3, 4, 6\}$ and midpoints works out to be $3(2) + 4 + 3 + 1.5(2) = 16$, which is the exact area.

18. Here is a graph of the function $y = r(x) = \tan(\cos(\pi x) + 0.5) + 2$:

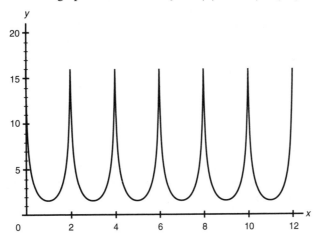

Estimate the total area A under this curve on the interval $[0, 12]$ with a Riemann sum using 36 equal subdivisions and *circumscribed rectangles*. (Hint: Use symmetry to make this problem easier.)

A. 57.340

B. 86.634

C. 14.439

D. 49.914

E. 28.044

Answer B. One way to solve this is by using symmetry. Subdividing the interval [0, 12] into 36 equal subintervals is the same as breaking the 6 smaller intervals [0, 2], [2, 4], [4, 6], [6, 8], [8, 10], and [10, 12] into 6 equal subintervals each. Considering only one such interval, [0, 2], compute the area under the curve using 6 equal subintervals and circumscribed rectangles, and then multiply this result by 6 to get your total. Here is the graph of the function $r(x)$ on the interval [0, 2], with the subdivisions drawn in.

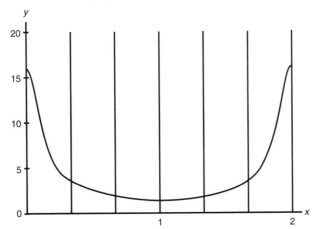

Since you want to use circumscribed rectangles, you must choose the largest function value on each subinterval. The Riemann sum on the interval $[0, 2]$ becomes

$$R = \frac{1}{3}\left(r(0) + r\left(\frac{1}{3}\right) + r\left(\frac{2}{3}\right) + r\left(\frac{4}{3}\right) + r\left(\frac{5}{3}\right) + r(2)\right).$$

Plugging in the function values gives $R = 14.439$. Remember that you must multiply this by 6 to get your total estimate, which is $14.439 \times 6 = 86.634$. You can make this problem even easier by using more symmetry. Do you see how?

---

19. The area under a curve, between $x = 0$ and $x = 2$, is estimated using Riemann sums with $n = 4$ subintervals. The left-hand Riemann sum gives an estimate of 10.25, and the right-hand Riemann sum gives an estimate of 8.25. What estimate would the trapezoid rule with $n = 4$ subintervals give?

A. 10

B. 9

C. 9.25

D. The correct area

E. There's no way to tell without knowing what the curve is.

Answer   C. The trapezoid rule gives the average of the left- and right-hand Riemann sums.

---

20. Estimate the area under the curve $f(x) = 6 - x^2$, between $x = 0$ and $x = 2$, using the trapezoid rule, with $n = 4$ subintervals.

A. 10

B. 11.25

C. 18.25

D. 9.25

E. 21.25

Answer   D. The trapezoid rule gives $\frac{2-0}{2\times4}(f(0) + 2f(0.5) + 2f(1) + 2f(1.5) + f(2)) = 9.25$.

21. We want to estimate the area under the curve shown in the graph. We want the best estimate we can get, so we draw trapezoids that seem to fit the curve nicely. Using the trapezoids in the picture, what is your estimate of the area under this curve?

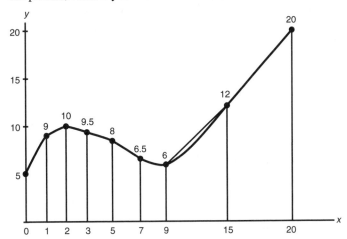

A. 168

B. 183.75

C. 192.5

D. 204.75

E. 249.5

Answer D. You can't use the trapezoidal *rule* here, because the trapezoids aren't all the same width. Find their areas individually and add them up. If you add the areas of all the trapezoids shown, you get $7 + 9.5 + 9.75 + 17.5 + 14.5 + 12.5 + 54 + 80 = 204.75$.

22. The table below shows the speedometer reading in miles per hour when a car tried to brake before an accident. Use the trapezoid rule to estimate how far the car traveled in this time by estimating the area under the curve.

time (sec)	0	1	2	3	4	5	6	7	8
speed (mph)	70	65	59	52	44	36	27	18	10

A. 341 feet

B. 500 feet

C. 682 feet

D. 232.5 feet

E. This isn't an area problem; the trapezoid rule doesn't apply.

Answer B. Notice that the different speeds of the car are given in miles per hour while the intervals are given in seconds. We want to work with consistent units. So let's first work with intervals whose lengths are in hours, so that the trapezoidal rule will give us an approximation for the distance traveled, which is in miles. Then, we will convert that estimated distance traveled, which is in miles, to feet. Since each time interval has a length of 1 second and there are 3,600 seconds in an hour, each time interval has a length of $\dfrac{1}{3,600}$ of an hour.

According to the trapezoidal rule, the car traveled this approximate distance.

$$\frac{1}{3,600 \times 2}(70 + 2(65) + 2(59) + 2(52) + 2(44) + 2((36) + 2(27) + 2(18) + 10)) =$$

$$\frac{1}{7,200}(682) = \frac{682}{7,200} \approx 0.09472 \text{ miles.}$$

All that remains to be done is to convert this distance to feet. There are 5,280 feet in a mile, so $0.09472 \text{ miles} = 0.09472 \text{ miles} \times 5,280 \dfrac{\text{feet}}{\text{mile}} \approx 500.1 \text{ feet.}$

This is closest to choice B, 500 feet.

---

23. If you use trapezoids to approximate the area A under the curve

$$y = f(x) = 5\sin(\pi x) + 2.5\cos(4\pi x)$$

on the interval $[0, 1]$ with 10 equal subdivisions you get $A \approx$

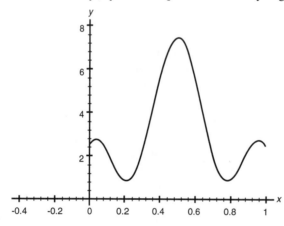

A. 3.157

B. 3.183

C. 6.314

D. 63.138

E. 3.078

Answer A. Since you're using subintervals of equal width, you can use the trapezoid rule:

$$A \approx T = \frac{1-0}{2(10)}(f(0) + 2f(0.1) + \times s + 2f(0.9) + f(1)) = 3.157.$$

This is the approximation of the area. Note, the actual area under the curve is 3.183, so the trapezoid rule gives you a pretty good approximation here.

---

24. Use the trapezoidal rule with 6 equal subdivisions to approximate the area beneath the curve $y = f(x) = \dfrac{1}{1+x^2}$ on the interval $[0, 6]$.

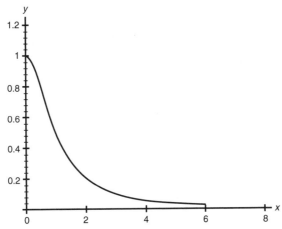

You get *Area* $\approx$

A. 1.406

B. 1.791

C. 0.896

D. 1.411

E. 2.822

Answer D. Since the subintervals are of equal width, you can use the trapezoid rule:

$$T = \frac{6}{2(6)}(f(0) + 2f(1) + 2f(2) + \times s + f(6)) = 1.411.$$

This is the approximation for the area under the curve. Note that the actual area under the curve is 1.406!

---

25. The table below gives data points for the continuous function $y = f(x)$

x	0	.2	.4	.6	.8	1.0	1.2	1.4	1.6	1.8	2.0
$f(x)$	30	44	55	68	66	54	39	37	26	25	40

Approximate the area A under the curve $y = f(x)$ on the interval $[0, 2]$ using trapezoids and 10 equal subdivisions.

A. 96.8

B. 898

C. 88.8

D. 89.8

E. 444

Answer   D. Since the subintervals are of equal width, you can use the trapezoidal rule:

$$T = \frac{2}{2(10)}(f(0) + 2f(0.2) + 2f(0.4) + \cdots + 2f(1.8) + 2f(2.0))$$

$$= \frac{1}{10}(30 + 88 + 110 + 136 + 132 + 108 + 78 + 74 + 52 + 50 + 40)$$

$$= 89.8.$$

This is the approximation for the area under the curve.

---

26.   Consider the curve $y = f(x) = \frac{1}{x}$, and the region under $f(x)$ between $x = 1$ and $x = 3$, which is graphed below:

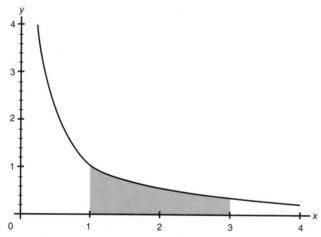

Suppose T is the approximation of this area using 15 trapezoids of equal width, R is the Riemann sum using right-hand endpoints and 15 equal subdivisions, and A is the true area of this region. Which of the following is correct?

A. $T = A = R$

B. $A < R < T$

C. $R < T < A$

D. $R < A < T$

E. $T < A < R$

Answer   D. Since the function is concave-up on the interval of concern, the trapezoid rule approximation will be greater than the true area. Furthermore, since the function is decreasing on the interval of concern, the Riemann sum using right-hand endpoints will be less than the true area.

That is, $R < A < T$.

---

27.   Consider the function $f(x) = \dfrac{5 \sin(\pi x - \frac{\pi}{2})}{(x^2 + 4x + 5)} + 5 \times 1.1^{-(x+2)}$, which is graphed below.

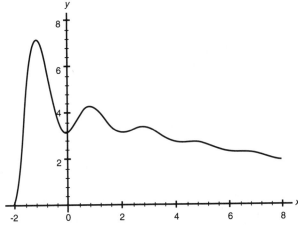

Estimate the area A under the curve from $x = 2$ to $x = 4$, using the trapezoidal rule and 3 equal subdivisions.

A.   12.414

B.   18.629

C.   6.514

D.   6.041

E.   6.207

Answer   E. The trapezoidal rule here gives you

$$T = \frac{4 - 2}{2(3)} \left(f(2) + 2f\left(\frac{8}{3}\right) + 2f\left(\frac{10}{3}\right) + f(4) \right) = 6.207.$$

This is the approximation for the area under the curve from $x = 2$ to $x = 4$.

B. Definite Integrals

▶Objective 1   Calculate definite integrals geometrically for circles, trapezoids, triangles, and rectangles.

Example 1   Given the graph of $h(x)$:

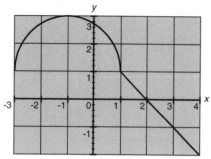

(where the curve is a semicircle connected to a line), find $\int_{-3}^{4} h(x)\,dx$.

Tip   Remember that area under the axis is interpreted as negative by a definite integral.

Answer   From $-3 \le x \le 1$, you have a semicircle on top of a rectangle:

$A = \frac{1}{2}\pi r^2 + bh = \frac{1}{2}\pi 2^2 + 4(1) = 2\pi + 4$. Then from $1 \le x \le 2$, you have a little triangle with $A = \frac{1}{2}$. Finally, from $2 \le x \le 4$, you're under the curve, so you get a signed area of $A = -2$. So,

$$\int_{-3}^{4} h(x)dx = 2\pi + 4 + \frac{1}{2} - 2 = \frac{5}{2} + 2\pi.$$

Example 2   Evaluate $\int_{-2}^{3} |x|\,dx$.

Tip   It will often help to visualize the area represented by a definite integral.

Answer   This is simply two triangles:

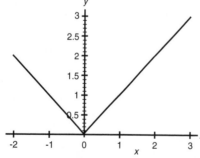

Using $A = \frac{1}{2}bh$.

$$\int_{-2}^{3} |x|dx = \frac{1}{2}2(2) + \frac{1}{2}3(3) = 2 + \frac{9}{2} = \frac{13}{2}.$$

▶Objective 2   Identify the area under a curve (and the definite integral) as a limit of a Riemann sum.

Example

A.   Convert the following limit of a Riemann sum into a definite integral:

$$\lim_{n \to \infty} \frac{2}{n} \sum_{k=1}^{n} \left(1 + \frac{2k}{n} \right)^2$$

B.   Convert the following integral into a limit of a Riemann sum:

$$\int_0^{\frac{\pi}{2}} \cos(x)\, dx$$

Tip   When you take the limit as $n \to \infty$ of $\sum_{k=1}^{n} f(x_k)\Delta x$, realize that the first rectangle ($k=1$) becomes $f(a)\, dx$ and the last rectangle ($k = \infty$) becomes $f(b)dx$, so you can use your limit skills to determine the limits of integration.

Answer

A.   When k is 1, and $n \to \infty$, then $1 + \dfrac{2k}{n} \to 1$.

When $k = n$, and $n \to \infty$, then $1 + \dfrac{2k}{n} \to 3$.

Also $\dfrac{b-a}{n} = \dfrac{3-1}{n} = \dfrac{2}{n}$, so that $\lim_{n \to \infty} \dfrac{2}{n} \sum_{k=1}^{n} (1 + \dfrac{2k}{n})^2 = \int_1^3 x^2\, dx$.

Another correct answer would be $\int_0^2 (1+x)^2\, dx$.

B. $\int_a^b f(x)\, dx = \lim_{\Delta x \to 0} \sum_{k=1}^{n} f(x_k)\Delta x = \lim_{n \to \infty} \dfrac{b-a}{n} \sum_{k=1}^{n} f(x_k)$, where $\Delta x = \dfrac{b-a}{n}$.

So $\int_0^{\frac{\pi}{2}} \cos(x)\, dx = \lim_{n \to \infty} \dfrac{\pi}{2n} \sum_{k=1}^{n} \cos(x_k)$, since $\dfrac{b-a}{n} = \dfrac{\frac{\pi}{2} - 0}{n} = \dfrac{\pi}{2n}$.

So finally, $\int_0^{\frac{\pi}{2}} \cos(x)\, dx = \lim_{n \to \infty} \dfrac{\pi}{2n} \sum_{k=1}^{n} \cos\left(\dfrac{k\pi}{2n} \right)$.

Note that instead of $x_k = \dfrac{k\pi}{2n}$ (right-hand endpoints), we could also have used either $x_k = \dfrac{(k-1)\pi}{2n}$ (left-hand endpoints) or $x_k = \dfrac{(k-\frac{1}{2})\pi}{2n}$ (midpoints).

The definition of $\int_a^b f(x)dx$ is that $\int_a^b f(x)dx = \lim_{n \to \infty} \sum_{k=1}^{n} f(x_k)(\Delta x_k)$ where the largest (Δx_k) approaches 0 as n goes to infinity. The (Δx_k)'s do not all have to be the same length.

▶Objective 3   Describe the difference between the area under a curve and a definite integral.

Example In your own words explain when $\int_a^b f(x)dx$ is **not** the area under the curve $f(x)$ for $a \leq x \leq b$.

Tip Remember that when $f(x) < 0$, the integral accumulates negative values, and strictly speaking area is always a positive quantity.

Answer If $f(x) < 0$ anywhere on the domain $[a, b]$.

▶**Objective 4** Approximate definite integrals using the same methods used for approximating area under a curve.

Example Approximate $\int_{-\frac{\pi}{3}}^{\frac{\pi}{3}} \tan(x)dx$ with $n = 4$ subintervals using:

   A. left endpoints

   B. right endpoints

   C. trapezoids

Answer

   A. left endpoints

$$L = \frac{\pi}{6}\left(\tan\left(-\frac{\pi}{3}\right) + \tan\left(-\frac{\pi}{6}\right) + \tan(0) + \tan\left(\frac{\pi}{6}\right)\right) = \frac{\pi\sqrt{3}}{6}$$

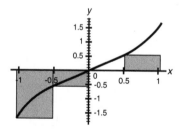

   B. right endpoints

$$R = \frac{\pi}{6}\left(\tan\left(-\frac{\pi}{6}\right) + \tan(0) + \tan\left(\frac{\pi}{6}\right) + \tan\left(\frac{\pi}{3}\right)\right)$$

$$= \frac{\pi}{6}\left(-\frac{\sqrt{3}}{3} + 0 + \frac{\sqrt{3}}{3} + \sqrt{3}\right) = \frac{\pi\sqrt{3}}{6}$$

C.   trapezoids

$$T = \frac{1}{2}\left(\frac{\pi}{6}\right)\left(\tan\left(-\frac{\pi}{3}\right) + 2\tan\left(-\frac{\pi}{6}\right) + 2\tan(0) + 2\tan\left(\frac{\pi}{6}\right) + \tan\left(\frac{\pi}{3}\right)\right)$$

$$= \frac{\pi}{12}\left(-\sqrt{3} - \frac{2\sqrt{3}}{3} + 0 + \frac{2\sqrt{3}}{3} + \sqrt{3}\right) = 0$$

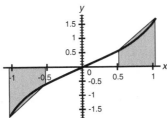

The fact that $\tan x$ is an odd function tells us that $\displaystyle\int_{-\frac{\pi}{3}}^{\frac{\pi}{3}} \tan(x)dx = 0$.

►Objective 5   Write the algebraic properties of the definite integral, and relate them to the geometric properties of area.

Properties of Definite Integrals

1. $\displaystyle\int_{a}^{a} f(x)dx = 0$. The area of a shape with width 0 is 0.

2. $\displaystyle\int_{a}^{b} f(x)dx = -\int_{b}^{a} f(x)dx$. Here the width changes signs, which makes the area change signs.

3. $\displaystyle\int_{a}^{b} kf(x)dx = k\int_{a}^{b} f(x)dx$. A shape that is k times as high will have k times the area.

4. $\displaystyle\int_{a}^{b} [f(x) + g(x)]dx = \int_{a}^{b} f(x)dx + \int_{a}^{b} g(x)dx$. Stacking one shape on top of another shape leads to an area that's the sum of the original two shapes.

5. $\displaystyle\int_{a}^{b} [f(x) - g(x)]dx = \int_{a}^{b} f(x)dx - \int_{a}^{b} g(x)dx$. The area between two curves is simply the larger area minus the smaller area (assuming that $f(x) \geq g(x)$).

6. If $f(x) \geq 0$ on $[a, b]$, then $\displaystyle\int_{a}^{b} f(x)dx \geq 0$. As long as the definite integral accumulates positive values, the total area will be positive.

7. If $f(x) \geq g(x)$ on $[a, b]$, then $\displaystyle\int_{a}^{b} f(x)dx \geq \int_{a}^{b} g(x)dx$. If one shape has the same width as another shape and one shape is always taller, the taller one will have more area.

8. $\min f \times (b - a) \leq \displaystyle\int_{a}^{b} f(x)dx \leq \max f \times (b - a)$, where $\min f$ and $\max f$ refer to the minimum and maximum values of f on $[a, b]$. On this one think of inscribed and circumscribed rectangles. Inscribed rectangles will always underestimate the area, and circumscribed rectangles will always overestimate the area.

9. For any a, b, and c, provided that f is continuous on the intervals joining them,

$\int_a^b f(x)dx = \int_a^c f(x)dx + \int_c^b f(x)dx$. If you split a shape into two parts, the total area is simply the sum of the two parts.

10. If $g(x)$ is odd, $\int_{-a}^a g(x)dx = 0$. The area under the axis will equal the area above the axis from $-a$ to a for an odd function.

11. If $g(x)$ is even, $\int_{-a}^a g(x)dx = 2\int_0^a g(x)dx$. The area from $-a$ to 0 will be the same as the area from 0 to a for an even function.

▶Objective 6 Use the properties of the definite integral to solve problems related to area.

Example If f is a continuous function, and $\int_a^b f(x)\,dx = 5$, $\int_0^b f(x)\,dx = 9$, what is $\int_0^a f(x)dx$? Justify your answer.

Answer While you might see this answer without any algebra or calculus, you've been asked to justify your answer so:

$$\int_a^b f(x)dx = \int_a^0 f(x)dx + \int_0^b f(x)dx, \text{ from property 9 above, so,}$$

$$5 = 9 + \int_a^0 f(x)\,dx \implies \int_a^0 f(x)\,dx = -4, \text{ and from property 2 above, we have}$$

$$\int_0^a f(x)\,dx = 4.$$

▶Objective 7 Identify the definite integral as an accumulator of values.

Example Evaluate $\int_0^{65\pi} \sin(x)dx$ given that $\int_0^{\frac{\pi}{2}} \sin(x)\,dx = 1$.

Tip Remember that the definite integral accumulates negative values when the function is below the x-axis.

Answer The answer comes from the symmetry of the sin function. After the integral has accumulated along the x-axis for an even number of π, the positive integral will be zero, since the area above the axis is equal to the area below the axis. Since the final bump from 64π to 65π is above the axis, the final answer will be 2:

$$\int_0^{65\pi} \sin(x)dx = 2.$$

▶Objective 8 Calculate the average value of a function for a given domain.

Example Find the average value of $\sin x$ on the domain $0 \le x \le \pi$.

Tip   Remember that the average value of a function is simply represented by a line that gives the same area under the curve as the function itself:

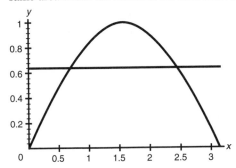

Answer

$$Avg = \frac{1}{b-a}\int_a^b f(x)dx = \frac{1}{\pi - 0}\int_0^\pi \sin(x)dx = \frac{2}{\pi} \approx 0.637$$

▶Objective 9   Calculate the change in position of an object from its velocity curve, using the definite integral.

Example   An object has a velocity given in the following graph of a semicircle connected to a straight line:

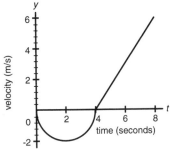

Where is the object relative to its original position at $t = 4$ and $t = 8$?

Tip   Remember that area under a velocity curve is accumulated change in position. Positive velocities are interpreted as motion to the right, and negative values are interpreted as motion to the left.

Answer   At $t = 4$, $\int_0^4 v(t)dt = -\frac{1}{2}\pi r^2 = -\frac{1}{2}\pi 2^2 = -2\pi$.

The object is 2π meters to the left of its original position.

At $t = 8$, $\int_4^8 v(t)dt = \frac{1}{2}bh = 12$.

The object moves back 12 meters toward the right from 4 to 8 seconds, to have a final position that is $(12 - 2\pi)$ meters to the right of the original position.

▶Objective 10   Calculate the net change in a quantity from the area under a rate of change function.

Example Water is pouring into a container at a rate given by $r(t) = 4t^2 - 2t^3$ in gallons per hour. How many gallons pour into the container from $t = 0$ hours to $t = 2$ hours?

Tip The definite integral of a rate of change accumulates all of the change to give the net change over the interval.

Answer The total water added will be $\int_0^2 r(t)dt = \left[\frac{4}{3}t^3 - \frac{1}{2}t^4\right]_0^2 = \frac{8}{3}$ gallons.

Multiple-Choice Questions on Definite Integrals

1. Think of estimating the area under the curve of $\sqrt{4-x^2}$ on $[-2,0]$ using right-hand endpoints. As you use larger and larger values for n, the approximated area will approach:

A. π

B. 2π

C. 4π

D. 8π

E. ∞

Answer A. As n goes to infinity the area approaches the true area under the curve, which is given by $Area = \frac{1}{4}$(area of circle) $= \frac{1}{4}\pi(2)^2 = \pi$.

2. Think of estimating the area under the curve of $\sqrt{4-x^2}$ on $[-2,0]$ with n going to infinity. What type of approximation should you use if you want to find the true area under the curve?

A. Right endpoints

B. Left endpoints

C. Midpoints

D. Trapezoids

E. Any of these

Answer E. As n goes to infinity, the area for any of the approximations approaches the true area under the curve.

3. Which of these is true?

I. $\int_a^b f(x)dx = \lim_{n\to\infty} \sum_{k=1}^n f(x_k)\Delta x$ where $\Delta x = \frac{b-a}{n}$

II. $\displaystyle\int_a^b f(x)dx = \lim_{n\to\infty} \frac{b-a}{n}\sum_{k=1}^{n} f(x_k)$ where $x_k = a + \dfrac{b-a}{n}k$ for $k = 1, \dots n$

III. $\displaystyle\int_a^b f(x)dx = \lim_{\Delta x\to 0} \sum_{k=1}^{n} f(x_k)\Delta x$ where $\Delta x = \dfrac{b-a}{n}$

A. I only

B. I and II only

C. I and III only

D. II only

E. I, II, and III

Answer   E. Since $\Delta x = \dfrac{b-a}{n}$ when you take equal subdivisions, and since $n \to \infty$ when $\Delta x \to 0$, all three of these limits give you the definite integral.

---

4. Given that $g(x) = \begin{cases} \frac{1}{2}x + \frac{1}{2} & \text{if } x < 1 \\ \sqrt{1-(x-1)^2} & \text{if } x \geq 1 \end{cases}$, evaluate $\displaystyle\int_{-1}^{2} g(x)dx$.

A. $\frac{3}{4} + \frac{1}{4}\pi$

B. π

C. $1 + \pi$

D. $1 + \frac{1}{4}\pi$

E. The area cannot be determined geometrically.

Answer   D. The area is simply a triangle with base 2 and height 1, plus $\frac{1}{4}$ of a circle:

$$\int_{-1}^{2} g(x)dx = Area = A_{triangle} + \frac{1}{4}A_{circle}$$

$$= \frac{1}{2}(2)(1) + \frac{1}{4}\pi(1)^2 = 1 + \frac{1}{4}\pi.$$

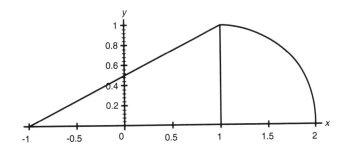

5. Given that $h(x)$ is a continuous function with the following values:

x	0	1	2	3	4	5	6	7	8
$h(x)$	2	2	8	26	62	122	212	338	506

approximate $\int_0^8 h(x)dx$ using midpoints with $n = 4$ subintervals.

A. 244

B. 488

C. 976

D. 1,278

E. More information is needed.

Answer   C. $A = \Delta x(h(1) + h(3) + h(5) + h(7)) = 2(2 + 26 + 122 + 338) = 976$.

6. $\int_1^3 \sin(x)dx$ is equivalent to which of the following:

A. $\lim\limits_{n \to \infty} \sum\limits_{k=1}^{n} \sin(1 + \frac{k}{n})$

B. $\lim\limits_{n \to \infty} \frac{2}{n}\sum\limits_{k=1}^{n} \sin(1 + \frac{2k}{n})$

C. $\lim\limits_{n \to \infty} \frac{2}{n}\sum\limits_{k=1}^{n} \sin(1 + \frac{k}{n})$

D. $\frac{2}{n}\sum\limits_{k=1}^{n} \sin(1 + \frac{k}{n})$

E. $\lim\limits_{n \to \infty} \frac{1}{n}\sum\limits_{k=1}^{n} \sin(1 + \frac{2k}{n})$

Answer   B. With $\Delta x = \dfrac{b - a}{n} = \dfrac{3 - 1}{n} = \dfrac{2}{n}$ and $x_k = 1 + \dfrac{2k}{n}$ we have:

$$a = \lim_{n \to \infty} x_{k=1} = \lim_{n \to \infty}\left(1 + \frac{2}{n}\right) = 1;$$

$$b = \lim_{n \to \infty} x_{k=n} = \lim_{n \to \infty}\left(1 + \frac{2n}{n}\right) = 3 \text{ and } f(x) = \sin x.$$

7. $\displaystyle\int_2^5 (2 + 3x)\, dx =$

A. 24

B. 37.5

C. 12.5

D. 75.0

E. 17.5

Answer   B. $\displaystyle\int_2^5 (2 + 3x)\, dx = 37.5$. Here is the graph of $y = 2 + 3x$ on the interval $[2, 5]$.

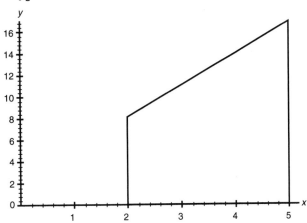

The area of this trapezoid is $\frac{1}{2}(8 + 17)(3) = 37.5$.

8. $\displaystyle\int_{-1}^2 \left(\sqrt{9 - (x + 1)^2} + 2 \right) dx =$

A. 7.0686

B. 8.3562

C. 34.274

D. 20.137

E. 13.069

Answer   E. $\displaystyle\int_{-1}^2 \left(\sqrt{9 - (x + 1)^2} + 2 \right) dx = 13.069$. Note that the graph of $y = \sqrt{9 - (x + 1)^2} + 2$ on the interval $[-1, 2]$ is one quarter of a circle of radius 3, shifted up and to the left.

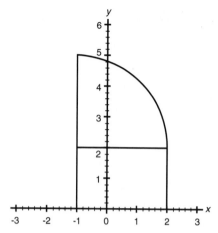

The area of the quarter circle is $\frac{1}{4}\pi(9) \approx 7.0686$, and the area of the rectangle below it is $2(3) = 6$. So the total area under the curve is about 13.069.

---

9. $\displaystyle\int_{-2}^{2}\left(2 - \sqrt{4 - x^2}\right)dx =$

A. 12.566

B. 14.2832

C. 7.1416

D. 1.717

E. 0.8585

Answer   D. $\displaystyle\int_{-2}^{2}(2 - \sqrt{4 - x^2})dx = 1.717$. Notice that the graph of $y = f(x)$ on the interval $[-2, 2]$ is a semicircle.

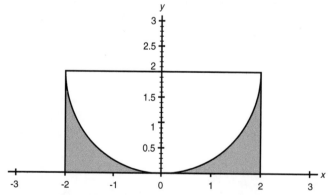

The semicircle is contained in the rectangle, which has vertices at $(-2, 0)$, $(-2, 2)$, $(2, 2)$, and $(2, 0)$, which has area $2 \times 4 = 8$. The semicircle has area $\frac{1}{2}\pi(4) = 6.2832$, which we must *subtract* from 8 to get the area below the curve. This gives us the answer.

The next two problems refer to the graph below:

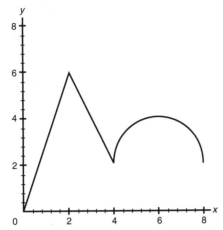

The graph of $y = f(x)$ consists of two straight lines and a semicircle.

---

10. $\displaystyle\int_0^4 f(x)\,dx =$

A. 28.283

B. 12.000

C. 14.000

D. 34.566

E. 20.283

Answer C. This is just the area under the curve $y = f(x)$ on the interval $[0, 4]$.

The area of the first triangle (on the interval $[0, 2]$) is $\frac{1}{2}(2)(6) = 6$, and the area of the trapezoid (on the interval $[2, 4]$) is $\frac{1}{2}(2)(6 + 2) = 8$. Adding these together gives you 14.000.

---

11. $\displaystyle\int_2^6 f(x)\,dx =$

A. 15.142

B. 22.283

C. 12.785

D. 28.283

E. 18.283

Answer A. $\int_2^6 f(x)\,dx = 15.142$, which is just the area under the curve $y = f(x)$ on the interval $[2, 6]$. The area under the trapezoid on the interval $[2, 4]$ is $\frac{1}{2}(2)(2 + 6) = 8.0$. The area under the curve on the interval $[4, 6]$ is the area of the quarter circle plus the area of the square beneath it; $\frac{1}{4}\pi(4) + 2 \times 2 = 7.142$. Adding these together you get 15.142.

---

12. From the definition of the definite integral, we have $\displaystyle\lim_{n\to\infty} \frac{2}{n} \sum_{k=1}^{n} \left(2 + \left(\frac{4k}{n} \right)^3 \right) =$

A. $\displaystyle\int_0^4 (2 + x^3)\,dx$

B. $\displaystyle\int_0^4 (2 + (2x)^3)\,dx$

C. $\displaystyle\int_0^2 (2 + x^3)\,dx$

D. $\displaystyle\int_0^2 (2 + (2x)^3)\,dx$

E. $\displaystyle\int_2^4 (2 + x^3)\,dx$

Answer D. The factor of $\dfrac{2}{n}$ to the left of the summation represents the size of the subintervals. With this in mind, you can rewrite

$$\lim_{n\to\infty} \frac{2}{n} \sum_{k=1}^{n} \left(2 + \left(\frac{4k}{n} \right)^3 \right) = \lim_{n\to\infty} \frac{2}{n} \sum_{k=1}^{n} \left(2 + \left(2 \times \frac{2k}{n} \right)^3 \right).$$

Now when $k = 1$, $\dfrac{2k}{n} = \dfrac{2}{n}$, and this goes to 0 as $n \to \infty$. When $k = n$, $\dfrac{2k}{n} = 2$. Thus, your x's range from 0 to 2, and

$$\lim_{n\to\infty} \frac{2}{n} \sum_{k=1}^{n} \left(2 + \left(\frac{4k}{n} \right)^3 \right) = \int_0^2 (2 + (2x)^3)\,dx.$$

---

13. From the definition of the definite integral, we have $\displaystyle\lim_{n\to\infty} \frac{3}{n} \sum_{k=1}^{n} \left(\frac{6k}{n} + \sin\left(\frac{6k\pi}{n} \right) \right) =$

A. $\displaystyle\int_0^6 (x + \sin(2\pi x))\,dx$

B. $\displaystyle\int_0^6 (x + \sin x)\,dx$

C. $\displaystyle\int_0^6 (2x + \sin(2\pi x))\,dx$

D. $\displaystyle\int_0^3 (2x + \sin(2\pi x))\,dx$

E. $\displaystyle\int_0^3 (x + \sin(2\pi x))\,dx$

Answer   D. In the sum, the factor of $\dfrac{3}{n}$ represents the width of the subintervals and also tells you that the width of the entire interval is 3 (recall that $\dfrac{3}{n} = \dfrac{b-a}{n}$). With this in mind, you can rewrite

$$\lim_{n\to\infty} \frac{3}{n} \sum_{k=1}^{n} \left(\frac{6k}{n} + \sin\left(\frac{6k\pi}{n}\right)\right)$$

$$= \lim_{n\to\infty} \frac{3}{n} \sum_{k=1}^{n} \left(2\left(\frac{3k}{n}\right) + \sin\left(2\pi\left(\frac{3k}{n}\right)\right)\right).$$

Now when $k = 1$, $\dfrac{3k}{n} = \dfrac{3}{n}$, which goes to 0 as $n \to \infty$. When $k = n$, $\dfrac{3k}{n} = 3$. Thus x goes from 0 to 3, and

$$\lim_{n\to\infty} \frac{3}{n} \sum_{k=1}^{n} \left(\frac{6k}{n} + \sin\left(\frac{6k\pi}{n}\right)\right) = \int_0^3 (2x + \sin(2\pi x))dx.$$

---

14.   For which of these functions $f(x)$ would $\displaystyle\int_0^5 f(x)\,dx$ give the area under the curve between $x = 0$ and $x = 5$?

I. $f(x) = 2$

II. $f(x) = x^2 + 4$

III. $f(x) = 4 - x^2$

A.   I

B.   I and II

C.   I and III

D.   All of them

E.   None of them

Answer   B. Both I: $f(x) = 2$ and II: $f(x) = x^2 + 4$ stay non-negative over the entire interval $[0, 5]$, so their definite integrals will give the true area under the curve. But III: $f(x) = 4 - x^2$ dips below the x-axis after $x = 2$, so the definite integral will include some negative area, and won't give the area under the curve.

15. $\int_0^\pi \sin x \, dx = 2$. Which of the following gives the correct answer for $\int_0^{4\pi} \sin x \, dx$?

A. 0

B. 2

C. 4

D. 8

E. There's not enough information.

Answer   0. The sine function is nicely balanced. Each hump above the x-axis is exactly the same size as each hump below the x-axis. Between 0 and 4π there are two humps above (whose area is counted as positive), and two humps below (whose area is counted as negative), for a net total of 0.

16. If $\int_0^4 h(x) \, dx = 12$ and $\int_0^3 h(x) \, dx = 6$, then $\int_3^4 h(x) \, dx =$

A. 2

B. 3

C. 4

D. 6

E. There's not enough information to decide.

Answer   D. The property that says $\int_a^b f(x) \, dx + \int_b^c f(x) \, dx = \int_a^c f(x) \, dx$ applies here.

$\int_0^3 h(x) \, dx + \int_3^4 h(x) \, dx = \int_0^4 h(x) \, dx$, so $6 + \int_3^4 h(x) \, dx = 12$, or $\int_3^4 h(x) \, dx = 6$.

17. Consider the area between the graphs of $y = 2 \sin x$ and $y = \sin x$ between $x = 0$ and $x = \pi$ as shown here:

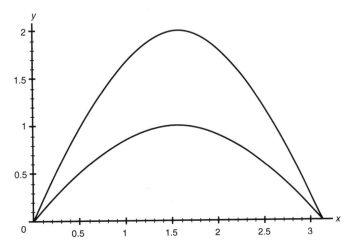

Given that $\int_0^{\pi} \sin x \, dx = 2$, what is the area of the boomerang shaped region?

A. 0

B. 2

C. 4

D. 6

E. Not enough information given.

Answer   B.

$$\int_0^{\pi} 2 \sin x \, dx - \int_0^{\pi} \sin x \, dx = \int_0^{\pi} (2 \sin x - \sin x) \, dx = \int_0^{\pi} \sin x \, dx = 2.$$

---

18.   Suppose that f and g are continuous functions, and that

$$\int_{-1}^{3} f(x) \, dx = 10, \quad \int_{-1}^{3} g(x) \, dx = 2, \quad \int_{1}^{3} f(x) \, dx = 1, \quad \text{and} \quad \int_{1}^{3} g(x) \, dx = 8.$$

$$\int_{-1}^{1} (2f(x) - g(x)) \, dx =$$

A.   18

B.   12

C.   2

D.   24

E.   Not enough information given.

Answer   D.

$$\int_{-1}^{1} (2f(x) - g(x))\, dx = \int_{-1}^{3} (2f(x) - g(x))\, dx - \int_{1}^{3} (2f(x) - g(x))\, dx.$$

Note that $\int_{-1}^{3} (2f(x) - g(x))\, dx = 2\int_{-1}^{3} f(x)\, dx - \int_{-1}^{3} g(x)\, dx = 18,$

and $\int_{1}^{3} (2f(x) - g(x))\, dx = 2\int_{1}^{3} f(x)\, dx - \int_{1}^{3} g(x)\, dx = -6.$

So $\int_{-1}^{1} (2f(x) - g(x))\, dx = 18 - (-6) = 24.$

---

19.   Given that $\int_{3}^{6} x^2\, dx = 63.0$ and $\int_{3}^{6} x\, dx = 13.5$ then $\int_{3}^{6} (3x^2 - 5x)\, dx =$

A.   256.5

B.   49.5

C.   274.5

D.   121.5

E.   Not enough information given.

Answer   D.

$$\int_{3}^{6} (3x^2 - 5x)\, dx = 3\int_{3}^{6} x^2\, dx - 5\int_{3}^{6} x\, dx$$

$$= 3(63) - 5(13.5) = 121.5.$$

---

20.   Given that $\int_{3}^{6} x^2\, dx = 63.0$ and $\int_{3}^{6} x\, dx = 13.5$, then $\int_{3}^{6} (2x - 3)^2\, dx =$

A.   324.0

B.   117.0

C.   99.0

D.   576

E.   Not enough information given.

Answer   B. To see this, rewrite

$$\int_3^6 (2x - 3)^2 \, dx = \int_3^6 (4x^2 - 12x + 9) \, dx$$

$$= 4\int_3^6 x^2 \, dx - 12 \int_3^6 x \, dx + 9 \int_3^6 1 \, dx.$$

The last integral $\int_3^6 1 \, dx$ is just the area under the curve $y = 1$ over the interval $[3, 6]$, and geometrically this is just a rectangle with area $3 \times 1 = 3$. So you get

$$4\int_3^6 x^2 \, dx - 12 \int_3^6 x \, dx + 9 \int_3^6 1 \, dx$$

$$= 4 \times 63 - 12 \times 13.5 + 9 \times 3 = 117.0.$$

---

21.   You're given that $\int_1^4 e^x \, dx = 51.880$ (to three decimal places).

Using properties of definite integrals and exponentials, find $\int_4^1 2e^{(x+3)} \, dx \approx$

A. -224.273

B. -143.931

C. -2084.075

D. 224.273

E.   Not enough information given.

Answer   C.

$$\int_4^1 2e^{(x+3)} \, dx = -2 \int_1^4 e^{(x+3)} \, dx = -2 \int_1^4 e^3 e^x \, dx$$

$$= -2e^3 \int_1^4 e^x \, dx = -2e^3 \times 51.88 = -2084.075.$$

---

22.   Suppose that f is a continuous function and

$$\int_{-3}^2 f(x) \, dx = 3 \qquad \int_{-3}^5 f(x) \, dx = -2 \qquad \int_3^5 f(x) \, dx = 1$$

Then $\int_2^3 f(x) \, dx =$

A. -3

B. -6

C. 3

D. 5

E. 6

Answer   B. To find this, rewrite

$$\int_{-3}^{5} f(x)\, dx = \int_{-3}^{2} f(x)\, dx + \int_{2}^{3} f(x)\, dx + \int_{3}^{5} f(x)\, dx.$$

Then, solving for $\int_{2}^{3} f(x)\, dx$, you get

$$\int_{2}^{3} f(x)\, dx = \int_{-3}^{5} f(x)\, dx - \int_{-3}^{2} f(x)\, dx - \int_{3}^{5} f(x)\, dx$$
$$= -2 - 3 - 1 = -6.$$

23.   Given that $f(x)$ is continuous and

$$\int_{-2}^{2} f(x)\, dx = 7 \qquad \int_{0}^{4} f(x)\, dx = -3 \qquad \int_{-2}^{4} f(x)\, dx = 2$$

Then $\int_{0}^{2} f(x)\, dx =$

A.   2

B.   4

C. -3

D. -6

E.   0

Answer   A. Using properties of definite integrals, you have

$$\int_{-2}^{2} f(x)\, dx + \int_{0}^{4} f(x)\, dx$$
$$= \left[\int_{-2}^{0} f(x)\, dx + \int_{0}^{2} f(x)\, dx \right] + \left[\int_{0}^{2} f(x)\, dx + \int_{2}^{4} f(x)\, dx \right]$$
$$= \int_{-2}^{4} f(x)\, dx + \int_{0}^{2} f(x)\, dx.$$

Then, filling in the given values, you get

$$7 + (-3) = 2 + \int_{0}^{2} f(x)\, dx \Rightarrow \int_{0}^{2} f(x)\, dx = 2.$$

24. Evaluate the definite integral $\int_{-2}^{6} (2 - |x|)\, dx$.

A. 12

B. 4

C. -8

D. -6

E. -4

Answer   E. If you graph the function $y = 2 - |x|$ on the interval $[-2, 6]$, you get the following:

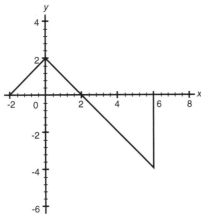

The area under the curve on the interval $[-2, 2]$ is 4. The "negative area" on the interval $[2, 6]$ is -8. Thus, $\int_{-2}^{6} (2 - |x|)\, dx = 4 - 8 = -4$.

---

25. Consider the function $f(x) = \begin{cases} -x - 2 & \text{if } -4 \le x < -2 \\ -\sqrt{4 - x^2} & \text{if } -2 \le x \le 2 \\ x - 2 & \text{if } 2 < x \le 4 \end{cases}$

The graph is below:

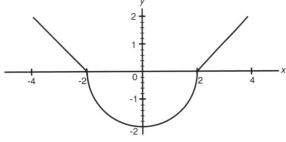

Evaluate $\int_{-4}^{4} f(x)\, dx$.

A. $2 - 2x$

B. $4 - \pi$

C. $2 - 4\pi$

D. $4 - 4\pi$

E. $4 - 2\pi$

Answer   E. The graph of $y = f(x)$ consists of two straight lines and a semicircle of radius 2.

The positive area is the area under the two straight lines, which totals 4. The negative area is the area above the semicircle, which totals $.5\pi(4) = 2\pi$.

So the value of the definite integral is $\displaystyle\int_{-4}^{4} f(x)\, dx = 4 - 2\pi$.

---

26.   The following graph of $f(x)$ consists of line segments and semicircles. $y = f(x)$

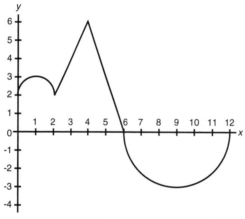

Use the graph to evaluate the definite integral $\displaystyle\int_{1}^{9} f(x)\, dx$.

A. $18 - 4\pi$

B. $16 - 2.5\pi$

C. $16 - 2\pi$

D. $14 - \pi$

E. $14 - 4\pi$

Answer   C. Below is a graph of the function on the interval $[1, 9]$, which we've partitioned into sections:

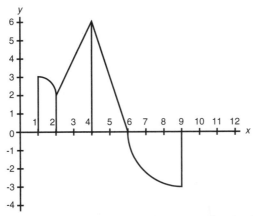

The positive area includes the area under the first quarter circle of radius 1, the area of the trapezoid beside it, and then the area of the triangle. The area under the first quarter circle is $\frac{1}{4}\pi + 2 \times 1$; the area under the trapezoid is $\frac{1}{2}2(2 + 6) = 8$; and the area under the triangle is $\frac{1}{2}2(6) = 6$. So the positive area involved is $16 + \frac{\pi}{4}$. The negative area here is just the area of the quarter circle of radius 3. That area is $\frac{9}{4}\pi$. So the definite integral becomes

$$\int_1^9 f(x)\,dx = 16 + \frac{\pi}{4} - \frac{9\pi}{4} = 16 - 2\pi.$$

---

27. The graph of $f(x)$ is shown below.

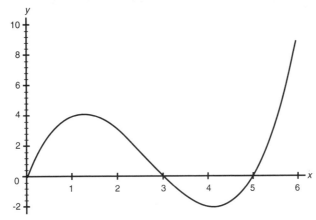

Which of the following statements is true?

I. $\displaystyle\int_0^4 f(x)\,dx > \int_0^5 f(x)\,dx$

II. $\displaystyle\int_0^6 f(x)\,dx > \int_0^5 f(x)\,dx$

III. $\displaystyle\int_0^5 f(x)\,dx > \int_0^3 f(x)\,dx.$

A. I only

B. II only

C.   III only

D.   II and III only

E.   I and II only

Answer E.

Statement I, $\int_0^4 f(x)\,dx > \int_0^5 f(x)\,dx$,

says that the signed area under the curve from 0 to 4 is greater than the signed area from 0 to 5. This is true, since the signed area under the curve from 4 to 5 is negative.

Statement II, $\int_0^6 f(x)\,dx > \int_0^5 f(x)\,dx$,

says that the signed area under the curve from 0 to 6 is greater than the signed area from 0 to 5. This is true, since the signed area under the curve from 5 to 6 is positive.

Statement III, $\int_0^5 f(x)\,dx > \int_0^3 f(x)\,dx$,

says that the signed area under the curve from 0 to 5 is greater than the signed area from 0 to 3. This is not true, since the signed area under the curve from 3 to 5 is negative!

Practice Using a Calculator to Compute Definite Integrals

The syntax for using **fnInt** on a TI-83 calculator to evaluate $\int_a^b f(x)\,dx$ is

fnInt$(f(x), x, a, b)$

In each of the following problems, use your calculator to estimate (to three decimal places) the value of the given definite integral.

---

1. $\int_{-3}^3 2\cos(x)\,dx =$

A.   0.564

B.   6.919

C.   3.459

D.   0.282

E.   3.736

Answer A. On the TI-83, the command to enter is fnInt$(2\char`\^\cos(X), X, -3, 3)$

2. $\int_0^6 \sin\left(\dfrac{\cos x}{x^2+1}\right) dx =$

A. 0.784

B. 1.236

C. 0.253

D. 0.495

E. 0.991

Answer   D. On the TI-83 the command to enter is fnInt$(\sin(\cos(X)/(X\char`\^2+1)), X, 0, 6)$.

---

3. $\int_{-1}^2 (e^{-x} - 1)\, dx =$

A. 3.315

B. -0.221

C. 4.021

D. -0.767

E. -0.417

Answer   E. On the TI-83, the command to enter is fnInt$(e\char`\^(-X) - 1, X, -1, 2)$.

---

4. $\int_0^{\sqrt{4}} \left(\sin(x^2) + \dfrac{x}{2}\right) dx =$

A. 3.628

B. 40.078

C. 4.734

D. 7.875

E. 4.320

Answer   A. On the TI-83, the command to enter is fnInt$(\sin(X\char`\^2) + X/2, X, 0, \sqrt{}(4))$.

---

5. $\int_1^e \dfrac{1}{x}\, dx =$

A. 1.000

B. -2.914

C. 2.000

D. 0.632

E. 0.693

Answer   A. The command to enter on the TI-83 is $\text{fnInt}(1/X, X, 1, e)$.

Chapter 7
Antiderivatives and the Fundamental Theorems of Calculus

In this chapter we review *antiderivatives* which are also known as *indefinite integrals*. As the name suggests, antidifferentiation reverses the process of differentiation. What is remarkable is that antiderivatives and definite integrals are closely related to each other. That's the subject of the *Fundamental Theorems of Calculus*, the most important mathematical results in the course, for they tie the two branches of calculus (differential and integral) together. This chapter assumes that you have experience with derivatives and integrals of polynomial and trigonometric functions.

A. Antiderivatives

▶Objective 1   Identify the antiderivative of a function as a family of functions.

Example   On your calculator graph three of the functions in the family of functions defined by:

$$\frac{dy}{dx} = 4x^2 - x$$

Tip   Realize that an antiderivative of $f(x)$ is of the form

$$F(x) + C$$

where $F'(x) = f(x)$ and C is an arbitrary constant.

The derivative of a constant is zero, so the derivative of $F(x) + C$ is $f(x)$ no matter what real number is chosen for the constant C.

Answer $y = \frac{4}{3}x^3 - \frac{1}{2}x^2 + C$

There are infinitely many correct answers. We chose:

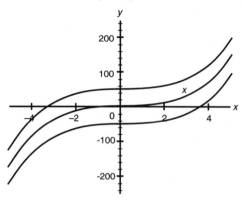

►Objective 2   Using the rules for differentiating basic functions, find antiderivatives of basic functions.

Example   Find the antiderivative for $g(x) = \dfrac{\pi}{2\sqrt{x}}$ and for $h(t) = \sec^2 t - t^{\frac{3}{4}}$.

Tip   Always check your answer by taking the derivative to see if you get your original function. And don't forget your constant of integration: $+C$.

Answer $G(x) = \pi\sqrt{x} + C$

$H(t) = \tan t - \frac{4}{7}t^{\frac{7}{4}} + C$

►Objective 3   Identify and solve simple differential equations.

Example   Given that the acceleration of an object is $a(t) = t - 8(\frac{\text{ft}}{\text{sec}}^2)$ and that $v(1) = 4\frac{\text{ft}}{\text{sec}}$ and $s(0) = -4$ ft, find $s(t)$, the position function.

Tip   The units can help you remember which way to go here. Each time you take the derivative you get one more *seconds* term in the denominator of the units:

$s(t)$ ft

$s'(t) = v(t)\frac{\text{ft}}{\text{sec}}$

$s''(t) = v'(t) = a(t)\frac{\text{ft}}{\text{sec}}^2$

Answer   Since $v'(t) = a(t)$, we get $v(t) = \frac{1}{2}t^2 - 8t + C$.

Using the fact that $v(1) = 4$, we find that $4 = \frac{1}{2}1^2 - 8 + C$.

Solution is $C = \frac{23}{2}$. Therefore $v(t) = \frac{1}{2}t^2 - 8t + \frac{23}{2}$.

Now $s'(t) = v(t)$, so that $s(t) = \frac{1}{6}t^3 - 4t^2 + \frac{23}{2}t + C$.

Using the fact that $s(0) = -4$, we find that $s(t) = \frac{1}{6}t^3 - 4t^2 + \frac{23}{2}t - 4$.

▶**Objective 4**   Identify when an antiderivative involves a composite function.

Example   Which of the following functions are derivatives of composite functions?

    A. $f(x) = 2x \sin(x^2)$

    B. $f(x) = \dfrac{\cos(\sqrt{x})}{2}$

    C. $f(x) = \dfrac{\cos(\sqrt{x})}{2\sqrt{x}}$

Tip $h'(x)g'(h(x))$ is the derivative of $g(h(x))$, from the chain rule. Think of it as the "chain rule backwards."

Answer

    A. $f(x) = 2x \sin(x^2)$. Yes, you can see that this is the derivative of $F(x) = -\cos(x^2) + C$.

    B. $f(x) = \dfrac{\cos(\sqrt{x})}{2}$. No, this doesn't work because the function $f(x)$ is not of the form $h'(x)g'(h(x))$, which is the result of a chain rule. You'd have to see a $\sqrt{x}$ in the denominator.

    C. $f(x) = \dfrac{\cos(\sqrt{x})}{2\sqrt{2}}$. Yes, this one does work because $\dfrac{1}{2\sqrt{x}}$ is the derivative of $\sqrt{x}$, and we find that $F(x) = \sin(\sqrt{x}) + C$.

▶**Objective 5**   Take derivatives of functions using differential notation.

Example   Assuming that $u = \sin(t)$, what is du?

Answer $du = \cos(t)\, dt$

▶**Objective 6**   Find antiderivatives for composite functions using substitution.

Example   Use substitution to help you find $\displaystyle\int \frac{\sec^2(\sqrt[3]{t})}{\sqrt[3]{t^2}}\, dt$.

Tip   Apply the following steps:

    1. Recognize the composite nature of the function.

    2. Set u equal to the "inside" of the function.

    3. Take the derivative of u using differential notation.

    4. Alter the original function to allow for substitution.

    5. Substitute and find the much easier antiderivative.

    6. Substitute back in terms of the original variable and check your answer.

Answer Let $u = \sqrt[3]{t} = t^{\frac{1}{3}}$, then $du = \frac{1}{3}t^{\frac{-2}{3}}dt = \dfrac{dt}{3\sqrt[3]{t^2}}$. Substituting, we get

$$\int \frac{\sec^2(\sqrt[3]{t})}{\sqrt[3]{t^2}}\, dt = 3 \int \frac{\sec^2(\sqrt[3]{t})}{3\sqrt[3]{t^2}}\, dt$$

$$= 3 \int \sec^2(u)\, du = 3\tan u + C = 3\tan(\sqrt[3]{t}) + C.$$

▶Objective 7 Find antiderivatives for composite functions, without using substitution.

Example Evaluate $\displaystyle\int \cos^4(3x)\sin(3x)\,dx$ without using substitution.

Tip Always check your final answer.

Answer Our first guess would be $\cos^5(3x)$, but its derivative is

$$\frac{d}{dx}(\cos^5(3x)) = 5\cos^4(3x)[-\sin(3x)3] = -15\cos^4(3x)\sin(3x).$$

Thus, we need to multiply by $-\frac{1}{15}$ to compensate:

$$\int \cos^4(3x)\sin(3x)\,dx = -\frac{1}{15}\cos^5(3x) + C.$$

Multiple-Choice Practice on Antiderivatives

---

1. A particle's acceleration along a straight line is given by $3 + 2t$ at any time t. At $t = 4$, the velocity is 4. Which of these gives the correct function for the velocity of this particle at any time t?

A. $v(t) = 3t + t^2 - 24$

B. $v(t) = 3$

C. $v(t) = 3t + t^2 + 4$

D. None of these

E. There's not enough information to decide

Answer A. Acceleration is the derivative of velocity, so you need to go backwards to find velocity:

$$\int (3 + 2t)\,dt = 3t + t^2 + C.$$

Then you need to solve to find C:

$$v(4) = 3(4) + 4^2 + C = 4, \text{ or } C = -24.$$

2. Which of these is an antiderivative of $f(x) = 3x^2$?

A. $F(x) = x^3$

B. $F(x) = x^3 + 102,000$

C. $F(x) = x^3 - 35$

D. All three of these

E. None of these

Answer   D. All three of these functions are antiderivatives of $f(x) = 3x^2$. The complete family of antiderivatives are given by $F(x) = x^3 + C$, where C is an arbitrary constant.

3. $\displaystyle\int 2\sin x \cos x\, dx =$

A. $\sin^2 x + C$

B. $-\cos^2 x + C$

C. $-\frac{\cos 2x}{2} + C$

D. All of these

E. None of these

Answer   D. The antiderivative is a family of functions, and the three functions

$$\sin^2 x, \qquad -\cos^2 x, \qquad \text{and} \qquad -\frac{\cos 2x}{2}$$

are just different members of the same family. You can convince yourself by using trigonometric identities or just looking at their graphs.

4. Evaluate $\displaystyle\int \left(\sin x + x^{\frac{1}{2}}\right) dx$.

A. $\cos x + \frac{1}{2}x^{-\frac{1}{2}} + C$

B. $-\cos x + \frac{1}{2}x^{-\frac{1}{2}} + C$

C. $-\cos x + \frac{3}{2}x^{\frac{3}{2}} + C$

D. $-\cos x + \frac{2}{3}x^{\frac{3}{2}} + C$

E. None of these

Answer D. Checking by taking its derivative:

$$\frac{d}{dx}(-\cos x + \frac{2}{3}x^{\frac{3}{2}} + C) = \sin x + x^{\frac{1}{2}}.$$

---

5. Evaluate $\int (x^{-4} + x^{-2} + x^{10})\, dx$.

A. $-x^{-5} - x^{-3} + x^{11} + C$

B. $-\dfrac{x^{-5}}{5} - \dfrac{x^{-3}}{3} + \dfrac{x^{11}}{11} + C$

C. $-\dfrac{x^{-3}}{3} - x^{-1} + \dfrac{x^{11}}{11} + C$

D. $\dfrac{x^5}{5} + \dfrac{x^3}{3} + x^{11} + C$

E. $\dfrac{x^{-5}}{5} + x^{-1} + \dfrac{x^{11}}{11} + C$

Answer C.

$$\int (x^{-4} + x^{-2} + x^{10})dx = -\frac{x^{-3}}{3} - x^{-1} + \frac{x^{11}}{11} + C.$$

---

6. Solve the differential equation $\dfrac{dy}{dx} = 3x^2 + \sin x + 2$, if $y = 2$ when $x = 0$.

A. $y = x^3 - \cos x + 2x + 3$

B. $y = x^3 - \cos x + 2x + 2$

C. $y = x^3 + \cos x + 2x + 1$

D. $y = x^3 - \cos x + 2x + C$

E. $y = x^3 + \cos x + 2x$

Answer A.

$$\int (3x^2 + \sin x + 2)dx = y = x^3 - \cos x + 2x + C.$$

Then use the other given information to solve for the constant C:

$$y(0) = 0^3 - \cos 0 + 2(0) + C = 2 \quad \Longrightarrow \quad C = 3.$$

---

7. $\int (-3x^3 + 4x + 2)dx =$

A. $-3x^4 + 4x^2 + 2x + C$

B. $-\frac{3}{4}x^4 + 2x^2 + 2x + C$

C. $-x^4 + 4x + 2x + C$

D. $-9x^2 + 4$

E. $-12x^3 + 8x^2 + 2x + C$

Answer   B. Use the rule $\int x^n dx = \dfrac{1}{n+1}x^{n+1} + C$ (for $n \neq -1$) and apply it term by term:

$$\int (-3x^3 + 4x + 2)dx = \int -3x^3 dx + \int 4x dx + \int 2 dx$$
$$= -\frac{3}{4}x^4 + 2x^2 + 2x + C.$$

Remember, you can always check your antiderivative by differentiating it!

---

8. $\int (x^2 - 3)^2 dx =$

A. $\frac{1}{5}x^5 - 9x + C$

B. $5x^5 - 9x + C$

C. $\frac{1}{5}x^5 - 2x^3 + 9x + C$

D. $4x^3 - 12x + C$

E. $5x^5 - 18x^3 + 9x + C$

Answer   C.

$$\int (x^2 - 3)^2 dx = \int (x^4 - 6x^2 + 9)dx$$
$$= \int x^4 dx - \int 6x^2 dx + \int 9 dx = \frac{1}{5}x^5 - 2x^3 + 9x + C.$$

The trick is to expand the square first, then antidifferentiate term by term.

---

9. $\int \left(\sqrt{t} + t^{-\frac{1}{2}}\right) dt =$

A. $\frac{3}{2}t^{\frac{3}{2}} - \frac{1}{2}t^{\frac{1}{2}} + C$

B. $\frac{3}{2}(\sqrt{t})^3 - 2\sqrt{t} + C$

C. $\frac{1}{2}t^{-\frac{1}{2}} - \frac{1}{2}t^{\frac{1}{2}} + C$

D. $\frac{2}{3}t^{\frac{3}{2}} - 2t^{\frac{1}{2}} + C$

E. $\frac{2}{3}(\sqrt{t})^3 + 2\sqrt{t} + C$

Answer   E. The antiderivative of $t^{\frac{1}{2}}$ is $\frac{1}{\frac{3}{2}}t^{\frac{3}{2}} + C = \frac{2}{3}(\sqrt{t})^3 + C$.

The antiderivative of $t^{-\frac{1}{2}}$ is $\frac{1}{\frac{1}{2}}t^{\frac{1}{2}} + C = 2\sqrt{t} + C$.

Combining these gives us

$$\int (\sqrt{t} + t^{-\frac{1}{2}})\, dt = \frac{2}{3}(\sqrt{t})^3 + 2\sqrt{t} + C.$$

---

10. $\displaystyle\int 4(\frac{1}{x} + x^{\frac{2}{5}})^2\, dx =$

A. $-\dfrac{4}{x} + 20x^{\frac{2}{5}} + \dfrac{20}{9}x^{\frac{9}{5}} + C$

B. $-\dfrac{4}{x} + \dfrac{40}{3}x^{\frac{3}{5}} + \dfrac{20}{9}x^{\frac{9}{5}} + C$

C. $-\dfrac{4}{x} + \dfrac{40}{3}x^{\frac{3}{5}} + \dfrac{100}{9}x^{\frac{29}{25}} + C$

D. $-\dfrac{1}{x} + \dfrac{10}{3}x^{\frac{3}{5}} + \dfrac{5}{9}x^{\frac{9}{5}} + C$

E. $-\dfrac{1}{x} + 5x^{\frac{2}{5}} + \dfrac{5}{9}x^{\frac{9}{5}} + C$

Answer   A. First rewrite so you can see all the exponents and expand to get

$$\int 4\left(\frac{1}{x} + x^{\frac{2}{5}}\right)^2 dx = 4\int \left(x^{-1} + x^{\frac{2}{5}}\right)^2 dx$$

$$= 4\int \left(x^{-2} + 2x^{-\frac{3}{5}} + x^{\frac{4}{5}}\right) dx$$

$$= 4\int x^{-2}\, dx + 8\int x^{-\frac{3}{5}}\, dx + 4\int x^{\frac{4}{5}}\, dx.$$

Now, the antiderivative of x^{-2} is $\dfrac{1}{-1}x^{-1} + C = -\dfrac{1}{x} + C$, the antiderivative of $x^{-\frac{3}{5}}$ is $\dfrac{1}{\frac{2}{5}}x^{\frac{2}{5}} + C = \dfrac{5}{2}x^{\frac{2}{5}} + C$, and the antiderivative of $x^{\frac{4}{5}}$ is $\dfrac{1}{\frac{9}{5}}x^{\frac{9}{5}} + C = \dfrac{5}{9}x^{\frac{9}{5}} + C$.

So

$$\int 4\left(\frac{1}{x} + x^{\frac{2}{5}}\right)^2 dx$$

$$= 4\left(-\frac{1}{x}\right) + 8\left(\frac{5}{2}x^{\frac{2}{5}}\right) + 4\left(\frac{5}{9}x^{\frac{9}{5}}\right) + C$$

$$= -\frac{4}{x} + 20x^{\frac{2}{5}} + \frac{20}{9}x^{\frac{9}{5}} + C.$$

11. $\int \dfrac{x^2 - 4}{x - 2}\, dx =$

A. $\dfrac{\frac{1}{3}x^3 - 4x}{\frac{1}{2}x^2 - 2x} + C$

B. $\dfrac{1}{3}x^3 - 4x + C$

C. $\dfrac{1}{2}x^2 + 2x + C$

D. $\dfrac{\frac{1}{3}x^3 - 4x + C}{\frac{1}{2}x^2 - 2x + C}$

E. None of the above

Answer C. First you must simplify the integrand by factoring. Notice that $x^2 - 4 = (x-2)(x+2)$. So you have

$$\int \frac{x^2 - 4}{x - 2}\, dx = \int \frac{(x-2)(x+2)}{x-2}\, dx$$
$$= \int (x+2)\, dx = \frac{1}{2}x^2 + 2x + C.$$

12. $\int (\sin x - 3\cot x \sin x)\, dx =$

A. $\cos x + 3\csc x + C$

B. $\cos x + 3\sin x + C$

C. $-\cos x - 3\sin x + C$

D. $-\cos x + 3\csc x + C$

E. None of the above.

Answer C. Using properties of antiderivatives and trig identities to simplify, you get

$$\int (\sin x - 3\cot x \sin x)\, dx = \int \sin x\, dx - 3\int \cot x \sin x\, dx$$
$$= \int \sin x\, dx - 3\int \frac{\cos x}{\sin x} \sin x\, dx = \int \sin x\, dx - 3\int \cos x\, dx.$$

The antiderivative of $\sin x$ is $-\cos x + C$, and the antiderivative of $\cos x$ is $\sin x + C$. Thus

$$\int (\sin x - 3\cot x \sin x)\, dx = -\cos x - 3\sin x + C.$$

13. $\displaystyle\int \left(2x^{-\frac{3}{7}} + \frac{5}{\sin^2 x}\right) dx =$

A. $\dfrac{7}{2}x^{\frac{4}{7}} - 5\tan x + C$

B. $\dfrac{7}{2}x^{\frac{4}{7}} - 5\cot x + C$

C. $-7x^{-\frac{2}{7}} + 5\tan x + C$

D. $-7x^{-\frac{2}{7}} - \cot x + C$

E. $\dfrac{7}{2}x^{\frac{4}{7}} + 5\cot x + C$

Answer   B. Rewriting and simplifying, you get

$$\int \left(2x^{-\frac{3}{7}} + \frac{5}{\sin^2 x}\right) dx = 2\int x^{-\frac{3}{7}}\, dx + 5\int \frac{1}{\sin^2 x}\, dx$$
$$= 2\int x^{-\frac{3}{7}} + 5\int \csc^2 x\, dx.$$

The antiderivative of $x^{-\frac{3}{7}}$ is $\dfrac{1}{\frac{4}{7}}x^{\frac{4}{7}} + C = \dfrac{7}{4}x^{\frac{4}{7}} + C$, and the antiderivative of $\csc x$ is $-\cot x + C$. Thus,

$$\int \left(2x^{-\frac{3}{7}} + \frac{5}{\sin^2 x}\right) dx = 2\left(\frac{7}{4}x^{\frac{4}{7}}\right) + 5(-\cot x) + C$$
$$= \frac{7}{2}x^{\frac{4}{7}} - 5\cot x + C.$$

---

14. $\displaystyle\int \csc^2 x \cos x\, dx =$

(Hint: Try using trigonometric identities to rewrite the integrand as something whose antiderivative you know.)

A. $-\csc x + C$

B. $-\cot x \csc x + C$

C. $-\csc^2 x + \sin x + C$

D. $-\cot^2 x + C$

E. $-\cos x\csc x + C$

Answer A. First rewrite the indefinite integral to simplify it to something you can antidifferentiate:

$$\int \csc^2 x \cos x \, dx = \int \frac{1}{\sin x \sin x} \cos x \, dx$$

$$= \int \frac{1}{\sin x} \frac{\cos x}{\sin x} \, dx = \int \csc x \cot x \, dx.$$

Since the antiderivative of $\csc x \cot x$ is $-\csc x + C$, we have

$$\int \csc^2 x \cos x \, dx = -\csc x + C.$$

---

15. $\displaystyle\int \frac{d}{dx}(3x^{-2} + \tan x - 4) \, dx =$

A. $-6x^{-3} + \sec^2 x + C$

B. $6x^{-2} + \tan x - 4$

C. $-\frac{3}{2}x^{-4} + \tan x - 4$

D. $3x^{-2} + \tan x + C$

E. $-3x^{-2} + \sec^2 x + C$

Answer D. You're looking for the antiderivative of $\dfrac{d}{dx}(3x^{-2} + \tan x - 4)$.

An antiderivative is a function $F(x)$, such that $\dfrac{d}{dx}F(x) = \dfrac{d}{dx}(3x^{-2} + \tan x - 4)$.

The family of functions that satisfies this is

$$F(x) = 3x^{-2} + \tan x - 4 + C, \quad \text{or more simply} \quad 3x^{-2} + \tan x + C.$$

The next two problems refer to the following situation:

A paperweight is thrown straight up with an initial upward velocity of 60 meters per second and an initial height of 6 meters. The acceleration (due to gravity) of the object is given by $a(t) = -9.8$ meters per second per second, where t is in seconds.

---

16. The equation that describes the paperweight's velocity as a function of time t is:

A. $v(t) = -9.8t + C$

B. $v(t) = -9.8t - 60$

C. $v(t) = -4.9t^2 + 60$

D. $v(t) = -9.8t + 60$

E. $v(t) = -4.9t^2 - 60$

Answer D. Remember that acceleration is the derivative of velocity, so $v'(t) = a(t)$.

That means $v(t)$ is an antiderivative for $a(t)$. But not just any antiderivative—it must also satisfy the condition that the initial velocity is 60 meters per second in an upward direction, or $v(0) = 60$. You have

$$v(t) = \int a(t)\, dt = \int -9.8\, dt = -9.8t + C \text{ with } v(0) = 60.$$

$$v(0) = -9.8(0) + C = 60 \Rightarrow C = 60.$$

The correct formula for the velocity is thus $v(t) = -9.8t + 60$.

---

17. The maximum height of the paperweight is:

A. 6.122 meters

B. 189.673 meters

C. $-4.9t^2 + 60t + 6$ meters

D. 3.06 meters

E. 144.692 meters

Answer B. The function that gives the height of the paperweight as a function of time is found by solving $h(t) = \int v(t)\, dt$, with $h(0) = 6$ (the initial height at time $t = 0$). You get

$$h(t) = \int (-9.8t + 60)\, dt = -4.9t^2 + 60t + C, \text{ with } h(0) = 6$$

$$h(0) = -4.9(0^2) + 60(0) + C = 6 \Rightarrow C = 6.$$

Thus $h(t) = -4.9t^2 + 60t + 6$.

The maximum height occurs when the velocity is 0: $-9.8t + 60 = 0 \Rightarrow t = \frac{60}{9.8}$.

Plugging this time into the equation for the height gives you $h(\frac{60}{9.8}) \approx 189.673$ meters.

---

18. If $u = f(t)$, $du =$

A. $f(t)\, dt$

B. $f'(t)\, dx$

C. $f'(t)$

D. $f'(t)\, dt$

E. $f(t)\, dx$

Answer D. Generally if $y = g(x)$, the differential $dy = g'(x)\, dx$, so if $u = f(t)$, then $du = f'(t)\, dt$.

19. If $u = \sin(t^2)$ then $du =$

A. $2t\,dt$

B. $\cos(t^2)2t\,dt$

C. $\cos(u)2t\,dt$

D. $\sin(u)\,dt$

E. None of these

Answer   B. Since $\dfrac{du}{dt} = \cos(t^2)2t\,dt$, we have $du = \cos(t^2)2t\,dt$. Generally if $y = g(x)$, the differential dy is given by $dy = g'(x)\,dx$.

---

20. In order to evaluate $\displaystyle\int t\sin(t^2)\,dt$ using substitution, you would probably use $u =$

A. t^2

B. $\cos(t^2)2t\,dt$

C. $2t\,dt$

D. $\sin(t^2)$

E. t

Answer   A. Set u equal to the "inside" of the composite function: $u = t^2$. Notice that the factor t that lets you make the $du = 2t\,dt$ substitution.

---

21. In order to evaluate $\displaystyle\int \frac{1}{\sqrt{x}}(\sqrt{x} - 5)^9\,dx$ using substitution, you might rearrange your indefinite integral to look like:

I. $2\displaystyle\int \frac{1}{2}x^{-\frac{1}{2}}(x^{\frac{1}{2}} - 5)^9\,dx$

II. $2\displaystyle\int \frac{1}{2}\sqrt{x}(\sqrt{x} - 5)^9\,dx$

III. $2\displaystyle\int (\sqrt{x} - 5)^9 \frac{1}{2\sqrt{x}}\,dx$

A. I only

B. II only

C. I and III only

D.   II and III only

E.   I, II, and III

Answer   C. You can use $u = \sqrt{x} - 5$, which gives $du = \dfrac{1}{2\sqrt{x}}\,dx$, or equivalently, $u = x^{\frac{1}{2}} - 5$, which gives $du = \frac{1}{2}x^{-\frac{1}{2}}\,dx$. In either case you need to put a factor 2 to compensate for the factor $\frac{1}{2}$ that arises.

---

22.   In order to evaluate $\displaystyle\int \dfrac{x}{\sqrt{1 - x^2}}\,dx$ using substitution, you would set $u =$

A. x

B. $1 - x^2$

C. $1 - x$

D. $\dfrac{1}{\sqrt{1 - x^2}}$

E.   Substitution won't work.

Answer   B. $u = 1 - x^2 \qquad du = -2x\,dx$

$$\int \frac{x}{\sqrt{1 - x^2}}\,dx = \frac{-1}{2}\int -2x(1 - x^2)^{-\frac{1}{2}}\,dx$$
$$= -\frac{1}{2}\int (u)^{-\frac{1}{2}}\,du = \frac{-1}{2}(2u)^{\frac{1}{2}} + C$$
$$= -\sqrt{u} + C = -\sqrt{1 - x^2} + C.$$

---

23.   If $g(x) = x\sqrt{9 - x^2}$, the antiderivative $G(x) =$

A. $\frac{3}{2}x^2 - \frac{1}{3}x^3 + C$

B. $\frac{3}{2}x^2 - x^3 + C$

C. $-\frac{1}{2}\left((9 - x^2)^{\frac{3}{2}}\right) + C$

D. $\dfrac{(9 - x^2)^{\frac{3}{2}}}{3} + C$

E. $-\frac{1}{3}\left(\sqrt{9 - x^2}\right)^3 + C$

Answer   E.

$$\int \left(x\sqrt{9-x^2} \right) dx = -\frac{1}{2} \int (9-x^2)^{\frac{1}{2}}(-2x)\, dx$$

$$u = 9 - x^2 \qquad du = -2x\, dx$$

$$= -\frac{1}{2} \int u^{\frac{1}{2}}\, du = -\frac{1}{2}\left(\frac{2}{3}u^{\frac{3}{2}} \right) + C = -\frac{1}{2}\left(\frac{2}{3}(9-x^2)^{\frac{3}{2}} \right) + C$$

$$= -\frac{1}{3}\left(\sqrt{(9-x^2)} \right)^3 + C.$$

---

24.   Given that $g'(t) = (\sin t)(5 + 5\cos t)^3$, find $g(t)$ if $g(0) = 0$.

A. $g(t) = -\frac{1}{4}(5 + 5\cos t)^4$

B. $g(t) = -\frac{1}{4}(5 + 5\cos t)^4 + 2,500$

C. $g(t) = \frac{1}{4}(5 + 5\cos t)^4 - 2,500$

D. $g(t) = -\frac{1}{20}(5 + 5\cos t)^4 + 500$

E. $g(t) = \frac{1}{20}(5 + 5\cos t)^4 - 500$

Answer   D. Using guess and check, your first guess might be $(5 + 5\cos t)^4$.

Then realize that you need to compensate for the factor 4 that comes down from the exponent and the factor -5 that comes from the derivative of $5 + 5\cos t$ when using the chain rule. Thus

$$g(t) = \frac{1}{4}\left(\frac{-1}{5} \right)(5 + 5\cos t)^4 + C = -\frac{1}{20}(5 + 5\cos t)^4 + C,$$

Differentiate to check your answer:

$$\frac{d}{dt}\left(-\frac{1}{20}(5 + 5\cos t)^4 + C \right)$$

$$= 4(-5\sin t)\left(-\frac{1}{20} \right)(5 + 5\cos t)^3 = (\sin t)(5 + 5\cos t)^3.$$

So $g(t) = -\frac{1}{20}(5 + 5\cos t)^4 + C$, and now we use the initial condition to solve for C:

$$g(0) = 0$$

$$-\frac{1}{20}(5 + 5\cos 0)^4 + C = 0$$

$$C = \frac{10^4}{20} = 500.$$

Finally, $g(t) = -\frac{1}{20}(5 + 5\cos t)^4 + 500$.

---

25. $\displaystyle\int 3x(x^2 + 7)^3\, dx =$

A. $\frac{3}{8}(x^2 + 7)^4 + C$

B. $\frac{3}{4}(x^2 + 7)^4 + C$

C. $\frac{1}{4}(x^2 + 7)^4 + C$

D. $3(x^2 + 7)^4 + C$

E. None of these.

Answer   A. First take $u = x^2 + 7$, which appears to be a good choice for u, since the derivative of u is a factor in the integrand. Then $du = 2x\,dx$.

Next, rewriting the integral you get

$$\int 3x(x^2 + 7)^3\,dx = \int \frac{3}{2}(x^2 + 7)^3 2x\,dx = \int \frac{3}{2}u^3\,du.$$

Then, finding the antiderivative for this simple indefinite integral gives you

$$\int \frac{3}{2}u^3\,du = \frac{3}{2 \cdot 4}u^4 + C = \frac{3}{8}u^4 + C.$$

Finally, substituting back with $u = x^2 + 7$, you get the answer

$$\int 3x(x^2 + 7)^3\,dx = \frac{3}{8}(x^2 + 7)^4 + C.$$

---

26. $\displaystyle\int \frac{x^2}{(2x^3 - 12)^4}\,dx =$

A. $-(2x^3 - 12)^{-3} + C$

B. $\dfrac{-1}{18(2x^3 - 12)^3} + C$

C. $\dfrac{-1}{3(2x^3 - 12)^3} + C$

D. $-\frac{1}{4}(2x^3 - 12)^{-3} + C$

E. None of these.

Answer   B. First take $u = 2x^3 - 12$, which appears to be a good choice for u, since the derivative of u is a factor in the integrand. Then $du = 6x^2\,dx$.

Next, rewriting the integral you get

$$\int \frac{x^2}{(2x^3 - 12)^4}\,dx = \int x^2(2x^3 - 12)^{-4}\,dx$$

$$= \int \frac{1}{6}(2x^3 - 12)^{-4}6x^2\,dx = \int \frac{1}{6}u^{-4}\,du.$$

Then, finding the antiderivative for this simple indefinite integral gives you

$$\int \frac{1}{6}u^{-4}\,du = \frac{1}{6(-3)}u^{-3} + C = \frac{-1}{18u^3} + C.$$

Finally, substituting back with $u = 2x^3 - 12$, you get the answer

$$\int \frac{x^2}{(2x^3 - 12)^4}\,dx = \frac{-1}{18u^3} + C = \frac{-1}{18(2x^3 - 12)^3} + C.$$

27. $\int \dfrac{(2 + \sqrt{x})^6}{\sqrt{x}}\, dx =$

A. $\frac{1}{14}\left(2 + x^{\frac{1}{2}}\right)^7 + C$

B. $\left(2 + x^{-\frac{1}{2}}\right)^7 + C$

C. $\frac{1}{7}\left(2 + x^{\frac{1}{2}}\right)^7 + C$

D. $\frac{2}{7}\left(2 + x^{\frac{1}{2}}\right)^7 + C$

E.   None of these.

Answer   D. First take $u = 2 + x^{\frac{1}{2}}$. Then $du = \frac{1}{2} x^{-\frac{1}{2}}\, dx$.

Next, rewriting the integral you get

$$\int \frac{(2 + \sqrt{x})^6}{\sqrt{x}}\, dx = \int \left(2 + x^{\frac{1}{2}}\right)^6 x^{-\frac{1}{2}}\, dx$$

$$= \int 2\left(2 + x^{\frac{1}{2}}\right)^6 \frac{1}{2} x^{-\frac{1}{2}}\, dx = \int 2u^6\, du = \frac{2}{7} u^7 + C.$$

Finally, substituting back with $u = 2 + x^{\frac{1}{2}}$, you get the answer

$$\int \frac{(2 + \sqrt{x})^6}{\sqrt{x}}\, dx = \frac{2}{7} u^7 + C = \frac{2}{7}\left(2 + x^{\frac{1}{2}}\right)^7 + C.$$

28. $\int \dfrac{4x - 8}{(x^2 - 4x + 7)^6}\, dx =$

A. $\dfrac{-4}{(x^2 - 4x + 7)^5} + C$

B. $\dfrac{-4}{5(x^2 - 4x + 7)^5} + C$

C. $\dfrac{-2}{(x^2 - 4x + 7)^5} + C$

D. $\dfrac{-2}{5(x^2 - 4x + 7)^5} + C$

E.   None of these.

Answer   D. First take $u = x^2 - 4x + 7$. Then $du = (2x - 4)\, dx$.

Next, rewriting the integral you get

$$\int \frac{4x - 8}{(x^2 - 4x + 7)^6}\, dx = \int (4x - 8)(x^2 - 4x + 7)^{-6}\, dx$$

$$= \int 2(x^2 - 4x + 7)^{-6}(2x - 4)\, dx = \int 2u^{-6}\, du.$$

Then, finding the antiderivative for this simple indefinite integral gives you

$$\int 2u^{-6}\, du = \frac{2}{-5} u^{-5} + C = \frac{-2}{5u^5} + C.$$

Finally, substituting back with $u = x^2 - 4x + 7$, you get the answer

$$\int \frac{4x - 8}{(x^2 - 4x + 7)^6}\, dx = \frac{-2}{5u^5} + C = \frac{-2}{5(x^2 - 4x + 7)^5} + C.$$

---

29. $\displaystyle \int (3 - \frac{1}{x})^{-2}(\frac{1}{x^2})\, dx =$

A. $-2(3 - \frac{1}{x})^{-1} + C$

B. $(3 - \frac{1}{x})^{-1} + C$

C. $2(3 - \frac{1}{x})^{-1} + C$

D. $-(3 - \frac{1}{x})^{-1} + C$

E.   None of these.

Answer   D. First, take $u = 3 - \frac{1}{x}$. Then $du = \frac{1}{x^2}\, dx$.

Next, rewriting the integral you get

$$\int \left(3 - \frac{1}{x}\right)^{-2} \left(\frac{1}{x^2}\right) dx = \int u^{-2}\, du = \int u^{-2}\, du = \frac{1}{-1} u^{-1} + C = -u^{-1} + C.$$

Finally, substituting back with $u = 3 - \frac{1}{x}$, you get the answer

$$\int \left(3 - \frac{1}{x}\right)^{-2} \left(\frac{1}{x^2}\right) dx = -\left(3 - \frac{1}{x}\right)^{-1} + C.$$

---

30. $\displaystyle \int \sec^2 x \tan x\, dx =$

A. $\frac{1}{3}\sec^3 x + C$

B. $\tan^2 x + C$

C. $\frac{1}{2}\tan^2 x + C$

D. $2\sec^2 x + C$

E. None of these.

Answer   C. Recognize first that the derivative of $\tan x$ is $\sec^2 x$. Since both these factors are in the integrand, it makes sense to take $u = \tan x$. Then $du = \sec^2 x\, dx$.

Next, rewriting the integral you get

$$\int \sec^2 x \tan x\, dx = \int \tan x \sec^2 x\, dx = \int u\, du = \frac{1}{2}u^2 + C.$$

Finally, substituting back with $u = \tan x$, you get the answer

$$\int \sec^2 x \tan x\, dx = \frac{1}{2}(\tan x)^2 + C = \frac{1}{2}\tan^2 x + C.$$

You could also solve this by using $u = \sec x$. Then the answer would be $\frac{1}{2}\sec^2 x + C$.
(Note that $\sec^2 x = \tan^2 x + 1$.)

---

31. $\int \cos(2x)\sqrt{\sin(2x)}\, dx =$

A. $\frac{1}{3}\sin^{3/2}(2x) + C$

B. $-2\sin^{-1/2}(2x) + C$

C. $\frac{2}{3}\sin^{3/2}(2x) + C$

D. $\frac{1}{6}\sin^{3/2}(2x) + C$

E. None of these.

Answer   A. Take $u = \sin(2x)$. Then $du = 2\cos(2x)\, dx$.

Next, rewriting the integral you get

$$\int \cos(2x)\sqrt{\sin(2x)}\, dx = \int \frac{1}{2}(\sin(2x))^{\frac{1}{2}} 2\cos(2x)\, dx$$

$$= \int \frac{1}{2}u^{\frac{1}{2}}\, du = \frac{1}{2(\frac{3}{2})}u^{\frac{3}{2}} + C = \frac{1}{3}u^{\frac{3}{2}} + C.$$

Finally, substituting back with $u = \sin(2x)$, you get the answer

$$\int \cos(2x)\sqrt{\sin(2x)}\, dx = \frac{1}{3}(\sin(2x))^{\frac{3}{2}} + C$$

$$= \frac{1}{3}\sin^{\frac{3}{2}}(2x) + C.$$

32. $\int \dfrac{\cos x}{(2 - 3\sin x)^4}\, dx =$

A. $\frac{1}{27}(2 - 3\sin x)^{-3} + C$

B. $\dfrac{1}{5(2 - 3\sin x)^5} + C$

C. $\frac{1}{9}(2 - 3\sin x)^{-3} + C$

D. $\dfrac{-1}{15(2 - 3\sin x)^5} + C$

E. None of these.

Answer   C. First take $u = 2 - 3\sin x$. Then $du = -3\cos x\, dx$.

Next, rewriting the integral you get

$$\int \dfrac{\cos x}{(2 - 3\sin x)^4}\, dx = \int (2 - 3\sin x)^{-4}\cos x\, dx$$
$$= \int \dfrac{-1}{3}(2 - 3\sin x)^{-4}(-3\cos x)\, dx = \int \dfrac{-1}{3}u^{-4}\, du.$$

Then, finding the antiderivative for this simple indefinite integral gives you

$$\int \dfrac{-1}{3}u^{-4}\, du = \dfrac{-1}{3(-3)}u^{-3} + C = \dfrac{1}{9}u^{-3} + C.$$

Finally, substituting back with $u = 2 - 3\sin x$, you get the answer

$$\int \dfrac{\cos x}{(2 - 3\sin x)^4}\, dx = \dfrac{1}{9}(2 - 3\sin x)^{-3} + C.$$

33. $\int \dfrac{x}{\cos^2(3x^2)}\, dx =$

A. $\frac{1}{3}\cos^{-1}(3x^2) + C$

B. $\frac{1}{6}\tan(3x^2) + C$

C. $-\frac{1}{6}\tan(3x^2) + C$

D. $\frac{1}{3}\sec(3x^2) + C$

E. None of these.

Answer B. One way to do this is to recognize that $\dfrac{1}{\cos^2(3x^2)} = \sec^2(3x^2)$.

Then you can rewrite the integral as $\displaystyle\int x\sec^2(3x^2)\,dx$.

Now take $u = 3x^2$. Then $du = 6x\,dx$.

Next, rewriting the integral you get

$$\int x\sec^2(3x^2)\,dx = \int \frac{1}{6}\sec^2(3x^2)6x\,dx = \int \frac{1}{6}\sec^2 u\,du = \frac{1}{6}\tan u + C.$$

Finally, substituting back with $u = 3x^2$, you get the answer

$$\int x\sec^2(3x^2)\,dx = \frac{1}{6}\tan(3x^2) + C.$$

---

34. $\displaystyle\int \sin(4x)\sec^6(4x)\,dx =$

A. $\frac{1}{7}\sec^7(4x) + C$

B. $\frac{1}{20}\sec^5(4x) + C$

C. $\frac{1}{28}\sec^7(4x) + C$

D. $\frac{1}{28}\sec^5(4x)\tan(4x) + C$

E. None of these.

Answer B. There are a few ways to do this one. The easiest way is to rewrite the integral as

$$\int \sin(4x)(\cos(4x))^{-6}\,dx.$$

Then, take $u = \cos(4x)$, and you get $du = -4\sin(4x)\,dx$.

Next, rewriting the integral you have

$$\int \sin(4x)(\cos(4x))^{-6}\,dx = \int \frac{-1}{4}(\cos(4x))^{-6}(-4\sin(4x))\,dx$$

$$= \int \frac{-1}{4}u^{-6}\,du = \frac{-1}{4(-5)}u^{-5} + C = \frac{1}{20}u^{-5} + C.$$

Finally, substituting back with $u = \cos(4x)$, you get the answer

$$\int \sin(4x)\sec^6(4x)\,dx = \frac{1}{20}u^{-5} + C$$

$$= \frac{1}{20}(\cos(4x))^{-5} + C = \frac{1}{20}\sec^5(4x) + C.$$

Practice Finding Antiderivatives by Substitution

Note: Whenever you find an antiderivative, you need to show your work.

This can be explicitly writing down the substitution (example: "$u = \sin x$, $du = \cos x \, dx$").

If you didn't use substitution, you should visibly check your answer by differentiating it.

---

1. Evaluate the following antiderivatives. Remember to show your substitution work OR show your derivative check if you use the "check and guess" method.

A. $\displaystyle\int \frac{\sin(\sqrt{x})}{\sqrt{x}} \, dx$

B. $\displaystyle\int \cos x \cos(\sin x) \, dx$

C. $\displaystyle\int \frac{4x^3 - 3}{(x^4 - 3x)^2} \, dx$

D. $\displaystyle\int 3x \tan(3x^2) \sec^2(3x^2) \, dx$

E. $\displaystyle\int (\sin^3 x \cos^2 x) \, dx$ (Hint: Use the identity $\sin^2 x = 1 - \cos^2 x$.)

---

2. Solve the differential equations subject to the given conditions.

A. An object's velocity (in meters per second) is given by the function $v(t) = 3t\sqrt{t^2 + 4}$. Find the position function $s(t)$ for the object if the object's position at time $t = 0$ is 25.

B. $f'(x) = 36x \sin(3x^2) \cos^2(3x^2)$, where $f(0) = 4$.

C. $f''(x) = 4\cos(2x) - \sin(3x)$, where $f(0) = 2$, and $f'(0) = \frac{4}{3}$.

---

3. During the summer, a certain city consumes water at a rate r given by the formula

$$r(t) = 7500 + 3{,}000 \sin(\frac{\pi}{90}t),$$

where r is in gallons per day and t (time) is in days, with $t = 0$ corresponding to the beginning of the first day of summer.

A. What is the city's rate of water consumption at the beginning of the 31st day of summer?

B. At what point during the summer is the city's rate of water consumption greatest?
(Note: There are 90 days of summer.)

C. Solve the differential equation $f'(t) = r(t)$ with $f(0) = 0$, and describe what the equation physically represents.

Answers

1.

A. $\int \frac{\sin(\sqrt{x})}{\sqrt{x}} dx = -2\cos\sqrt{x} + C.$ Using substitution:

$$u = x^{\frac{1}{2}}, \; du = \frac{1}{2}x^{-\frac{1}{2}} dx, \int \frac{\sin(\sqrt{x})}{\sqrt{x}} dx$$

$$= \int 2\sin u \, du = -2\cos u + C = -2\cos\sqrt{x} + C.$$

B. $\int \cos x \cos(\sin x) \, dx = \sin(\sin x) + C.$ Using substitution:

$$u = \sin x, \; du = \cos x \, dx, \int \cos x \cos(\sin x) \, dx$$

$$= \int \cos u \, du = \sin u + C = \sin(\sin x) + C.$$

C. $\int \frac{4x^3 - 3}{(x^4 - 3x)^2} dx = \frac{-1}{(x^4 - 3x)} + C.$ Using substitution:

$$u = x^4 - 3x, \; du = (4x^3 - 3) \, dx, \int \frac{4x^3 - 3}{(x^4 - 3x)^2} dx$$

$$= \int u^{-2} \, du = -u^{-1} + C = \frac{-1}{(x^4 - 3x)} + C.$$

D. $\int 3x \tan(3x^2) \sec^2(3x^2) \, dx = \frac{1}{4}\tan^2(3x^2) + C$ (or $\frac{1}{4}\sec^2(3x^2) + C$).

Using substitution:

$$u = \tan(3x^2), \; du = \sec^2(3x^2)6x \, dx,$$

$$\int 3x \tan(3x^2) \sec^2(3x^2) \, dx = \int \frac{1}{2}u \, du = \frac{1}{4}u^2 + C = \frac{1}{4}\tan^2(3x^2) + C.$$

Or, alternatively:

$$u = \sec(3x^2), \; du = \sec(3x^2) \tan(3x^2)6x \, dx,$$

$$\int 3x \tan(3x^2) \sec^2(3x^2) \, dx = \int \frac{1}{2}u \, du = \frac{1}{4}u^2 + C = \frac{1}{4}\sec^2(3x^2) + C.$$

Notice that although these two answers look different, they're actually equivalent because of the trig identity $1 + \tan^2 x = \sec^2 x$.

E. $\displaystyle\int (\sin^3 x \cos^2 x)\, dx = \frac{-1}{3}\cos^3 x + \frac{1}{5}\cos^5 x + C.$ Substituting $1 - \cos^2 x$ for $\sin^2 x$:

$$\int (\sin^3 x \cos^2 x)\, dx = \int (\sin^2 x \sin x \cos^2 x)\, dx$$

$$= \int (1 - \cos^2 x)\sin x \cos^2 x\, dx = \int \sin x \cos^2 x\, dx - \int \sin x \cos^4 x\, dx.$$

Then, with $u = \cos x$, $du = -\sin x\, dx$ for each integral:

$$\int \sin x \cos^2 x\, dx - \int \sin x \cos^4 x\, dx = \int -u^2\, du + \int u^4\, du$$

$$= \frac{-1}{3}u^3 + \frac{1}{5}u^5 + C$$

$$= \frac{-1}{3}\cos^3 x + \frac{1}{5}\cos^5 x + C.$$

2.

A. $s(t) = (t^2 + 4)^{3/2} + 17.$

First you need to find the antiderivative: $\displaystyle\int 3t\sqrt{t^2 + 4}\, dt = (t^2 + 4)^{3/2} + C.$

Then the condition must be used: $s(0) = 25$ yields

$$25 = (0^2 + 4)^{3/2} + C = 8 + C \Rightarrow C = 17.$$

B. $f(x) = -2\cos^3(3x^2) + 6.$

First you need to find the antiderivative:

$$\int 36x \sin(3x^2)\cos^2(3x^2)\, dx = -2\cos^3(3x^2) + C.$$

Then, the condition must be dealt with: $f(0) = 4$ yields

$$4 = -2\cos^3(3(0)^2) + C = -2 + C \Rightarrow C = 6.$$

C. $f(x) = -\cos 2x + \frac{1}{9}\sin 3x + x + 3.$

First you must find $f'(x)$:

$$\int (4\cos(2x) - \sin(3x))\, dx = 2\sin 2x + \frac{1}{3}\cos 3x + C.$$

Dealing with the condition for f' : $f'(0) = \frac{4}{3}$ yields

$$\frac{4}{3} = 2\sin(2 \cdot 0) + \frac{1}{3}\cos(3 \cdot 0) + C = \frac{1}{3} + C \Rightarrow C = 1.$$

So, you get $f'(x) = 2\sin 2x + \frac{1}{3}\cos 3x + 1.$

Next, to find $f(x)$, you need to get the antiderivative of $f'(x)$:

$$\int \left(2\sin 2x + \frac{1}{3}\cos 3x + 1\right) dx = -\cos 2x + \frac{1}{9}\sin 3x + x + C.$$

The condition $f(0) = 2$ must be used, which yields

$$2 = -\cos(2 \cdot 0) + \frac{1}{9}\sin(3 \cdot 0) + (0) + C = -1 + C \Rightarrow C = 3.$$

3.

A. This is just $r(30) = 7,500 + 3,000 \sin(\frac{\pi}{3}) = 7,500 + 1,500\sqrt{3} \approx 10,098$ gallons per day.

B. The rate is maximum where the sine function is maximum, that is when $\frac{\pi}{90}t = \frac{\pi}{2}$, so at $t = 45$, which is at the beginning of the 46th day of summer, or half-way through summer.

A.

$$f(t) = 7,500t - \frac{270,000}{\pi} \cos\left(\frac{\pi}{90}t\right) + \frac{270,000}{\pi}.$$

The function $f(t)$ represents the number of gallons of water consumed by the city from the beginning of the summer until time t.

First, the antiderivative:

$$\int \left(7,500 + 3,000 \sin\left(\frac{\pi}{90}t\right)\right) dt = 7,500t - \frac{270,000}{\pi} \cos\left(\frac{\pi}{90}t\right) + C.$$

Then, the condition $f(0) = 0$ yields

$$0 = 7,500(0) - \frac{270,000}{\pi} \cos\left(\frac{\pi}{90} \cdot 0\right) + C = -\frac{270,000}{\pi} + C \Rightarrow C = \frac{270,000}{\pi}.$$

B. The Fundamental Theorems of Calculus

▶Objective 1   Write both Fundamental Theorems of Calculus from memory.

The First Fundamental Theorem of Calculus: If f is continuous on $[a, b]$, then

$$F(x) = \int_a^x f(t)\, dt$$

is a differentiable function of x on $[a, b]$, and its derivative is

$$F'(x) = \frac{d}{dx} \int_a^x f(t)dt = f(x).$$

The Second Fundamental Theorem of Calculus: If f is continuous on $[a, b]$, and F is any antiderivative of f, then

$$\int_a^b f(x)\, dx = F(x)\Big|_a^b = F(b) - F(a).$$

▶Objective 2 Use the First Fundamental Theorem to find derivatives of functions that are defined as integrals.

Example Given that $g(x) = \int_3^x e^{\sin t} dt$, find $g'(x)$.

Tip When a function is defined by an integral, you can find its derivative directly. You don't need to actually integrate or differentiate.

Answer $g'(x) = e^{\sin x}$

▶Objective 3 Use the Second Fundamental Theorem to evaluate definite integrals.

Example Evaluate $\int_{-5}^{-1} \frac{1}{x^2} dx$.

Tip $\int_a^b f(x)\, dx = F(b) - F(a)$ is an amazingly powerful tool.

Answer

$$\int_{-5}^{-1} x^{-2}\, dx = -x^{-1}\Big|_{-5}^{-1} = -\frac{1}{-1} - \left(-\frac{1}{-5}\right) = \frac{4}{5}$$

▶Objective 4 Use substitution to change the form of a definite integral.

Example Is $\int_2^3 x\cos(x^2)\, dx$ equal to $\int_4^9 \cos(u)\, du$.

Tip Make sure you understand the idea:

$$\int_a^b f(g(x))g'(x)\, dx = \int_{g(a)}^{g(b)} f(u)\, du,$$

where $u = g(x)$ and $du = g'(x)dx$.

Answer No. It's missing a factor of 2. Substitute $u = x^2$, so that $du = 2x\, dx$:

$$\int_2^3 x\cos(x^2)\, dx = \frac{1}{2}\int_2^3 2x\cos(x^2)\, dx = \frac{1}{2}\int_4^9 \cos(u)\, du.$$

▶Objective 5 Adjust the limits of integration when using substitution to solve definite integrals.

Example   If $\displaystyle\int_{\frac{\pi}{4}}^{\pi} \cos^2(x)(-\sin(x))\,dx = \int_a^b u^2\,du$ find a and b.

Tip   Don't worry if the limits seem a little strange after a substitution.

Answer   Let $u = \cos x$, so $x = \pi \implies u = \cos \pi = -1$, and $x = \frac{\pi}{4} \implies u = \cos\frac{\pi}{4} = \frac{\sqrt{2}}{2}$. Thus,

$$\int_{\frac{\pi}{4}}^{\pi} \cos^2(x)(-\sin(x))\,dx = \int_{\frac{\sqrt{2}}{2}}^{-1} u^2\,du.$$

Thus, $a = \dfrac{\sqrt{2}}{2}$ and $b = -1$.

▶**Objective 6**   Use substitution to identify equivalent definite integrals.

Example   Justify the statement $\displaystyle\int_{-1}^{3} 2x f(x^2)\,dx = \int_1^9 f(x)\,dx$, and then show that they're equivalent if $f(x) = |x|$.

Tip   Try this for $f(x) = x$; $f(x) = \sin x$, and other examples. You should also play with your graphing calculator to see that the (signed) areas represented by these integrals are equivalent.

Answer   Let $u = x^2$, so $du = 2x\,dx$. Then $x = 3 \implies u = 9$ and $x = -1 \implies u = 1$.

$$\int_{-1}^{3} 2x f(x^2)\,dx = \int_1^9 f(u)\,du,$$

but realize that u is just a dummy variable and can be any letter (including x):

$$\int_{-1}^{3} 2x f(x^2)\,dx = \int_1^9 f(x)\,dx.$$

With $f(x) = |x|$, realize that $f(x^2) = |x^2| = x^2$.

$$\int_{-1}^{3} 2x|x^2|\,dx = \int_{-1}^{3} 2x^3\,dx = \left[\frac{1}{2}x^4\right]_{-1}^{3} = \frac{1}{2}3^4 - \frac{1}{2} = 40.$$

With $\displaystyle\int_1^9 |x|\,dx$, realize that $|x| = x$ for the stated domain $[1, 9]$.

$$\int_1^9 |x|\,dx = \int_1^9 x\,dx = \left[\frac{1}{2}x^2\right]_1^9 = 40.$$

▶**Objective 7**   Solve definite integrals involving composite functions, with or without using substitution.

Example Find $\displaystyle\int_1^9 \frac{1}{\sqrt{x}(\sqrt{x}+2)^3}\,dx$.

Answer The solution here is done with substitution. First we rewrite the integral using algebra:

$$\int_1^9 \frac{1}{\sqrt{x}(\sqrt{x}+2)^3}\,dx = \int_1^9 \frac{(\sqrt{x}+2)^{-3}}{\sqrt{x}}\,dx = 2\int_1^9 \frac{(\sqrt{x}+2)^{-3}}{2\sqrt{x}}\,dx.$$

Let $u = \sqrt{x}+2$ and $du = \dfrac{1}{2\sqrt{x}}\,dx.$

$$2\int_1^9 \frac{(\sqrt{x}+2)^{-3}}{2\sqrt{x}}\,dx = 2\int_3^5 u^{-3}\,du = 2\left[\frac{-1}{2}u^{-2}\right]_3^5 = \left[-u^{-2}\right]_3^5$$
$$= \frac{-1}{5^2} - \frac{-1}{3^2} = \frac{16}{225}.$$

►**Objective 8** Analyze functions defined by definite integrals.

Example Find the maximum value of $g(x)$ on $-\pi \le x \le \pi$, for $g(x) = \displaystyle\int_{\frac{\pi}{2}}^x \cos(t)\,dt$. Justify your answer.

Tip It really helps to visualize the (signed) area represented by a function defined in terms of a definite integral.

Answer

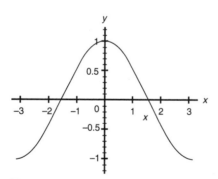

From the graph of cosine shown here you should be able to see that

$$\int_{\frac{\pi}{2}}^x \cos(t)\,dt$$

will always be less than or equal to zero:

To the right of $\frac{\pi}{2}$ you're under the axis, and to the left of $\frac{\pi}{2}$ you have a smaller limit on top, so the integral is negative. Therefore the maximum value for $g(x)$ is zero. Since you're asked to "justify your answer" let's do a quick number line:

$$g(x) = \int_{\frac{\pi}{2}}^x \cos(t)\,dt, \qquad g'(x) = \cos(x).$$

	$-\pi \leq x < -\frac{\pi}{2}$	$-\frac{\pi}{2} < x < \frac{\pi}{2}$	$\frac{\pi}{2} < x \leq \pi$
$g'(x) = \cos x$	*Negative*	*Positive*	*Negative*
$g(x) = \displaystyle\int_{\frac{\pi}{2}}^{x} \cos(t)\, dt$	Decreasing	Increasing	Decreasing

Possible maximum values for $g(x)$ are:

$$g(-\pi) = \int_{\frac{\pi}{2}}^{-\pi} \cos(t)\, dt = -1 \text{ and } g(\tfrac{\pi}{2}) = \int_{\frac{\pi}{2}}^{\frac{\pi}{2}} \cos(t)\, dt = 0.$$

So, $g(x)$ has a maximum value of 0 on the domain $-\pi \leq x \leq \pi$.

▶Objective 9   Use the chain rule and the First Fundamental Theorem of Calculus to analyze functions defined by definite integrals with functions in the limits.

Example   The function f has a graph shown in the illustration below:

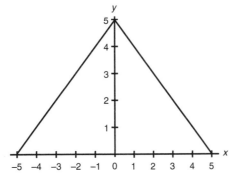

What is the minimum value of $g(x) = \displaystyle\int_{1}^{x^2} f(t)\, dt$ on the domain $-2 \leq x \leq 2$?

Tip   If you saw the answer intuitively without going through the numberline test that's OK, but be prepared to do an analysis like this if you're asked to "justify your answer."

Answer $g'(x) = 2x f(x^2)$.

Notice that $f(x^2)$ is always positive, so that $g'(x)$ changes sign when $x = 0$.

x	$-2 \leq x < 0$	0	$0 < x \leq 2$
$g'(x)$	*Negative*	0	*Positive*
$g(x)$	Decreasing	Relative *Min*	Increasing

So, $g(0)$ represents the minimum value of $g(x)$ on the domain $-2 \leq x \leq 2$.

$$g(0) = \int_{1}^{0} f(t)\, dt = -\int_{0}^{1} f(t)\, dt = -\left(\frac{5+4}{2}\right) = -\frac{9}{2}.$$

Multiple-Choice Questions on the Fundamental Theorems of Calculus

1. Which of these is the function that gives the signed area under the curve $y = t^2$ between $t = 2$ and $t = x$ as a function of x?

   A. $\quad a = \displaystyle\int_0^x t^2 \, dt$

   B. $\quad a = \displaystyle\int_2^x x^2 \, dx$

   C. $\quad a = \displaystyle\int_2^x t^2 \, dt$

   D. $\quad a = \displaystyle\int_x^2 t^2 \, dt$

   E. None of these

Answer   C. This definite integral gives the signed area between 2 and x under the curve $y = t^2$.

2. Which of these has x^2 as its derivative?

   A. $\quad \dfrac{x^3}{3} + 75$

   B. $\quad \displaystyle\int_{-386}^x t^2 \, dt$

   C. $\quad \displaystyle\int_0^x u^2 \, du$

   D. All of these

   E. None of these

Answer   D. They all have x^2 as their derivative with respect to x. When you use the integral to write down an antiderivative, it doesn't matter what the lower limit of the integral is, and it also doesn't matter what variable you use in the integrand (as long as it's different from the variable in the limit).

3. The graph of f is shown in the illustration below:

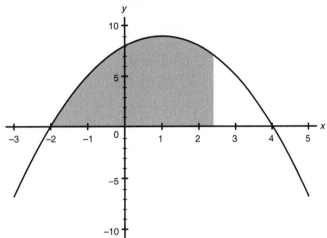

Which of the following are true about $A(x) = \displaystyle\int_{-2}^{x} f(t)\, dt$?

I. $A'(-1) = A'(3)$

II. The maximum value of A occurs at $x = 1$.

III. The maximum value of A' occurs at $x = 1$.

A. III only

B. I and II

C. I and III

D. I, II, and III

E. None of these is true.

Answer C. $A'(x)$ is just the value of the curve, so I says that $f(-1) = f(3)$, which is true.

III says that the maximum value of this curve is at $x = 1$, which is also true.

But the maximum value of A doesn't happen until $x = 4$, so II is not true.

4. Evaluate $\displaystyle\int_{0}^{1} \sqrt{1 - x^2}\, dx$.

A. $\dfrac{\sqrt{2}}{2}$

B. $\dfrac{\pi}{4}$

C. 1

D. 0

E. None of these

Answer   B. This is actually the area inside the unit circle, in the first quadrant, so its area $= \frac{\pi}{4}$.

---

5. $\dfrac{d}{dx} \displaystyle\int_{-17}^{x} t\cos(t^2)\, dt =$

A. $\frac{1}{2}\sin x^2 - \frac{1}{2}\sin(-17^2)$

B. $x\cos(x^2) - (-17)\cos(-17^2)$

C. $\frac{1}{2}\sin x^2$

D. $x\cos(x^2)$

E. None of the above.

Answer   D. The First Fundamental Theorem tells you directly that

$$\frac{d}{dx} \int_{-17}^{x} t\cos(t^2)\, dt = x\cos(x^2).$$

---

6. If $F(x) = \displaystyle\int_{\pi}^{x} e^{2t}\sin^2(3t)\, dt$, then $F'(x) =$

A. $e^{2x}\sin^2(3x)$

B. $\displaystyle\int_{\pi}^{x} e^{2t}\sin^2(3t)\, dt - \int_{\pi}^{x} e^{2\pi}\sin^2(3\pi)\, dt$

C. $e^{2\pi}\sin^2(3x)$

D. 0

E. None of the above.

Answer   A. You have that $F(x)$ is a function defined by an integral,

$$F(x) = \int_{\pi}^{x} e^{2t}\sin^2(3t)\, dt.$$

The First Fundamental Theorem tells you that when you have a function defined by an integral of this form, it's easy to find that function's derivative. You get

$$F'(x) = \frac{d}{dx}F(x) = \frac{d}{dx}\int_{\pi}^{x} e^{2t}\sin^2(3t)\, dt = e^{2x}\sin^2(3x).$$

---

7. If $F(x) = \displaystyle\int_{7}^{x} t^2\sin(t)\, dt$, then $F'(\frac{5\pi}{2}) =$

A. 0

B. 29.493

C. $\frac{25\pi^2}{4}$

D. $x^2 \sin(x)$

E. None of the above

Answer   C. The First Fundamental Theorem will give you $F'(x) = x^2 \sin(x)$. Thus,

$$F'\left(\frac{5\pi}{2}\right) = \left(\frac{5\pi}{2}\right)^2 \sin\left(\frac{5\pi}{2}\right) = \frac{25\pi^2}{4} \cdot 1 = \frac{25\pi^2}{4}.$$

---

8. $\dfrac{d}{dx} \displaystyle\int_{-3}^{1} (2t^3 + 3)\, dt =$

A. $2t^3 + 3$

B. 56.0

C. 5

D. -28.0

E. 0

Answer   E. The First Fundamental Theorem will give you that $\dfrac{d}{dx} \displaystyle\int_{-3}^{x} (2t^3 + 3)\, dt = 2x^3 + 3.$

But we do not have the x in the upper limit of integration here. Our problem is to find $\dfrac{d}{dx} \displaystyle\int_{-3}^{1} (2t^3 + 3)\, dt$. Remember that $\displaystyle\int_{-3}^{1} (2t^3 + 3)\, dt$, being a definite integral with constant limits of integration, is just a constant itself! The derivative with respect to x of any constant is 0, because constants don't change. This is an important case to watch out for!

---

9.   The graph of the function f is illustrated below:

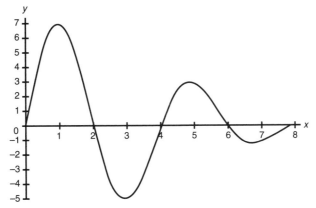

If we define $h(x) = \displaystyle\int_{0}^{x} f(t)\, dt$, then which of the following statements are true?

I. $h'(3) > 0$

II. $h'(x)$ has a relative maximum at $x = 2$.

III. $h(x)$ has a relative maximum at $x = 5$.

A.   I only

B.   II only

C.   III only

D.   II and III only

E.   None of these statements is true.

Answer   E. The First Fundamental Theorem tells you that $h'(x) = f(x)$ and therefore

$$h'(3) = f(3) = -5$$

so statement I is false.

But $f(x)$ does not have a relative maximum at $x = 2$, so neither does $h'(x)$. Thus statement II is false.

Finally, $h(x)$ does not have a relative maximum at $x = 5$, since the signed area under the f is still increasing as you increase x past 5. Thus statement III is false.

---

10.   Use the Second Fundamental Theorem of calculus to evaluate $\int_1^3 \left(6x^2 + \frac{6}{x^2} \right) dx$.

A. $\frac{160}{3}$

B.   52

C. $\frac{128}{3}$

D.   56

E.   None of the above.

Answer   D. First note that an antiderivative of $6x^2 + \frac{6}{x^2}$ is $F(x) = 2x^3 - \frac{6}{x}$.

Then, by the Second Fundamental Theorem, you have

$$\int_1^3 \left(6x^2 + \frac{6}{x^2} \right) dx = F(3) - F(1)$$
$$= \left[2(3)^3 - \frac{6}{3} \right] - \left[2(1)^3 - \frac{6}{1} \right]$$
$$= 52 - (-4) = 56.$$

---

11. Use the Second Fundamental Theorem of calculus to evaluate $\int_{-\pi}^{\pi/4} \cos(2x)\,dx$.

   A. -1

   B. $\frac{1}{2}$

   C. 0

   D. 1

   E. None of the above

Answer   B. First note that an antiderivative of $\cos(2x)$ is $\frac{1}{2}\sin(2x)$.

   Then, by the Second Fundamental Theorem, you have

$$\int_{-\pi}^{\frac{\pi}{4}} \cos(2x)\,dx = \left[\frac{1}{2}\sin(2x)\right]_{-\pi}^{\frac{\pi}{4}}$$
$$= \frac{1}{2}\sin(2(\frac{\pi}{4})) - \frac{1}{2}\sin(2(-\pi))$$
$$= \frac{1}{2}\sin(\frac{\pi}{2}) - \frac{1}{2}\sin(-2\pi) = \frac{1}{2}.$$

---

12. $\int_{0}^{\pi} \sec^2 x\,dx =$

   A. $\frac{-1}{\sqrt{3}}$

   B. -1

   C. 0

   D. $-\sqrt{3}$

   E. None of the above

Answer   E. If you try to use the Second Fundamental Theorem you get

$$\int_{0}^{\pi} \sec^2 x\,dx = [\tan x]_0^{\pi} = \tan\pi - \tan 0 = 0 - 0 = 0.$$

However, you CANNOT use the Second Fundamental Theorem because the function $\sec^2 x$ is not continous on the interval $[0, \pi]$.

This is a very important thing to look out for! In fact, since $\sec^2 x$ is always positive where it's defined, the definite integral $\int_{0}^{\pi} \sec^2 x\,dx$ could not be 0 or negative.

Note: This is called an *improper integral* (a BC topic) and it is possible to show that $\int_{0}^{\pi} \sec^2 x\,dx = \infty$!

13. The velocity of a particle moving along a line is $t^2 - 2t\sin(t^2)$ meters per second. Then, the net change in position during the time interval $0 \le t \le 4$ is:

A. 21.4 meters

B. $\frac{61}{3} + \cos 16$ meters

C. $\frac{67}{3} + \cos 16$ meters

D. 16.45 meters

E. 20.376 meters

Answer   B. The net change in position during the time interval $0 \le t \le 4$ is given by

$$\int_0^4 (t^2 - 2t\sin(t^2))\, dt.$$

An antiderivative of $t^2 - 2t\sin(t^2)$ is $\frac{1}{3}t^3 + \cos(t^2)$. Then by the Second Fundamental Theorem we have that

$$\int_0^4 (t^2 - 2t\sin(t^2))\, dt = \left[\frac{1}{3}t^3 + \cos(t^2)\right]_0^4$$
$$= \frac{1}{3}(4)^3 + \cos(4^2) - \frac{1}{3}0^3 - \cos(0^2)$$
$$= \frac{64}{3} + \cos 16 - 1 = \frac{61}{3} + \cos 16$$

14. Which of the following substitutions will help you solve $\displaystyle\int_a^b x(\sqrt[3]{x^2 + 3})\, dx$?

I. $u = x^2$

II. $u = x^2 + 3$

III. $u = x$

A. I only

B. II only

C. I and II only

D. I and III only

E. II and III only

Answer   C.

I. With $u = x^2$ $du = 2x\,dx$

$$\int_a^b x\left(\sqrt[3]{x^2+3}\right)dx = \frac{1}{2}\int_a^b 2x\left(\sqrt[3]{x^2+3}\right)dx = \frac{1}{2}\int_{a^2}^{b^2}\sqrt[3]{u+3}\,du,$$

which you can solve.

II. With $u = x^2 + 3$ $du = 2x\,dx$

$$\int_a^b x\left(\sqrt[3]{x^2+3}\right)dx = \frac{1}{2}\int_a^b 2x\left(\sqrt[3]{x^2+3}\right)dx = \frac{1}{2}\int_{a^2+3}^{b^2+3}\sqrt[3]{u}\,du,$$

which you can solve.

III. Setting $u = x$ merely amounts to relabeling the variables.

---

15.   Which of the following integrals is equivalent to $\displaystyle\int_a^b \frac{\sin\sqrt{x}}{\sqrt{x}}\,dx$?

A. $\displaystyle 2\int_{\sqrt{a}}^{\sqrt{b}} \sin(u)\,du$

B. $\displaystyle 2\int_a^b \sin(u)\,du$

C. $\displaystyle \frac{1}{2}\int_{a^2}^{b^2} \sin(u)\,du$

D. $\displaystyle \int_{\sqrt{a}}^{\sqrt{b}} -\cos(u)\,du$

E. $\displaystyle \frac{1}{2}\int_{\sqrt{a}}^{\sqrt{b}} \sin(u)\,du$

Answer   A. Using $u = \sqrt{x}$ $du = \dfrac{1}{2\sqrt{x}}\,dx$,

$$\int_a^b \frac{\sin\left(\sqrt{x}\right)}{\sqrt{x}}\,dx = 2\int_a^b \frac{\sin\left(\sqrt{x}\right)}{2\sqrt{x}}\,dx = 2\int_{\sqrt{a}}^{\sqrt{b}} \sin(u)\,du.$$

---

16.   Which of the following statements are true?

I. $\displaystyle\int_2^3 x f(x^2)\,dx = \int_4^9 f(u)\,du.$

II. $\displaystyle\int_a^b g'(t)\sin(g(t))\,dt = \int_{g(a)}^{g(b)} \sin(x)\,dx.$

III. $\displaystyle\int_0^3 x\sqrt{25 - x^2}\, dx = \frac{-1}{2}\int_{25}^{16}\sqrt{u}\, du.$

A. I only

B. II only

C. III only

D. I and II only

E. II and III only

Answer E.

I. if false. With $u = x^2$ and $du = 2x\, dx$.

$$\int_2^3 xf(x^2)\, dx = \frac{1}{2}\int_2^3 2xf(x^2)\, dx = \frac{1}{2}\int_4^9 f(u)\, du,$$

II. is true, since all you need to do is let $x = g(t)$, $dx = g'(t)\, dt$, and by substituting,

$$\int_a^b g'(t)\sin(g(t))\, dt = \int_{g(a)}^{g(b)} \sin(x)\, dx.$$

The limits also have to change; using $x = g(t)$, $t = a$ means $x = g(a)$, and $t = b$ means $x = g(b)$.

III. is true. Let $u = 25 - x^2$ and $du = -2x\, dx$, so

$$\int_0^3 x\sqrt{25 - x^2}\, dx = \frac{-1}{2}\int_0^3 -2x\sqrt{25 - x^2}\, dx = \frac{-1}{2}\int_{25}^{16}\sqrt{u}\, du,$$

since $25 - 3^2 = 16$ and $25 - 0^2 = 25$.

---

17. The area under the curve $y = 2x\tan(x^2)$ for $0 \le x \le \frac{1}{4}$ is equal to the area under the curve $y = \tan x$ for $0 \le x \le a$. What does a equal?

A. $a = 1$

B. $a = \frac{1}{2}$

C. $a = \frac{1}{4}$

D. $a = \frac{1}{8}$

E. $a = \frac{1}{16}$

Answer E. If you think of two equivalent integrals,

$$\int_0^{\frac{1}{4}} 2x\tan(x^2)\, dx = \int_0^{\frac{1}{16}} \tan(u)\, du,$$

where $u = x^2$, $x = \frac{1}{4}$ means $u = \left(\frac{1}{4}\right)^2 = \frac{1}{16}$.

Notice that you know the two are equivalent even though you don't know what the area actually is!

18. Which of the following integrals is equal to $\int_1^5 \dfrac{x}{\sqrt{2x-1}}\,dx$, given that $u = \sqrt{2x-1}$?

A. $\int_1^3 (u^2 + 1)\,du$

B. $\dfrac{1}{2}\int_1^3 \dfrac{u^2 + 1}{u}\,du$

C. $\int_1^3 \dfrac{u^2 + 1}{2}\,du$

D. $\int_1^3 \dfrac{u^2 + 1}{u}\,du$

E. $\int_1^5 \dfrac{u^2 + 1}{u}\,du$

Answer   C. Let $u = \sqrt{2x-1}$, so $x = \dfrac{u^2 + 1}{2}$ and $dx = u\,du$.

$$\int_1^5 \dfrac{x}{\sqrt{2x-1}}\,dx = \int_1^5 x\dfrac{1}{\sqrt{2x-1}}\,dx = \int_1^3 \dfrac{u^2 + 1}{2}\left(\dfrac{1}{u}\right)u\,du = \int_1^3 \dfrac{u^2 + 1}{2}\,du.$$

19. Which of these is the derivative with respect to t of $\int_{t^2}^{t^3} \cos(x^2)\,dx$?

A. $\cos t^3 (3t^2) - \cos t^2 (2t)$

B. $[\cos t^3 - \cos t^2](-2x)$

C. $[\sin t^6 - \sin t^4](3t^2 - 2t)$

D. $\sin t^6 (3t^2) - \sin t^4 (2t)$

E. $3t^2 \cos(t^6) - 2t\cos(t^4)$

Answer   E. By the FTC, $\dfrac{d}{dx}\int_{u(x)}^{v(x)} f(t)\,dt = f(v(x))\dfrac{dv}{dx} - f(u(x))\dfrac{du}{dx}$

In this case, then,

$$\dfrac{d}{dt}\left(\int_{t^2}^{t^3} \cos(x^2)\,dx\right)$$
$$= \cos((t^3)^2)(3t^2) - \cos((t^2)^2)(2t) = 3t^2\cos(t^6) - 2t\cos(t^4).$$

20. The volume of a container at a certain height h is given by $\int_0^h \pi \left(\tan \left(\frac{\pi y}{20} \right) \right)^2 dy$.

    Liquid is being poured into the container in such a way that the height (depth) is increasing at a constant rate of 0.2 feet per second. At what rate (in cubic feet per second) is the liquid coming into the container when the height (depth) of the liquid is 5 feet?

    A. Not enough information

    B. 4.292 cubic feet per second

    C. 0.628 cubic feet per second

    D. 15.708 cubic feet per second

    E. 0.858 cubic feet per second

Answer C. This is a related-rates problem. You want $\frac{dV}{dt}$; you know $\frac{dh}{dt}$; and you can figure $\frac{dV}{dh}$ from the Fundamental Theorem. So,

$$\frac{dV}{dt} = \frac{dV}{dh}\frac{dh}{dt} = \pi \left(\tan \left(\frac{\pi h}{20} \right) \right)^2 (0.2).$$

When $h = 5$, you get $0.2\pi \approx 0.628$ cubic feet per second. Notice that you didn't have to figure out this integral at all, either as an antiderivative or with your calculator.

21. The distance in meters traveled after T seconds by a pebble stuck in the rim of a bicycle tire that rolls off a roof is given by

$$\int_0^T \sqrt{(6 + 6\cos(30t))^2 + (-9.8t - 6\sin(30t))^2} \, dt.$$

About how far does it travel between 3 and 3.1 seconds?

    A. Not enough information

    B. $\sqrt{(6 + 6\cos(30(3)))^2 + (-9.8(3) - 6\sin(30(3)))^2} \cdot ((0.1) \text{ meters})$

    C. $(\sqrt{(6 - 6\sin(30(3)))^2 + (-9.8t - 6\cos(30(3)))^2}) \cdot ((0.1) \text{ meters})$

    D. $\sqrt{(6 + 6\cos(30(3)))^2 + (-9.8t - 6\sin(30(3)))^2} \text{ meters}$

    E. $\sqrt{(6 + 6\cos(30(3.1)))^2 + (-9.8t - 6\sin(30(3.1)))^2} \text{ meters}$

Answer B. This is the same situation you've been in before. In this setting, you don't even really know what's being accumulated here. But you do know that the integral is accumulating something, so the rule $\int_a^{a+h} f(x)\,dx$ is approximately equal to $f(a) \cdot h$ applies. Here, the approximation would be

$$\sqrt{(6 + 6\cos(30(3)))^2 + (-9.8(3) - 6\sin(30(3)))^2} \cdot ((0.1) \text{ meters}).$$

ANTIDERIVATIVES AND THE FUNDAMENTAL THEOREMS OF CALCULUS

Practice Using Substitution and the Fundamental Theorems

1. Use the method of substitution to evaluate each of the following definite integrals. That is, evaluate each definite integral by utilizing the fact that

$$\int_a^b f(g(x))g'(x)\,dx = \int_{g(a)}^{g(b)} f(u)\,du,$$

where $u = g(x)$ and $du = g'(x)\,dx$.

A. $\displaystyle\int_1^2 x^3(4-x^4)^3\,dx$

B. $\displaystyle\int_0^{\frac{\pi}{4}} \cos(2x)\sqrt{2-\sin(2x)}\,dx$

2. Evaluate the following definite integrals as directed:

A. $\displaystyle\int_{-3}^3 (x\sqrt{9-x^2}+2\sqrt{9-x^2})\,dx$. (Rewrite it as a sum of two integrals, one that can be interpreted in terms of area and the other that you can solve by substitution.)

B. $\displaystyle\int_0^{\sqrt{2}} x\sqrt{4-x^4}\,dx$. Use the substitution $u = x^2$. (Note: This is an example of how a clever substitution can be used to transform a problem that looks impossible to solve into one whose solution is known.)

3. Use substitution to prove the following properties. Explain geometrically.

A. If f is continuous for all real numbers, then $\displaystyle\int_a^b f(-x)\,dx = \int_{-b}^{-a} f(x)\,dx$.

B. If f is continuous for all real numbers, then $\displaystyle\int_a^b f(x+c)\,dx = \int_{a+c}^{b+c} f(x)\,dx$.

4. You're given that $\displaystyle\int_0^8 e^x\,dx = 2979.958$ (to three decimal places).

Use substitution to find $\displaystyle\int_0^{-2} 2x^2 e^{-x^3}\,dx$.

Apex Learning   **323**

Answers

1.

A. $\frac{-20655}{16} \approx -1290.9375.$

Work: The most obvious substitution is $u = 4 - x^4$. Then you have

$$du = -4x^3\,dx, u(1) = 4 - 1^4 = 3, u(2) = 4 - 2^4 = -12.$$

$$\int_1^2 x^3(4 - x^4)^3\,dx = \int_1^2 \frac{-1}{4}(4 - x^4)^3(-4x^3)\,dx = \int_3^{-12} \frac{-1}{4}u^3\,du$$

$$= \left[\frac{-1}{16}u^4\right]_3^{-12} = \frac{-1}{16}(-12^4) - \frac{-1}{16}(3^4) = \frac{-20655}{16} \approx -1290.9375.$$

B. $-\frac{1}{3} + \frac{2}{3}\sqrt{2} \approx 0.6095.$

Work: $u = 2 - \sin(2x)$, which then gives you

$$du = -2\cos(2x)\,dx, u(0) = 2 - \sin(0) = 2, u(\frac{\pi}{4}) = 2 - \sin(\frac{\pi}{2}) = 1.$$

$$\int_0^{\frac{\pi}{4}} \cos(2x)\sqrt{2 - \sin(2x)}\,dx = \int_0^{\frac{\pi}{4}} \frac{-1}{2}\sqrt{2 - \sin(2x)}(-2\cos(2x))\,dx$$

$$= \int_2^1 \frac{-1}{2}u^{\frac{1}{2}}\,du = \left[\frac{-1}{2} \times \frac{2}{3}u^{\frac{3}{2}}\right]_2^1 = \frac{-1}{3} \times 1 - \frac{-1}{3} \times 2^{\frac{3}{2}} = -\frac{1}{3} + \frac{2}{3}\sqrt{2}.$$

2.

A. $9\pi \approx 28.274.$

Work: Following the instructions, the integral should be split up as

$$\int_{-3}^3 \left(x\sqrt{9 - x^2} + 2\sqrt{9 - x^2}\right)\,dx = \int_{-3}^3 x\sqrt{9 - x^2}\,dx + \int_{-3}^3 2\sqrt{9 - x^2}\,dx.$$

The first one: $u = 9 - x^2$, $du = -2x\,dx$, $u(-3) = 0, u(3) = 0.$

So, $\int_{-3}^3 x\sqrt{9 - x^2}\,dx = \int_0^0 \frac{-1}{2}u^{\frac{1}{2}}\,du = 0.$

The second one: $\int_{-3}^3 2\sqrt{9 - x^2}\,dx = 2\int_{-3}^3 \sqrt{9 - x^2}\,dx.$ The integral here is just the area of the top half of a circle of radius 3, so $2\int_{-3}^3 \sqrt{9 - x^2}\,dx = 2 \cdot \frac{1}{2}9\pi = 9\pi \approx 28.274.$

B. $\frac{\pi}{2} = 1.5708.$

Work: $u = x^2$, $du = 2x\,dx$. $u(0) = 0, u(\sqrt{2}) = 2.$

$\int_0^{\sqrt{2}} x\sqrt{4 - x^4}\,dx = \int_0^2 \frac{1}{2}\sqrt{4 - u^2}\,du = \frac{1}{2}\int_0^2 \sqrt{4 - u^2}\,du.$ Again, notice that this integral is just a quarter of a circle of radius 2, so $\frac{1}{2}\int_0^2 \sqrt{4 - u^2}\,du = \frac{1}{2}\frac{4\pi}{4} = \frac{\pi}{2} = 1.5708.$

3.

A. Substitute with $u = -x$. Then $du = -dx$, $u(a) = -a$, $u(b) = -b$.

So, $\int_a^b f(-x)\,dx = \int_{-a}^{-b} -f(u)\,du = \int_{-b}^{-a} f(u)\,du = \int_{-b}^{-a} f(x)\,dx$.

Geometrically: Changing $-x$ to x flips the graph across the y-axis, giving a mirror image of the original graph. So, the part of the graph that was between $x = a$ and $x = b$ is now between $x = -b$ and $x = -a$.

B. Substitute with $u = x + c$.

Then $du = dx$, $u(a) = a + c$, $u(b) = b + c$.

So $\int_a^b f(x+c)\,dx = \int_{a+c}^{b+c} f(u)\,du = \int_{a+c}^{b+c} f(x)\,dx$.

Geometrically: Changing $x + c$ into x shifts the graph c units to the right. So, the part of the graph that was between $x = a$ and $x = b$ is now between $x = a + c$ and $x = b + c$.

---

4. -1986.639.

Work: $u = -x^3$, then you get $du = -3x^2\,dx$. $u(0) = 0$, $u(-2) = 8$. So, $\int_0^{-2} 2x^2 e^{-x^3}\,dx$

$= \int_0^8 \frac{-2}{3} e^u\,du = \frac{-2}{3} \int_0^8 e^u\,du = \frac{-2}{3}(e^8 - 1) \approx \frac{-2}{3} \times 2979.958 = -1986.639$.

Chapter 8   *Applications of the Integral*

In this chapter we review objectives related to applications of definite integrals. Since we can think of definite integrals in terms of measuring area, it is not surprising that area calculation is one such application. Integrals can also be used to find volumes, distance traveled, and many other quantities. In general, the definite integral can measure the total accumulated change from a given rate of change.

A. Area

►Objective 1   Use the definite integral to calculate the area between two curves.

Example   Find the area bounded by the curves $y = x^3 - ax^2$ and $y = ax^2 - x^3$, where $a > 0$.

Tip   When you're given an unspecified constant such as a, you can investigate the behavior of the function by putting in some specific values for a. **But be very careful!** Your analysis has to be valid for **ALL** a. This means that you must never use specific chosen values for a on which to base your justifications.

Answer   First find the intersections:

$$x^3 - ax^2 = ax^2 - x^3$$
$$2x^3 - 2ax^2 = 0$$
$$2x^2(x - a) = 0$$
$$x = 0, a$$

So, the region defined is on the domain $0 \le x \le a$.

Next, recognize that $y = x^3 - ax^2 = x^2(x - a)$ is negative when $0 \le x \le a$, and that $y = ax^2 - x^3 = x^2(a - x)$ is positive when $0 \le x \le a$. Thus the area is

$$\text{Area} = \int_0^a (ax^2 - x^3) - (x^3 - ax^2)dx = \int_0^a (2ax^2 - 2x^3)dx$$

$$= \left[\frac{2}{3}ax^3 - \frac{1}{2}x^4\right]_0^a = \frac{1}{6}a^4.$$

▶**Objective 2**   Calculate the area of regions bounded by multiple curves and/or axis lines.

Example   Using your calculator, find the area of a region bounded by the curves $y = \sin x$, $y = 3x$, and $y = 30 - 3x$.

Tip   Remember to keep all digits during the problem and only round to three digits at the end of the problem.

Answer

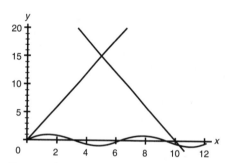

Intersections are at $x = 0$, 5, and 10.243402.

$$Area = \int_0^5 (3x - \sin x)dx + \int_5^{10.243402} (30 - 3x - \sin x)dx \approx 73.228.$$

▶**Objective 3**   Calculate areas by accumulation along the y-axis.

Example   Find the area bounded by $x = (y - 2)^2$ and $y = 4 - x$.

Tip   Sometimes it's easier to add the areas by slicing horizontally.

Answer   Find the intersections of $x = (y - 2)^2$, and $x = 4 - y$:

$$(y - 2)^2 = 4 - y$$
$$y^2 - 4y + 4 = 4 - y$$
$$y^2 - 3y = 0$$
$$y = 0, 3.$$

Draw a rough sketch or simply note that $4 - y$ is larger than $(y - 2)^2$ when $0 < y < 3$.

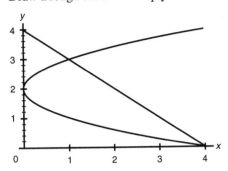

$$Area = \int_0^3 ((4 - y) - (y - 2)^2)dy = \int_0^3 ((4 - y) - (y^2 - 4y + 4))dy$$

$$= \int_0^3 (3y - y^2)dy = \left[\frac{3}{2}y^2 - \frac{1}{3}y^3\right]_0^3 = \frac{9}{2}$$

► **Objective 4**   Use the definite integral to find the average value of a function.

Example   The figure below is a square of base 4 meters topped by a semicircle.

What is the average height of this figure?

Tips   Remember that the area under an average value line is the same as the area under the original function curve.

Also, this figure has a shape with a well-known area. But if you were trying to find the area of an unfamiliar function, you could always determine the integral using your calculator.

Answer   You can think of a rectangle with this average height:

and realize that it will have the same area as the original figure.

The area of the original figure is $Area = A_{semicircle} + A_{square} = \frac{1}{2}\pi(2)^2 + 4^2 = 2\pi + 16$.

The area of the rectangle with average height is $Area = bh_{avg} = 4h_{avg}$.

Setting the two equal to each other, you find $4h_{avg} = 2\pi + 16$.

$h_{avg} = \frac{2\pi+16}{4} = \frac{1}{2}\pi + 4 \approx 5.571$, which gives the average height.

►Objective 5   Use numerical integration to estimate the average value of a function given as a table of data.

Example   The temperature over a given period is found to be:

time (hrs)	0	2	4	6	8	10	12
Temp °F	70	71	75	80	82	79	74

a)   Estimate the average temperature from $0 \leq t \leq 12$ using trapezoids.

B.   Estimate the average temperature from $0 \leq t \leq 12$ using midpoints.

Tips   This should be old hat by now. Notice that your subintervals will have to be bigger to use midpoints, and so the trapezoidal approximation is probably better in this case.

Remember that you don't have to use the trapezoid rule. You can figure the areas of the trapezoids directly.

Answer

a) $Avg = \dfrac{1}{b-a} \displaystyle\int_a^b f(x)\,dx$

$= \frac{1}{12}(2\frac{70+71}{2} + 2\frac{71+75}{2} + 2\frac{75+80}{2} + 2\frac{80+82}{2} + 2\frac{82+79}{2} + 2\frac{79+74}{2}) = \frac{153}{2} = 76.5$.

B. $Avg = \dfrac{1}{b-a} \displaystyle\int_a^b f(x)\,dx$

$= \frac{1}{12}(\Delta x(71 + 80 + 79)) = \frac{1}{12}(4(71 + 80 + 79)) = 76.667$.

Notice that using midpoints forced you to use subintervals with a width of 4 hours.

►Objective 6   Given an area and a function, find the correct limits of integration for a definite integral to yield that given area.

Example   The area bounded by the curve $y = b - x^2$ and the x-axis is 288. What is the value of b?

Tip   In this problem use symmetry to change $\displaystyle\int_{-\sqrt{b}}^{\sqrt{b}} (b - x^2)\,dx = 2\int_0^{\sqrt{b}} (b - x^2)\,dx$, which cuts down considerably on the algebra.

Answer Since the curve $y = b - x^2$ intersects the x-axis at $x = \pm\sqrt{b}$, half of the area is

$$\int_0^{\sqrt{b}} (b - x^2)\, dx = \frac{288}{2} = 144$$

$$\left[bx - \tfrac{1}{3}x^3 \right]_0^{\sqrt{b}} = 144$$

$$(b(\sqrt{b}) - \tfrac{1}{3}(\sqrt{b})^3) - (0 - 0) = \tfrac{2}{3}(\sqrt{b})^3 = 144$$

$(\sqrt{b})^3 = 216$, so the solution is $b = 36$.

Multiple-Choice Questions on Area

1. Which of the following statements about the region bounded by the curves $y = |x|$ and $y = 2 - x^2$ are true?

A. The region is symmetrical about the y-axis.

B. The area of the region is $\displaystyle\int_{-1}^{1} (2 - x^2 - |x|)\, dx$.

C. The area of the region is $\displaystyle 2\int_{0}^{1} (2 - x^2 - x)\, dx$.

D. The area of the region is $\frac{7}{3}$.

E. All of these.

Answer E. Make sure you understand why each statement is true:

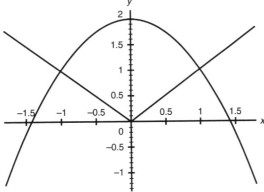

A. The region is symmetrical about the y-axis, since both $y = |x|$ and $y = 2 - x^2$ are **even**.

B. $A = \displaystyle\int_a^b L(x)\, dx = \int_{-1}^{1} (2 - x^2 - |x|)\, dx$, since $2 - x^2 \geq |x|$ on the interval $[-1, 1]$.

C. If you only think about the area to the right of the y-axis, you can remove the absolute value sign, $L(x) = 2 - x^2 - x$, on the interval $[0, 1]$. You can then use the symmetry to show that $A = \displaystyle 2\int_0^1 (2 - x^2 - x)\, dx$.

D. Evaluating: $2\int_0^1 (2 - x^2 - x)\,dx = 2[2x - \frac{1}{3}x^3 - \frac{1}{2}x^2]_0^1 = \frac{7}{3}$.

---

2. The area bounded by $x = 2y^2 - 5$ and $x = y^2 + 4$ is:

A. 48

B. 36

C. 24

D. 18

E. 9

Answer   B. Find the intersections: $2y^2 - 5 = y^2 + 4$, so $y^2 = 9$; intersection at $y = -3, 3$. Then find the length of each differential rectangular element: $L(y) = (y^2 + 4) - (2y^2 - 5) = 9 - y^2$, since $y^2 + 4 \geq 2y^2 - 5$ when $-3 \leq y \leq 3$.

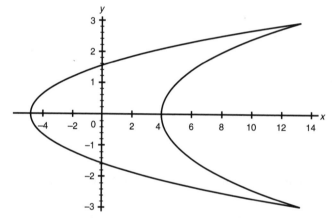

$$A = \int_{-3}^{3} ((y^2 + 4) - (2y^2 - 5))\,dy = \int_{-3}^{3} (9 - y^2)\,dy$$

$$= \left[9y - \frac{1}{3}y^3\right]_{-3}^{3} = 36.$$

---

3. The area bounded by $y = x$, $y = -\dfrac{x}{2}$, and $y = 5$ is equal to:

I. $\displaystyle\int_{-10}^{0} (5 + \frac{x}{2})\,dx + \int_0^5 (5 - x)\,dx$

II. $\displaystyle\int_{-10}^{0} (5 - \frac{x}{2})\,dx + \int_0^5 (5 - x)\,dx$

III. $\displaystyle\int_0^5 (y+2y)\,dy$

A. I only

B. II only

C. III only

D. I and III

E. II and III

Answer   D. The area can be cut vertically to give $A = \displaystyle\int_{-10}^0 \left(5+\frac{x}{2}\right)dx + \int_0^5 (5-x)\,dx = 25 + \frac{25}{2} = \frac{75}{2}$, or horizontally to give $A = \displaystyle\int_0^5 (y+2y)\,dy = \frac{75}{2}$.

---

4.   The area between the curve $y = \dfrac{x^2}{2} + 2$ and $y = -x - 3$ on the interval $-4 \le x \le 4$ is:

A. $\dfrac{40}{3}$

B. 16

C. 24

D. $\dfrac{184}{3}$

E. $\dfrac{212}{6}$

Answer   D. The graph below depicts $y = f(x)$, as the top line, and $y = g(x)$, as the bottom line.

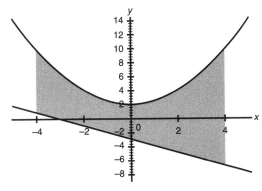

The function $f(x)$ is greater than the function $g(x)$ on the interval of concern, so the length of the rectangles that you're using to sum up the area is given by $L(x) = f(x) - g(x)$.

So the area between the curves is given by

$$Area = \int_{-4}^{4} L(x)\,dx = \int_{-4}^{4} (f(x) - g(x))\,dx$$
$$= \int_{-4}^{4} \left(\frac{x^2}{2} + 2 - (-x - 3)\right) dx = \int_{-4}^{4} \left(\frac{x^2}{2} + x + 5\right) dx.$$

So $Area = \left[\frac{1}{6}x^3 + \frac{1}{2}x^2 + 5x\right]_{-4}^{4}$
$$= \frac{1}{6}(4)^3 + \frac{1}{2}(4)^2 + 5(4) - \left(\frac{1}{6}(-4)^3 + \frac{1}{2}(-4)^2 + 5(-4)\right)$$
$$= \frac{64}{6} + 8 + 20 - \left(\frac{-64}{6} + 8 - 20\right) = \frac{116}{3} - \left(-\frac{68}{3}\right) = \frac{184}{3}.$$

---

5. Find the area between the curves $y = f(x) = x^3 - 3x$ and $y = g(x) = x\sqrt{4 - x^2} + x$ on the interval $0 \le x \le 2$.

A. $\dfrac{8}{3}$

B. $\dfrac{20}{3}$

C. $\dfrac{4}{3}$

D. $\dfrac{44}{3}$

E. $\dfrac{14}{3}$

Answer   B. The graph below depicts the two curves, $y = f(x)$ and $y = g(x)$.

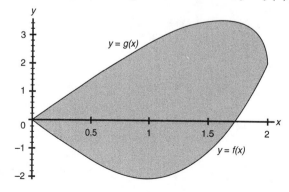

The expression that represents the area between the two curves from $x = 0$ to $x = 2$ is $Area = \int_0^2 (g(x) - f(x))\, dx$. Using this expression, you get

$$Area = \int_0^2 (x\sqrt{4 - x^2} + x - x^3 + 3x)\, dx$$

$$= \int_0^2 (x\sqrt{4 - x^2} + 4x - x^3)\, dx$$

$$= \int_0^2 (4x - x^3)\, dx + \int_0^2 x\sqrt{4 - x^2}\, dx$$

$$= \left[2x^2 - \frac{1}{4}x^4\right]_0^2 + \left[\frac{-1}{3}(4 - x^2)^{\frac{3}{2}}\right]_0^2$$

$$= [4 - 0] + \left[0 - \frac{-1}{3}(8)\right] = 4 + \frac{8}{3} = \frac{20}{3}.$$

6. The area of the region bounded by the curves $y = f(x) = 2x^3 - 6x^2 - 2x + 6$ and $y = g(x) = -x^3 + 3x^2 + x - 3$ is given by:

A. $\displaystyle\int_{-1}^1 (3x^3 - 3x^2 - x + 3)\, dx - \int_1^3 (-3x^3 - 3x^2 - x + 3)\, dx$

B. $\displaystyle\int_{-1}^1 (-3x^3 + 9x^2 + 3x - 9)\, dx + \int_1^3 (3x^3 - 9x^2 - 3x + 9)\, dx$

C. $\displaystyle\int_{-1}^3 (3x^3 - 9x^2 - 3x + 9)\, dx$

D. $\displaystyle\int_{-1}^1 (3x^3 - 9x^2 - x + 3)\, dx - \int_1^3 (3x^3 - 9x^2 - x + 3)\, dx$

E. $\displaystyle\int_{-1}^1 (3x^3 - 9x^2 - 3x + 9)\, dx - \int_1^3 (3x^3 - 9x^2 - 3x + 9)\, dx$

Answer   E. The graph below depicts $y = f(x)$ and $y = g(x)$.

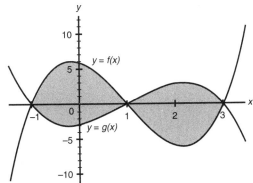

Cutting the area vertically into very narrow rectangles, with $L(x)$ representing the height of the rectangles and dx the width, you get that the area of the region bounded by the two curves is given by $A(x) = \int_{-1}^{3} L(x)\,dx$. Since $f(x) \geq g(x)$ on the interval $[-1, 1]$ and $g(x) \geq f(x)$ on the interval $[1, 3]$, the function $L(x)$ is given by

$$L(x) = \begin{cases} f(x) - g(x) & \text{if} \quad -1 \leq x < 1 \\ -(f(x) - g(x)) & \text{if} \quad 1 \leq x \leq 3. \end{cases}$$

So the area between the curves is given by

$$Area = \int_{-1}^{1} (f(x) - g(x))\,dx + \int_{1}^{3} -(f(x) - g(x))\,dx.$$

Then, since $f(x) - g(x) = 2x^3 - 6x^2 - 2x + 6 - (-x^3 + 3x^2 + x - 3) = 3x^3 - 9x^2 - 3x + 9$, you get

$$\begin{aligned} Area &= \int_{-1}^{1} (3x^3 - 9x^2 - 3x + 9)\,dx \\ &+ \int_{1}^{3} -(3x^3 - 9x^2 - 3x + 9)\,dx \\ &= \int_{-1}^{1} (3x^3 - 9x^2 - 3x + 9)\,dx \\ &- \int_{1}^{3} (3x^3 - 9x^2 - 3x + 9)\,dx. \end{aligned}$$

---

7. Find the area of the region bounded by the curves $y^2 = x$, $y - 4 = x$, $y = -2$, and $y = 1$. (Hint: Sketch this one on paper first.)

A. $\dfrac{27}{2}$

B. $\dfrac{22}{3}$

C. $\dfrac{33}{2}$

D. $\dfrac{34}{3}$

E. 14

Answer   C. The graph below shows all four curves:

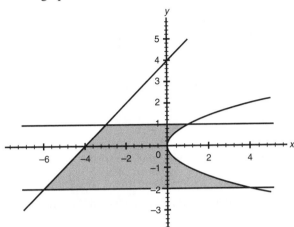

It appears that the easiest way to find the area of the region bounded by these four curves is to cut the area horizontally, evaluating an integral of the form $\int_{-2}^{1} L(y)\,dy$, where $L(y)$ represents the length (horizontal) of the rectangles and dy represents the width. The proper expression for the lengths of the rectangles is $L(y) = y^2 - (y-4) = y^2 - y + 4$, which on the graph above represents the width of the shaded region at any particular y value. Thus you have

$$Area = \int_{-2}^{1} L(y)\,dy = \int_{-2}^{1} (y^2 - y + 4)\,dy$$

$$= \left[\frac{1}{3}y^3 - \frac{1}{2}y^2 + 4y \right]_{-2}^{1}$$

$$= \left(\frac{1}{3} - \frac{1}{2} + 4 \right) - \left(\frac{-8}{3} - 2 - 8 \right) = \frac{33}{2}.$$

---

8. Find the area of the region bounded by $y = \sin x$, $y = \csc^2 x$, $x = \frac{\pi}{4}$, and $x = \frac{3\pi}{4}$.

A. $2 - \sqrt{2}$

B. $2 + \dfrac{\sqrt{2}}{2}$

C. $-2 + \sqrt{2}$

D. $2 - \dfrac{\sqrt{2}}{2}$

E. $2 + \sqrt{2}$

Answer A. Below is a graph of the four curves.

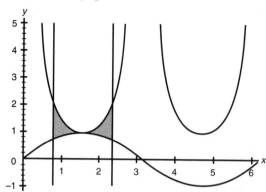

From this, it appears that the area of the region bounded by the four curves is best expressed as $Area = \int_{\frac{\pi}{4}}^{\frac{3\pi}{4}} L(x)\,dx$, where $L(x)$ is the distance between $\csc^2 x$ and $\sin x$.

That is, $Area = \int_{\frac{\pi}{4}}^{\frac{3\pi}{4}} (\csc^2 x - \sin x)\,dx$.

Thus, $Area = \left[-\cot x + \cos x\right]_{\frac{\pi}{4}}^{\frac{3\pi}{4}} = \left(1 - \frac{\sqrt{2}}{2}\right) - \left(-1 + \frac{\sqrt{2}}{2}\right) = 2 - \sqrt{2}.$

---

9. Find the area of the region bounded by the curves $y = x^{-\frac{1}{2}}$, $y = x^{-2}$, $y = 1$, and $y = 3$.

A. $\dfrac{1}{2}\sqrt{3} + \dfrac{4}{3}$

B. $2\sqrt{3} - \dfrac{8}{3}$

C. $\dfrac{1}{2}\sqrt{3} - \dfrac{32}{3}$

D. $2\sqrt{3} - \dfrac{32}{3}$

E. $\dfrac{8}{3} - 2\sqrt{3}$

Answer B. A graph of the four curves involved follows.

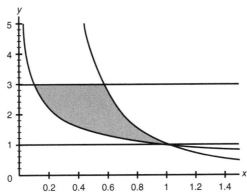

The best way to find the area of the region bounded by these curves is to cut the region horizontally and thus to evaluate an expression of the form $\int_1^3 L(y)\,dy$, where $L(y)$ represents the length (horizontal) of the rectangles to be summed. Since the length L of the rectangles is a function of y, you must rewrite the curves $y = x^{-\frac{1}{2}}$ and $y = x^{-2}$ as functions of y.

You get $y = x^{-\frac{1}{2}} \Rightarrow x = y^{-2}$ and $y = x^{-2} \Rightarrow x = y^{-\frac{1}{2}}$, so $L(y) = y^{-\frac{1}{2}} - y^{-2}$, and

$$
\begin{aligned}
Area &= \int_1^3 (y^{-\frac{1}{2}} - y^{-2})\,dy \\
&= \left[2y^{\frac{1}{2}} + y^{-1} \right]_1^3 \\
&= 2\sqrt{3} + \frac{1}{3} - 2 - 1 \\
&= 2\sqrt{3} - \frac{8}{3}.
\end{aligned}
$$

---

10. Find the area of the region IN THE FIRST QUADRANT (upper right quadrant) bounded by the curves $y = \sin x \cos^2 x$, $y = 2x \cos(x^2)$, and $y = 4 - 4x$.

A. 1.8467

B. 0.16165

C. 0.36974

D. 1.7281

E. 0.37859

Answer   E. Below is the graph of the three curves in the first quadrant with the region shaded:

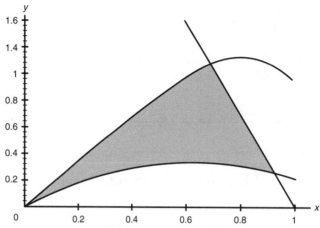

The easiest way to compute the area of the region is by cutting the region vertically, thereby coming up with a definite integral for area of the form $\int_0^b L(x)\, dx$.

The first thing to do is find where the curve $y = 4 - 4x$ intersects the other two curves. Using your calculator, you should find that the curve $y = 4 - 4x$ intersects the curve $y = \sin x \cos^2 x$ at approximately $x = .928113$. Also using your calculator, you should find that the curve $y = 4 - 4x$ intersects the curve $y = 2x \cos(x^2)$ at approximately $x = .692751$.

So you have now that $Area = \int_0^{.928113} L(x)\, dx$, where

$$L(x) = \begin{cases} 2x \cos(x^2) - \sin x \cos^2 x & \text{if} \quad 0 \le x < .692751 \\ 4 - 4x - \sin x \cos^2 x & \text{if} \quad .692751 \le .928113. \end{cases}$$

In other words,

$$Area = \int_0^{.692751} \left(2x \cos(x^2) - \sin x \cos^2 x\right) dx$$
$$+ \int_{.692751}^{.928113} \left(4 - 4x - \sin x \cos^2 x\right) dx = .37859$$

(using your calculator to evaluate).

_____

11.   The region R is the region in the first quadrant bounded by the curves $y = x^2 - 4x + 4$, $x = 0$, and $y = 0$, as pictured below:

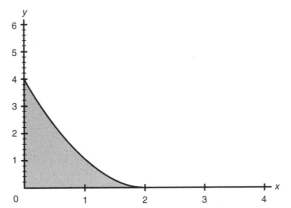

If the vertical line $x = h$ divides the region R into two regions of equal area, then $h =$

A. 0.39987

B. 0.41260

C. 0.42739

D. 0.44123

E. 0.43120

Answer B. The area of the region R is given by $\displaystyle\int_0^2 (x^2 - 4x + 4)\,dx$, while the areas of the regions

that the vertical line $x = h$ creates are $\displaystyle\int_0^h (x^2 - 4x + 4)\,dx$ (left part of the region) and

$\displaystyle\int_h^2 (x^2 - 4x + 4)\,dx$ (right part of the region). You can solve the problem two ways. The
first way is to find the area of the region R, and then solve for the value of h that makes
either of the other two areas (left part or right part) equal to half that.

The second way to solve the problem is to set the two areas (left part and right part) equal
to each other which is what we'll do:

$$\int_0^h (x^2 - 4x + 4)\,dx = \int_h^2 (x^2 - 4x + 4)\,dx$$

Solving:

$$\int_0^h (x^2 - 4x + 4)\,dx = \int_h^2 (x^2 - 4x + 4)\,dx$$

$$\Rightarrow \left[\frac{1}{3}x^3 - 2x^2 + 4x\right]_0^h = \left[\frac{1}{3}x^3 - 2x^2 + 4x\right]_h^2$$

$$\Rightarrow \frac{1}{3}h^3 - 2h^2 + 4h = \frac{1}{3}2^3 - 2(2)^2 + 4(2)$$

$$-\left(\frac{1}{3}h^3 - 2h^2 + 4h\right) \Rightarrow \frac{2}{3}h^3 - 4h^2 + 8h = \frac{8}{3}$$

$\Rightarrow h = 0.41260$ (to five decimal places, using graphing calculator).

So the line $x = 0.41260$ divides the region R into two regions of equal area.

12. Which of these would correctly give the area of the shaded region if the two unshaded regions are squares and the region consisting of the two unshaded regions and the shaded region is a square?

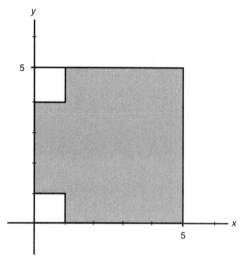

I. $\displaystyle\int_0^1 3\,dx + \int_1^5 5\,dx$

II. $\displaystyle\int_0^1 4dy + \int_1^4 5dy + \int_4^5 4dy$

III. $5^2 - 1^2 - 1^2$

A. I and II

B. III

C. I, II, and III

D. I and III

E. None of these.

Answer   C. The first integral ($\displaystyle\int_0^1 3\,dx + \int_1^5 5\,dx$) is how you'd set it up to integrate with respect to x, the second integral ($\displaystyle\int_0^1 4\,dy + \int_1^4 5dy + \int_4^5 4\,dy$) is how you'd set it up to integrate with respect to y, and the string of numbers ($5^2 - 1^2 - 1^2$) is how you'd figure the area as just squares.

13. Find the average value of $y = 1 - \sqrt{1 - x^2}$ for $-1 \leq x \leq 1$.

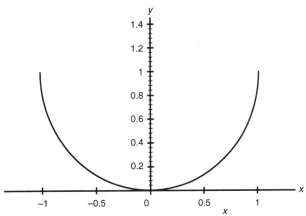

A. $\dfrac{2 - \pi}{2}$

B. $\dfrac{2 - \pi}{4}$

C. $\dfrac{4 - \pi}{4}$

D. $\dfrac{1 - \pi}{2}$

E. None of the above.

Answer   C. The average value here is

$$\frac{1}{1 - (-1)} \int_{-1}^{1} \left(1 - \sqrt{1 - x^2}\right) dx,$$

which is just $\frac{1}{2}$ times the area under the curve. You can calculate the area directly (it's the area of the rectangle minus the area of the semicircle) as $2 - \dfrac{\pi}{2}$. So the average value is

$\dfrac{1}{2}\left(2 - \dfrac{\pi}{2}\right) = \dfrac{4 - \pi}{4}$.

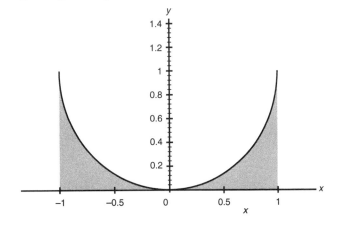

14. Find b so that $\int_0^b x^3 \, dx = 100$.

   A. 10

   B. 5.774

   C. 4.642

   D. 3.162

   E. None of these is correct

Answer E. Integrate to get $\dfrac{x^4}{4}\Big|_0^b = \dfrac{b^4}{4}$, then solve $\dfrac{b^4}{4} = 100$ for b. The correct answer is ≈ 4.472.

---

15. Find the number a so that $\int_2^5 x^2 \, dx$ is the same as $\int_2^5 a \, dx$.

   A. 13

   B. 8.33

   C. 19.5

   D. 39

   E. 117

Answer A. $\int_2^5 x^2 \, dx = 39$, and the length of this interval is 3, so $\int_2^5 13 \, dx = 39$ also.
(This is just another average value problem.)

B. Volume

▶**Objective 1**   Use the definite integral to find volumes by accumulating cross-sectional area.

Example   A solid is formed by revolving about the x-axis the region bounded by the x-axis and the curve $y = \sqrt{\sin x}$ for $0 \le x \le \pi$.

   A. Write an expression that represents the area of a cross section perpendicular to the x-axis.

   B. Find the volume of the solid.

    C.   What does this solid look like?

Answer

    A. $A(x) = \pi r^2 = \pi(\sqrt{\sin x})^2 = \pi \sin x.$

    B. $V = \displaystyle\int_0^\pi \pi \sin x \, dx = \left[-(\cos x)\pi \right]_0^\pi = 2\pi.$

    C.   A rugby ball, maybe?

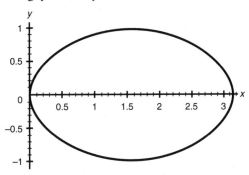

▶**Objective 2**   Use the definite integral to find the volume of a solid of revolution.

Example   A region is bounded by the y-axis, $y = 5$, and the curve $y = \begin{cases} x & \text{if } 0 \le x < 1 \\ \frac{1}{2}x^2 + \frac{1}{2} & \text{if } x \ge 1. \end{cases}$

A solid is formed by revolving the region about the y-axis. Determine the volume of the solid formed.

Tip   Notice that the first integral is just a cone with base π and height 1.

Answer

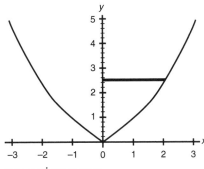

$y = x$ gives $x = y$.

$y = \frac{1}{2}x^2 + \frac{1}{2}$, gives $x = \sqrt{(2y - 1)}$.

Because of the piecewise function, it will help to divide this into two integrals.

From $y = 0$ to $y = 1$ you have $A(y) = \pi r^2 = \pi y^2$, and from $y = 1$ to $y = 5$ you have

$$A(y) = \pi r^2 = \pi(\sqrt{(2y-1)})^2 = \pi(2y-1).$$

$$V = \int_a^b A(y)dy = \int_0^1 \pi y^2 dy + \int_1^5 \pi(2y-1)dy.$$

$$V = \frac{1}{3}\pi + 20\pi = \frac{61}{3}\pi.$$

▶Objective 3   Calculate volumes of solids of revolution created by rotating curves about lines that are not the x-axis or the y-axis.

Example   Find the volume of the solid formed when the region enclosed by $x^2 + y^2 = 4$ is revolved about the line $x = 3$.

Tip   Draw a rough picture and then determine the area of an arbitrary slice.

Answer   First realize that this doughnut will need to be cut horizontally.

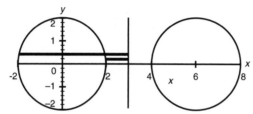

This gives washers with an outer radius of $R = 3 + \sqrt{4 - y^2}$ and an inner radius of $r = 3 - \sqrt{4 - y^2}$.

The area of one arbitrary slice will be given by

$$A(y) = \pi R^2 - \pi r^2 = \pi(3 + \sqrt{4 - y^2})^2 - \pi(3 - \sqrt{4 - y^2})^2 = 12\pi\sqrt{4 - y^2}.$$

Using the calculator:

$$\int_{-2}^2 12\pi\sqrt{4 - y^2}\,dy \approx 236.871.$$

▶Objective 4   Calculate volumes of solids that are created with well-defined bases and cross-sectional shapes.

Example   Let R be the region in the first quadrant under the graph of $y = \frac{1}{x}$ for $4 \le x \le 9$. Find the volume of the solid whose base is the region R and whose cross sections (cut by planes perpendicular to the x-axis) are squares.

Answer

$$A(x) = \left(\frac{1}{x}\right)^2 = \frac{1}{x^2}.$$

$$V = \int_a^b A(x)dx = \int_4^9 \frac{1}{x^2}dx = \left[-\frac{1}{x}\right]_4^9 = -\left(\frac{1}{9} - \frac{1}{4}\right) = \frac{5}{36}.$$

▶Objective 5 Given the shape of a solid (described with a function or a set of functions), determine the limits of integration needed to create a specific volume.

Example Let R be the region in the first quadrant under the graph of $y = \frac{1}{x}$ for $k \le x \le 9$.

The volume of the solid whose base is the region R and whose cross sections cut by planes perpendicular to the x-axis are squares is $\frac{1}{2}$. What is the value of k?

Answer

$$A(x) = \left(\frac{1}{x}\right)^2 = \frac{1}{x^2}.$$

$$V = \int_a^b A(x)dx = \int_k^9 \frac{1}{x^2}dx = \left[-\frac{1}{x}\right]_k^9 = -\left(\frac{1}{9} - \frac{1}{k}\right) = \frac{1}{k} - \frac{1}{9}.$$

Setting this equal to the known volume gives

$$\frac{1}{2} = \frac{1}{k} - \frac{1}{9}$$
$$\frac{11}{18} = \frac{1}{k}$$
$$k = \frac{18}{11}.$$

Multiple-Choice Questions on Volumes

1. The region bounded by the x-axis, the y-axis, and the portion of the curve $y = 4 - x^2$ in the first quadrant is revolved around the y-axis. Which of these integrals correctly gives the volume of this solid of revolution?

A. $\int_0^2 \pi(4 - x^2)^2\, dx$

B. $\int_0^4 \pi(4 - y)\, dy$

C. $\displaystyle\int_0^4 \pi(4-y^2)^2\,dy$

D. $\displaystyle\int_0^2 (4-x^2)^2\,dx$

E. $\displaystyle\int_{-2}^2 \pi(4-x^2)^2\,dx$

Answer B. Since you're revolving around the y-axis, you want to make horizontal slices.

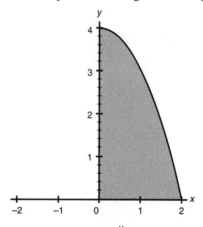

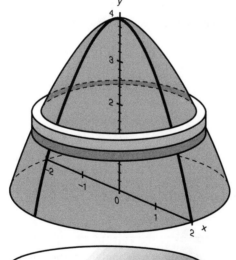

The cross section is a circle, and its radius is actually x. Because the curve is given in the form $y = f(x)$, you have to do a little algebra to find x in terms of y.

The radius is $x = \sqrt{4-y}$. So the area of your cross section is $\pi(\sqrt{4-y})^2 = \pi(4-y)$, and the integral is $\displaystyle\int_0^4 \pi(4-y)\,dy$.

2. The curve $y = \sqrt{1-x^2}$ between $x = -1$ and $x = 1$ is revolved around the x-axis. Which of these gives the correct volume of the resulting solid?

I. $\displaystyle\int_{-1}^{1} \sqrt{1-x^2}\, dx$

II. $\displaystyle\int_{-1}^{1} (\sqrt{1-x^2})^2\, dx$

III. $\dfrac{4\pi}{3}$

A. I only

B. II only

C. III only

D. I and II

E. None of these is correct.

Answer C. Neither integral is correct. But this curve is a semicircle. When we revolve it around the x-axis, we end up with a sphere of radius 1, and we can figure its volume the old-fashioned way: $V = \frac{4}{3}\pi r^3$, so the volume is $\frac{4\pi}{3}$. (By the way, the correct integral for this volume would be $\displaystyle\int_{-1}^{1} \pi(\sqrt{1-x^2})^2\, dx$.)

3. The region bounded by $y = \sin x$, $y = \cos x$, $x = 0$, and $x = \frac{\pi}{4}$ is revolved around the x-axis. Which of these integrals correctly gives the volume of this solid of revolution?

A. $\displaystyle\int_{0}^{\pi/4} \pi(\cos^2 x - \sin^2 x)\, dx$

B. $\displaystyle\int_{0}^{1} \pi(\cos^2 x - \sin^2 x)\, dx$

C. $\displaystyle\int_{0}^{\pi/4} \pi(\sin^2 x - \cos^2 x)\, dx$

D. $\displaystyle\int_{0}^{1} \pi(\sin^2 x + \cos^2 x)\, dx$

E. $\displaystyle\int_{0}^{\pi/4} \pi(\cos x - \sin x)^2\, dx$

Answer A. Here's the picture of the solid:

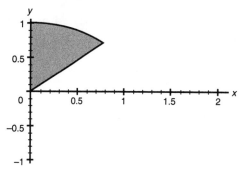

Note that for small values of x, $\sin x$ is approximately equal to x.

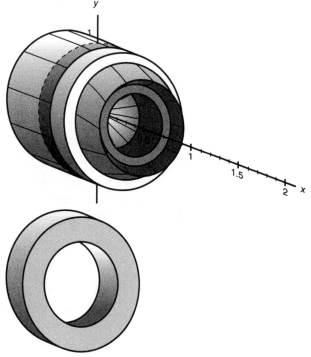

The cross sectional area is $\pi R^2 - \pi r^2 = \pi(\cos^2 x - \sin^2 x)$ for x-values between 0 and $\frac{\pi}{4}$.

So the integral that gives the volume is $\displaystyle\int_0^{\pi/4} \pi(\cos^2 x - \sin^2 x)\,dx$.

---

4. A region R is bounded by the curves $y = \sqrt{x}$ and $y = x^3$. The solid formed by revolving region R around the y-axis has a volume of:

I. $\displaystyle\pi \int_0^1 ((y^3)^2 - (\sqrt{y})^2)\,dy$

II. $\displaystyle\pi \int_0^1 ((x^3)^2 - (\sqrt{x})^2)\,dx$

III. $\pi \displaystyle\int_0^1 (y^{\frac{2}{3}} - y^4)dy$

A. I only

B. II only

C. III only

D. I and II only

E. I and III only

Answer C. Our slice is a washer with outer radius $y^{1/3}$, inner radius y^2, and width dy. Thus, the volume

$$V = \int_0^1 A(y)dy = \int_0^1 \left(\pi(y^{1/3})^2 - \pi(y^2)^2\right) dy$$

$$= \pi \int_0^1 \left(y^{2/3} - y^4\right) dy.$$

---

5. The base of a solid is the region in the second quadrant enclosed by the graph of $y = x^3 + 7x^2$ and the x-axis. If every cross section perpendicular to the x-axis is a square, then the volume of the solid is:

A. 24,640.349

B. 7,843.267

C. 200.083

D. -200.083

E. $-7,843.266$

Answer B.

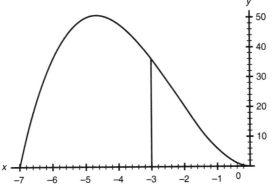

The area of each cross section is $A(x) = (x^3 + 7x^2)^2$.

The volume of one slice is $dV = (x^3 + 7x^2)^2 \, dx$.

The total volume of the solid is $V = \displaystyle\int_{-7}^{0} (x^3 + 7x^2)^2 \, dx \approx 7843.267$.

---

6. Which of the following solids have a volume of $V = \displaystyle\int_{0}^{b} x^2 \, dx$?

   I. A solid whose base is a region bounded by $y = -\frac{x}{2}$; $y = \frac{x}{2}$ and $x = b$, and whose cross sections perpendicular to the x-axis are squares

   II. A four-sided pyramid whose base is a square with area b^2 and whose height is b

   III. A solid formed when the region bounded by the x-axis, $y = \frac{x}{2}$, and $x = b$ is revolved about the x-axis

   A. I only

   B. II only

   C. II and III only

   D. I and II only

   E. I, II, and III

Answer
D. If you think of the slices like a stack of cards, you should see that I is just II with the cards forming the volume shifted over to one side. In both cases the cross-sectional area, $A(x) = x^2$, leads to a volume of $V = \displaystyle\int_{0}^{b} x^2 \, dx = \frac{b^3}{3}$.

The solid formed in III will have cross-sectional areas of $A(x) = \pi \left(\frac{x}{2}\right)^2$, which will lead to a volume of $V = \frac{\pi}{4} \displaystyle\int_{0}^{b} x^2 \, dx = \frac{1}{12}\pi b^3$.

---

7. The base of a solid is bounded by $y = x^2$ and the line $y = d$. Every cross section of the solid perpendicular to the y-axis is an isosceles triangle with a height that is three times its base (with the short leg lying in the xy plane). If the volume of the solid is 75, what is the value of d?

   A. 15

   B. 9

   C. 8.660

   D. 5

   E. 2.5

Answer   D. For each triangle:

$$b = 2x = 2\sqrt{y}$$
$$h = 3b = 6\sqrt{y}$$
$$A(y) = \frac{1}{2}bh = \frac{1}{2}(2\sqrt{y})(6\sqrt{y}) = 6y$$
$$V = \int_0^d 6y\,dy = \left[3y^2\right]_0^d = 3d^2$$
$$3d^2 = 75$$
$$d = 5.$$

C.   Other Applications of the Definite Integral

▶**Objective 1**   Calculate net and total changes from rates of change presented numerically, analytically, or graphically.

Example 1   The rate at which water is dripping into a tub of water is given by $r(t) = \dfrac{t+2}{t+1}$ (gal/hour). Find how much water entered the tub from $t = 1$ hour to $t = 3$ hours.

Tip   This is really just the Fundamental Theorem of Calculus: $f(b) - f(a) = \displaystyle\int_a^b f'(t)\,dt.$

Answer

$$\int_1^3 \frac{t+2}{t+1}\,dt \approx 2.693 \text{ gallons}$$

Example 2   The birth rate for a population of animals is given by

$$b(t) = 30\cos(\frac{\pi t}{6}) + 45 \text{ (births/year)},$$

and the death rate is given by

$$d(t) = 45 - 15\cos(\frac{\pi t}{3}) \text{ (deaths/year)}.$$

A.   About how many total births occurred in the years $t = 0$ to $t = 6$?

B.   What was the net change in the population from in the years $t = 0$ to $t = 10$?

Tip   In both of these examples, the rates of change are presented algebraically (with a formula). You may also need to work from a table of data or from a graph.

Answer

A. Total births $= \int_0^6 b(t)\, dt = \int_0^6 (30\cos(\frac{\pi t}{6}) + 45)\, dt = 270.$

Note: You need to round to a whole number.

B. The rate of change in the population is $r(t) = b(t) - d(t)$.

Net change $= \int_0^{10} b(t) - d(t)\, dt = \int_0^{10} (30\cos(\frac{\pi t}{6}) + 15\cos(\frac{\pi t}{3}))\, dt \approx -62,$

which means that the population is 62 members lower at year 10 than at year 0.

▶Objective 2   Use definite integrals to solve problems in applications where a quantity accumulates or is averaged.

Example 1   An economy car gets mileage (in miles per gallon) $m(v)$ approximated by $m = .5v + 20$ (for speeds under 60 mph). The car has a velocity given by $v = 14 + 24t - 4t^2$ for $t = 0$ to $t = 4$ hours. Set up an integral (in terms of t) that gives the amount of gas used during those four hours.

Tip   You should think of a small interval because then you can treat the varying quantities as approximately constant for that small interval of time.

Answer   If we think of a very small segment of time dt, we need to ask ourselves how much gas will be used during that interval.

Gas used $= \dfrac{\text{distance traveled}}{\text{mileage}}$. Notice the units $\dfrac{\text{miles}}{\text{miles/gallon}} = $ gallons.

So, the gas used during the interval dt is $\dfrac{\text{distance traveled}}{\text{mileage}} = \dfrac{v\, dt}{.5v + 20}$.

Thus, the total gas used over the entire time interval from $t = 0$ to $t = 4$ hours is

$$\int_0^4 \frac{v\, dt}{.5v + 20} = \int_0^4 \frac{14 + 24t - 4t^2}{.5(14 + 24t - 4t^2) + 20}\, dt,$$

which is about 3.954 gallons, from a calculator.

Example 2   The temperature outside a house on a given day is

Time	6 AM	9 AM	Noon	3 PM	6 PM	9 PM	Midnight
Temp	25°F	45°F	60°F	65°F	60°F	47°F	38°F

A. Estimate (using trapezoids) the average temperature outside the house from 6 AM to midnight.

B. Heating the house costs 10 cents per hour for each degree the house is below 60°F.

Estimate (using trapezoids) the total cost of heating the house from 6 AM to midnight.

Tip Note that you can't use the average in part A to help you do part b because any temp $\geq 60°F$ leads to a cost equal to zero.

Answer

A. $Avg = \frac{1}{18}(3\frac{25+45}{2} + 3\frac{45+60}{2} + 3\frac{60+65}{2} + 3\frac{65+60}{2} + 3\frac{60+47}{2} + 3\frac{47+38}{2}) = \frac{617}{12} \approx 51.417$

B.

Time	6 AM	9 AM	*Noon*	3 PM	6 PM	9 PM	*Midnight*
Temp	25°F	45°F	60°F	65°F	60°F	47°F	38°F
°F below 60°F	35	15	0	0	0	13	22

$$\textbf{Total cost} = \int_{6am}^{Midnight} Cost(t)\, dt \approx (10)(3\frac{35+15}{2} + 3\frac{15+0}{2} + 3\frac{0+13}{2} + 3\frac{13+22}{2})$$
$$= 1,695 \text{ cents or } \$16.95$$

Free-Response Questions on Applications

---

1. Using your calculator, estimate the **total** distance traveled from $t = 0$ to $t = 2.5$ sec by an object whose velocity is given by

$$v(t) = \cos(e^t)\left(\frac{cm}{sec}\right).$$

---

2. Consider a tidal flat that's connected to the ocean by a narrow canal. As the ocean rises and falls with the local tides, the amount of water flowing through the canal during a 12-hour period is given by:

Time of Day	12 M	2 AM	4 AM	
Flow (gal/sec)	4	20	14	

Time of Day	6 AM	8 AM	10 AM	12 N
Flow (gal/sec)	0	−15	−16	0

Positive water flow is taken to be into the tidal flat, and negative values represent flow out of the tidal flat. Estimate the total amount of water that moved through the canal (irrespective of direction) in the 12-hour period, using trapezoids.

---

3. The velocity function of an object moving in a straight line is given by $v(t) = t\cos(\pi t^2)$, where v is in meters per second and t is in seconds.

A. What is the net distance traveled by the object on the interval $0 \leq t \leq 4$?

(Do **not** use a calculator here.)

B. What is the total distance traveled by the object on the interval $0 \le t \le 4$? (You may use a calculator here.)

C. What is the average speed of the object on the interval $0 \le t \le 4$? What is the average velocity of the object on that interval?

D. Find, algebraically, the first three times after time $t = 0$ that the object returns again to its initial position at time 0.

---

4. The acceleration function (in meters per second) and initial velocity are given for an object moving along a straight line: $a(t) = 4t - 1$, $v(0) = -6$.

A. What is the total distance traveled by the object in the first 5 seconds? (Do **not** use a calculator except to simplify your answer at the end of the problem).

B. What is the maximum speed the object attains within the first 5 seconds? Justify your answer, and note that graphical justification is **not** sufficient.

C. What is the minimum velocity that the object attains within the first 5 seconds?

D. What is the average acceleration of the object over the first 5 seconds?

---

5. A rocket propulsion system is designed so that its overall vertical acceleration at t seconds after launch will be given by $a(t) = t^K$ meters per second per second, where K is some positive constant determined before launch by adjustments made by well-trained engineers.

A. Suppose that the rocket is required to have a vertical velocity of 200 meters per second at precisely 10 seconds after launch. Set up an equation that the engineers must solve to determine the constant K that should be used when setting up the propulsion system. Then solve for the constant K by simplifying your equation as far as you can and then using your graphing calculator. (Assume that the rocket's vertical velocity at launch is 0 meters per second).

B. Suppose instead that the engineers set $K = 2.0$. Set up an equation that could be solved to find the time T (seconds after launch) at which the vertical velocity of the rocket will be 400 meters per second. Solve your equation for T exactly. (Again assume that the rocket's vertical velocity at launch is 0 meters per second.)

---

6. Show that the average velocity of an object moving in a straight line over a time interval $[a, b]$ is equal to the average value of its velocity function on that interval.

Answers

---

1. The correct answer is 1.394 cm. Use the absolute value function on your calculator to approximate $\int_0^{2.5} |\cos(e^t)| \, dt \approx 1.394$.

2. The correct answer is 482,400 gal. The estimate is

$$\int_a^b |r(t)|dt \approx \frac{b-a}{2n}(|r(a)| + 2|r(t_1)| + 2|r(t_2)| + \cdots + 2|r(t_{n-1})| + |r(b)|)$$

$$n = 6$$

$$a = 0 \text{ (midnight)}$$

$$b = (12)(60)(60) = 43,200 \text{ seconds (noon)}$$

$$\int_0^{43200} |r(t)|dt$$

$$\approx \frac{43,200}{2(6)}(4 + 2(20) + 2(14) + 2(0) + 2(15) + 2(16) + 0)$$

$$= 482,400 \text{ gallons.}$$

(You may have used actual trapezoids instead of the trapezoidal rule.)

3.

A. The net distance traveled by the object on the interval $0 \le t \le 4$ is 0 meters.

Work: The net distance traveled by the object on the interval $0 \le t \le 4$ is

$$\int_0^4 t\cos(\pi t^2)dt = \left[\frac{1}{2\pi}\sin(\pi t^2)\right]_0^4 = \frac{1}{2\pi}(\sin(16\pi) - \sin(0)) = 0.$$

B. The total distance traveled by the object on the interval $0 \le t \le 4$ is $\int_0^4 |t\cos(\pi t^2)|\,dt \approx$ 5.093 meters.

C. The average speed is total distance traveled divided by time:

Average speed $\approx \dfrac{5.093}{4} = 1.27325$ meters per second.

The average velocity is net distance traveled divided by time:

Average Velocity $= \dfrac{0}{4} = 0$ meters per second.

D. The first three times after time $t = 0$ that the object returns again to its initial position are: $T = 1$, $T = \sqrt{2}$, and $T = \sqrt{3}$ seconds.

Work: You need to solve the following for T:

$$\int_0^T t\cos(\pi t^2)dt = 0$$

$$\Rightarrow \left[\frac{1}{2\pi}\sin(\pi t^2)\right]_0^T = 0$$

$$\frac{1}{2\pi}\sin(\pi T^2) = 0.$$

The first 3 times after $x = 0$ that $\sin(x)$ is 0 are at $x = \pi$, $x = 2\pi$, and $x = 3\pi$.

So the three values for T are:

$$\pi T^2 = \pi \Rightarrow T = 1 \text{ seconds}$$
$$\pi T^2 = 2\pi \Rightarrow T = \sqrt{2} \text{ seconds}$$
$$\pi T^2 = 3\pi \Rightarrow T = \sqrt{3} \text{ seconds}$$

4.

A. The total distance traveled by the object in the first 5 seconds is $\frac{349}{6} \approx 58.167$ meters.

First, you need to solve the differential equation to find the velocity.

Integrate $a(t) = 4t - 1$ to get $v(t) = 2t^2 - t + C$.

Then use the initial condition $v(0) = -6$ to solve for $C = -6$.

The velocity of the object is given by $v(t) = 2t^2 - t - 6$.

The total distance traveled by the object in the first 5 seconds is $\int_0^5 |2t^2 - t - 6| dt$.

Now, $2t^2 - t - 6 = 0$ has roots $t = 2$ and $-\frac{3}{2}$ (which can be obtained by factoring or quadratic formula). The function is negative for $-\frac{3}{2} < t < 2$ and positive for $t > 2$. So the total distance traveled by the object in the first 5 seconds is

$$\int_0^5 |2t^2 - t - 6| dt = -\int_0^2 (2t^2 - t - 6) dt + \int_2^5 (2t^2 - t - 6) dt$$
$$= -\left[\frac{2}{3}t^3 - \frac{1}{2}t^2 - 6t\right]_0^2 + \left[\frac{2}{3}t^3 - \frac{1}{2}t^2 - 6t\right]_2^5 = \frac{26}{3} + \frac{99}{2} = \frac{349}{6}$$
$$\approx 58.167 \text{ meters.}$$

B. The maximum speed attained within the first 5 seconds is 39 meters per second.

Justification: To get this algebraically, you must check relative extrema and endpoint values of velocity and determine which one has the greatest absolute value. Remember how to find relative extrema? Set the derivative equal to 0 to find the critical points (and remember also to pick up those points where the derivative is undefined).

Here, $v'(t) = a(t) = 0$ when $t = \frac{1}{4}$ and is never undefined.

Checking the values of v at the critical point and at the endpoints ($t = 0$ and $t = 5$), we find that $v(t)$ attains a relative minimum at $t = \frac{1}{4}$, at which time

$$v\left(\frac{1}{4}\right) = 2\left(\frac{1}{4}\right)^2 - \frac{1}{4} - 6 = -\frac{49}{8} = -6.125.$$

At $t = 0$, the velocity is -6, and at $t = 5$, the velocity is $2(5)^2 - (5) - 6 = 39$.

So the maximum speed on the interval $0 \le t \le 5$ is 39 meters per second.

C. The minimum velocity on the interval $0 \leq t \leq 5$ is -6.125 meters per second.

Method: We already did most of the work for this in part b.

$v(t)$ has a minimum at $t = \frac{1}{4}$. To see this, set the first derivative of velocity (acceleration) equal to 0 and solve.

At this time, $v(\frac{1}{4}) = 2(\frac{1}{4})^2 - (\frac{1}{4}) - 6 = -\frac{49}{8} = -6.125$. The one thing we didn't do yet is check that this was a relative minimum. To check, you'd use the first or second derivative test.

D. The average acceleration on the interval is 9 meters per second per second.

Just as average velocity is change in distance divided by elapsed time, average acceleration is change in velocity divided by elapsed time. Here, the velocity at $t = 0$ is -6 meters per second, and the velocity at $t = 5$ is 39 meters per second, so the average acceleration is
$$\frac{39 - (-6)}{5} = 9 \frac{\text{meters}}{\text{second}^2}$$
You could also figure the average value of the acceleration function:

$$\frac{1}{5} \int_0^5 a(t)\, dt = \frac{1}{5} \int_0^5 (4t - 1)dt = \frac{1}{5} \left[2t^2 - t \right]_0^5$$

$$= 9 \frac{\text{meters}}{\text{second}^2}.$$

5.

A. Equation is $\int_0^{10} t^K\, dt = 200$. Answer is $K \approx 1.738551$ because

$$\int_0^{10} t^K\, dt = 200$$

$$\Rightarrow \left[\frac{1}{K+1} t^{(K+1)} \right]_0^{10} = 200$$

$$\Rightarrow \frac{1}{K+1} 10^{(K+1)} = 200$$

$$\Rightarrow 10^{(K+1)} = 200(K+1)$$

You can use your calculator at either of the last two lines.

Using your calculator, you should get $K \approx 1.738551$.

B. Equation is $\int_0^T t^2\, dt = 400$. The answer is $T = \sqrt[3]{1,200}$ seconds after launch.

$$\int_0^T t^2\, dt = 400$$

$$\Rightarrow \left[\frac{1}{3} t^3 \right]_0^T = 400$$

$$\Rightarrow \frac{1}{3} T^3 = 400$$

$$\Rightarrow T^3 = 1,200$$

$$\Rightarrow T = \sqrt[3]{1,200} \approx 10.62659 \text{ seconds.}$$

6. The average velocity on the interval $[a, b]$ is given by $\dfrac{s(b) - s(a)}{b - a}$, where $s(t)$ represents the position function for the object.

The average value of the velocity function on the interval $[a, b]$ is

$$\frac{1}{b-a} \int_a^b v(t)\, dt = \frac{1}{b-a}(s(b) - s(a)) = \frac{s(b) - s(a)}{b - a}$$

So they are equal. (Which is good, since the language is the same.)

Chapter 9   *Inverse Functions and Transcendental Functions*

This chapter reviews the basic objectives of differential and integral calculus all over again, but with an expanded cast of characters—transcendental functions. Transcendental functions are so named for their descriptions *transcend* algebraic formulas.

Algebraic functions include polynomials (like $x^3 - 15x$), rational functions (like $\dfrac{x}{x^2 - 4}$) and those involving radicals (like $\sqrt[3]{5 - x}$).

Transcendental functions include trigonometric functions (like $\sin x$ and $\cos x$), inverse trigonometric functions (like $\arctan x$), exponentials (like 2^x), and logarithms (like $\log_{10} x$).

A. Inverse Functions

▶Objective 1   Find an inverse function from a given algebraic or trigonometric function.

Example   Suppose $f^{-1}(x)$ is the inverse function of $f(x) = \dfrac{x}{x - 3}$. Write an expression for $f^{-1}(x)$ in terms of x.

Tip   The notation for an inverse function is $f^{-1}(x)$. Realize that this is **not** $\dfrac{1}{f(x)}$.

Answer   If $y = f(x)$ gives us $y = \dfrac{x}{x-3}$, then $f^{-1}(x)$ gives $x = \dfrac{y}{y-3}$. Solving for y:

$$xy - 3x = y$$
$$xy - y = 3x$$
$$y(x-1) = 3x$$
$$y = \dfrac{3x}{x-1}.$$

So, $f^{-1}(x) = \dfrac{3x}{x-1}.$

▶Objective 2   Find the derivative of an inverse algebraic or trigonometric function, using implicit differentiation.

Example   Given that $g(x) = x^5 + x^3 + x$, find an expression that gives $\dfrac{dy}{dx}$ for the slope of the graph of the inverse function $y = g^{-1}(x)$.

Tips   Use implicit differentiation after swapping x and y values.

Your answer may include x or y (or both).

Answer   For the inverse function, simply exchange y and x values, so that for $g^{-1}(x)$ you get $x = y^5 + y^3 + y$.

Taking the derivative implicity:

$$1 = 5y^4 \dfrac{dy}{dx} + 3y^2 \dfrac{dy}{dx} + \dfrac{dy}{dx}, \text{ or}$$
$$1 = (5y^4 + 3y^2 + 1)\dfrac{dy}{dx}, \text{ so}$$
$$\dfrac{dy}{dx} = \dfrac{1}{5y^4 + 3y^2 + 1}.$$

▶Objective 3   Exploit the graphical symmetry of inverse functions to analyze functions.

Example 1   Find the slope of the tangent line to the curve $y = g^{-1}(x)$ at point $(3, 1)$ if $g(x) = x^5 + x^3 + x$.

Tips   Notice that if point (a, b) is on the curve $y = g(x)$, the point (b, a) is on the curve $y = g^{-1}(x)$.

Remember that the slope of $g(x)$ at point (a, b) is the reciprocal of the slope of $g^{-1}(x)$ at point (b, a).

If g is the inverse of f, then $g'(x) = \dfrac{1}{f'(g(x))}.$

Answer   There are two separate ways to attack this problem. The first is just to use the expression $\dfrac{dy}{dx} = \dfrac{1}{5y^4 + 3y^2 + 1}$ that you found in the last problem and evaluate for $y = 1$:

$$\frac{dy}{dx} = \frac{1}{5 + 3 + 1} = \frac{1}{9} \text{ at the point } x = 3, y = 1.$$

The alternative method is to find the slope at point $(1, 3)$ on $g(x) = x^5 + x^3 + x$:

$$g'(x) = 5x^4 + 3x^2 + 1$$
$$g'(1) = 5 + 3 + 1 = 9.$$

Then realize that the slope of the inverse function at any point (b, a) is the reciprocal of the slope of the original at the corresponding point (a, b):

$$\frac{dy}{dx} = \frac{1}{9}.$$

Example 2   Given that $g(x)$ and $f(x)$ are inverse functions of each other, write an equation for the tangent line to $g(x)$ at $x = 2$ if $f(x) = 3x^5 - x$.

Answer   First realize that the point $(1, 2)$ on the curve of $f(x) = 3x^5 - x$ corresponds to the point $(2, 1)$ on the curve $g(x)$. Since

$$f'(x) = 15x^2 - 1$$
$$f'(1) = 15 - 1 = 14.$$

Therefore

$$g'(2) = \frac{1}{f'(1)} = \frac{1}{14}.$$

Then, using the fact that the point $(2, 1)$ is on the curve of $g(x)$:

$$y - 1 = \frac{1}{14}(x - 2), \text{ or}$$
$$y = \frac{1}{14}(x - 2) + 1.$$

▶**Objective 4**   Identify the domain restrictions of the inverse trigonometric functions.

The inverse sine function, $\sin^{-1}(x)$ or $\arcsin(x)$, is defined by

$$y = \arcsin(x) \quad \text{if and only if} \quad x = \sin y$$

for $-1 \le x \le 1$ and $-\frac{\pi}{2} \le y \le \frac{\pi}{2}$.

The inverse sine function, $\cos^{-1}(x)$ or $\arccos(x)$, is defined by

$$y = \arccos(x) \quad \text{if and only if} \quad x = \cos y$$

for $-1 \le x \le 1$ and $0 \le y \le \pi$.

The inverse tangent function, $\tan^{-1}(x)$ or $\arctan(x)$, is defined by

$$y = \arctan(x) \quad \text{if and only if} \quad x = \tan y$$

for $-\infty < x < \infty$ and $-\frac{\pi}{2} < y < \frac{\pi}{2}$.

▶Objective 5   Memorize the derivatives of the inverse trigonometric functions and use them when asked to find derivatives.

$$\frac{d}{dx}(\arcsin u) = \frac{d}{dx}(\sin^{-1} u) = \frac{1}{\sqrt{1-u^2}}\left(\frac{du}{dx}\right).$$

$$\frac{d}{dx}(\arccos u) = \frac{d}{dx}(\cos^{-1} u) = -\frac{1}{\sqrt{1-u^2}}\left(\frac{du}{dx}\right).$$

$$\frac{d}{dx}(\arctan u) = \frac{d}{dx}(\tan^{-1} u) = \frac{1}{1+u^2}\left(\frac{du}{dx}\right).$$

▶Objective 6   Find derivatives for composite functions involving inverse trigonometric functions.

Example   Find the derivatives for:

$$h(t) = \arctan(t^2) \quad g(x) = x^2 \arcsin x \quad f(x) = \frac{\arccos x}{x}$$

Answers

$$h(t) = \arctan(t^2) \Rightarrow h'(t) = \frac{2t}{1+t^4}.$$

$$g(x) = x^2 \arcsin x \Rightarrow g'(x) = 2x(\arcsin x) + \frac{x^2}{\sqrt{1-x^2}}.$$

$$f(x) = \frac{\arccos x}{x} \Rightarrow f'(x) = \frac{-\dfrac{1}{\sqrt{1-x^2}}x - \arccos x}{x^2}.$$

▶Objective 7   Use inverse trigonometric functions to model situations.

Example   A lighthouse light is revolving such that its beam shines on a wall that is perpendicular to the line of sight, as shown below, where θ is the angle (in radians) from the normal line and y is the distance (in meters) along the wall from the normal line. The distance from the lighthouse to the wall is 20 m.

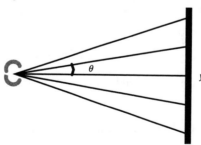

A.   Write an equation that gives θ as a function of y.

B.   Find the rate of change of θ with respect to y.

C. Given that the lighthouse is rotating at 2 rev/sec, find the rate at which the beam is moving on the wall when it's at $y = 3$ m.

Tip Drawing a picture can often help you visualize situation problems.

Answers

A. Since we have a right triangle:

$$\tan \theta = \frac{y}{20}, \text{ and}$$

$$\theta = \arctan\left(\frac{y}{20}\right).$$

B. $\theta = \arctan\left(\frac{y}{20}\right)$, so

$$\frac{d\theta}{dy} = \left(\frac{1}{20}\right)\left(\frac{1}{\left(\frac{y}{20}\right)^2 + 1}\right) = \frac{1}{20\left(\frac{1}{400}y^2 + 1\right)}$$

$$= \frac{1}{\frac{1}{20}y^2 + 20} = \frac{20}{y^2 + 400}.$$

C. You'll need to convert the angular velocity into units of radians/second. Using $\theta = \arctan\left(\frac{y}{20}\right)$, we find:

$$\frac{d\theta}{dt} = \frac{20}{y^2 + 400} \times \frac{dy}{dt}.$$

Since the lighthouse light moves at $2\frac{\text{rev}}{\text{sec}} = 4\pi\frac{\text{radians}}{\text{sec}}$,

$$4\pi = \frac{20}{(3)^2 + 400} \times \frac{dy}{dt}, \text{ and}$$

$$\frac{dy}{dt} = 4\pi\frac{409}{20} = \frac{409}{5}\pi = 256.982\,\frac{\text{m}}{\text{s}}.$$

Alternative solution:

$$y = 20\tan\theta$$

$$\frac{dy}{dt} = 20\sec^2\theta \times \frac{d\theta}{dt}$$

$$\theta = \arctan\left(\frac{3}{20}\right) \approx 0.14889$$

$$\frac{dy}{dt} = 20(4\pi)\sec^2(0.14889) = 256.982\,\frac{\text{m}}{\text{s}}.$$

Yet another solution:

$$y = 20\tan\theta$$

$$\frac{dy}{dt} = 20\sec^2\theta \times \frac{d\theta}{dt}$$

Then use a right triangle, with legs 20, 3, and $\sqrt{409}$ to find $\sec^2\theta$ directly:

$$\frac{dy}{dt} = 20(4\pi)\frac{409}{400} = 256.982\,\frac{\text{m}}{\text{s}}.$$

Apex Learning

Multiple-Choice Questions on Inverse Functions and Inverse Trigonometric Functions

1. Which of these expresses the inverse function for the function $y = x^2$?

    A. $y = \sqrt{x}$

    B. $x = \pm\sqrt{x}$

    C. $x = y^2$

    D. $y = x^{-2}$

    E. None of these.

Answer   E. The function $y = x^2$ is not one-to-one so its inverse isn't a function at all. You'd have to restrict its domain to get an inverse function. For example, if you restrict yourself to just $x \geq 0$, you can write its inverse as $y = \sqrt{x}$.

2. Which of these functions has an inverse function?

    I.

x	0	1	2	3
$y = f(x)$	0	1	2	3

    II.

x	0	1	2	3
$y = g(x)$	3	1	2	3

    III.

x	0	1	2	3
$y = h(x)$	0	0	0	0

    A. I only

    B. II only

    C. III only

    D. I and II

    E. Neither I, II, nor III

Answer A. The first function, $y = f(x)$, is one-to-one (no two x's have the same y's), so its inverse is a function. Neither of the other functions is one-to-one, so they don't have inverse functions.

---

3. Consider the function $x^3 + 2x + 1$. Its derivative is always positive, so it's one-to-one, and it has an inverse function. Let's call that inverse function g.

   Which of these is a correct expression for g'?

   A. $g'(x) = 3x^2 + 2$

   B. $g'(x) = \dfrac{1}{3x^2 + 2}$

   C. $g'(x) = 3y^2 + 2$

   D. $g'(x) = \dfrac{1}{3y^2 + 2}$

   E. $g'(x) = \dfrac{1}{x^3 + 2x + 1}$

Answer D. You didn't get an explicit definition for g here, and the expression for the derivative is also implicit. You get it by swapping the x's and y's, and then differentiating implicitly:

$$x = y^3 + 2y + 1$$
$$1 = 3y^2 y' + 2y', \text{ or } y' = \frac{1}{3y^2 + 2}.$$

---

4. Which of the following statements are true?

   I. If a function is always increasing, so is its inverse function.

   II. If L is the tangent line to $y = f(x)$ at the point (a, b) and M is the tangent line to $y = f^{-1}(x)$ at the point (b, a), then L and M are perpendicular.

   III. If a function is odd, then its inverse is even.

   A. I only

   B. II only

   C. III only

   D. I, II, and III

   E. None of these.

Answer   A. I is true. Flipping an increasing graph across the line $y = x$ gives you another increasing graph. Or you can work it out algebraically. If f is always increasing, f' is always positive, and f' ends up in the denominator when you work out what the derivative of f^{-1} is. So the derivative of the inverse is always positive, too.

II is false. The slopes of this pair of lines are reciprocals, but they're not negative reciprocals, which is what you'd need for them to be perpendicular. (This could be true in the special case where one line is horizontal and the other is vertical.)

III is false. For example, $f(x) = x^3$ is odd and so is its inverse $f^{-1}(x) = \sqrt[3]{x}$.

---

5. Here is the graph of $y = f(x)$ and its tangent line at $(-4, -8)$. The equation for the tangent line is $y = 3x + 4$. Find the equation of the tangent line to the graph of $y = f^{-1}(x)$ at the point $(-8, -4)$.

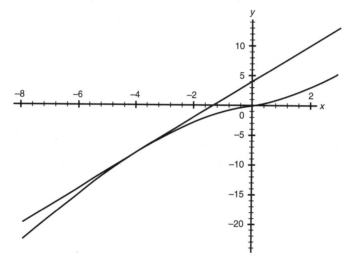

A. $y = 3x + 20$

B. $y = \frac{1}{3}x + 4$

C. $y = -\frac{1}{3}x - \frac{20}{3}$

D. $y = \frac{1}{3}x - \frac{4}{3}$

E.   There's not enough information to answer this question.

Answer   D. The slope of the line you're looking for is just the reciprocal of the slope of the line you have. Then you have a point $(-8, -4)$ and a slope, so it's easy to write down the equation of this line: $y = \frac{1}{3}x - \frac{4}{3}$.

---

6. Find the exact value of $\cos(\arcsin \frac{5}{13})$.

A.   0.9231

B.   0.6985

C. $\frac{12}{13}$

D. $\frac{3\pi}{8}$

E. $-\frac{5}{13}$

Answer C. You can find the exact value by drawing a right triangle and filling in the pieces. Remember that $\sin\theta$ is opposite over hypotenuse, and you can find the length of the third leg from the Pythagorean Theorem. Then you can read off the cosine of your angle: It's $\frac{12}{13}$!

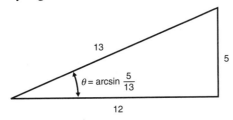

---

7. If $y = f(x) = 3x^3 - 5$, then the inverse function $f^{-1}(x) =$

A. $3y^3 - 5$

B. $\left(\dfrac{x-5}{3}\right)^3$

C. $\dfrac{1}{3}x^{\frac{1}{5}} + 5$

D. $\left(\dfrac{x+5}{3}\right)^{\frac{1}{3}}$

E. $(3x^3 - 5)^{\frac{1}{3}} + 5$

Answer D. To find the inverse function $f^{-1}(x)$, you can swap x's with the y's and solve for y:

$$x = 3y^3 - 5 \Rightarrow x + 5 = 3y^3 \Rightarrow \frac{x+5}{3} = y^3 \Rightarrow \left(\frac{x+5}{3}\right)^{\frac{1}{3}} = y.$$

So

$$f^{-1}(x) = \left(\frac{x+5}{3}\right)^{\frac{1}{3}}.$$

---

8. If $f(x) = x^3 + \frac{3}{2}x^2 - 18x + 4$, then over which of these domains could you define $f^{-1}(x)$?

I. $(-\infty, -3]$

II. $(-3, \infty)$

III. $(2, \infty)$

A. I

B. II

C. III

D. Either I or II

E. Either I or III

Answer E. Here's the graph of $y = f(x) = x^3 + \frac{3}{2}x^2 - 18x + 4$.

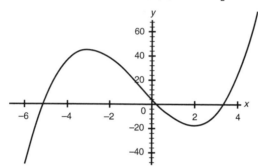

The function f is one-to-one wherever it is monotonic ("monotonic" simply means "increasing or decreasing").

To find these intervals, take the first derivative, which is $f'(x) = 3x^2 + 3x - 18$, and set it equal to zero. You find that the derivative is 0 at $x = -3$ and $x = 2$.

To the left of $x = -3$ $f(x)$ is increasing and therefore one-to-one, while between $x = -3$ and $x = 2$ the function is decreasing and one-to-one, and finally to the right of $x = 2$ the function is increasing and one-to-one.

So, on each of these intervals $(-\infty, -3]$, $[-3, 2]$, and $[2, \infty)$ the function is one-to-one and has an inverse.

---

9. Suppose $f(x) = \sqrt{9 - x}$. On any interval where the inverse function $y = f^{-1}(x)$ exists, then $\dfrac{d}{dx}(f^{-1}(x)) =$

A. $-2x$

B. $9 - x^2$

C. $2x$

D. $\dfrac{1}{\frac{1}{2}(9 - x)^{-\frac{1}{2}}}$

E. $2(9 - x)^{\frac{1}{2}}$

Answer A. First write the inverse function implicitly as $x = f(y)$: $x = \sqrt{9-y}$.

Then solve for y explicitly in terms of x if possible:

$$x^2 = 9 - y \Rightarrow y = 9 - x^2.$$

So,

$$y = f^{-1}(x) = 9 - x^2, \text{ and thus } \frac{d}{dx}(f^{-1}(x)) = -2x.$$

---

10. Consider the function $f(x) = x^2 - 4x + 5$, for $x \geq 2$. Then $\dfrac{d}{dx}(f^{-1}(x)) =$

   A. $\dfrac{1}{2y - 4}$, where x and y are related by the equation $x = y^2 - 4y + 5$ for $x \geq 1$.

   B. $2y - 4$, where x and y are related by the equation $y = x^2 - 4x + 5$ for $x \geq 2$.

   C. $\dfrac{1}{2x - 4}$, for $x \geq 1$.

   D. $\dfrac{1}{2x - 4}$, for $x \geq 2$.

   E. $\dfrac{1}{2y - 4}$, where x and y are related by the equation $y = x^2 - 4x + 5$ for $x \geq 2$.

Answer A. First, you need to find an equation for $y = f^{-1}(x)$ which is either explicit or implicit.

Notice that the domain of $y = f(x)$ is $x \geq 2$. The range is $y \geq 1$.

So $y = f^{-1}(x)$ has domain $x \geq 1$ and range $y \geq 2$, and the equation describing $y = f^{-1}(x)$ implicitly is produced by swapping the x's and y's in the original equation for $y = f(x)$. You get that $x = y^2 - 4y + 5$, $x \geq 1$ is the implicit equation for $y = f^{-1}(x)$.

To find $\dfrac{d}{dx}(f^{-1}(x))$ differentiate implicitly: $1 = (2y - 4)y' \Rightarrow y' = \dfrac{1}{2y - 4}$.

Thus,

$$\frac{d}{dx}(f^{-1}(x)) = \frac{1}{2y - 4},$$

where x and y are related by the equation $x = y^2 - 4y + 5$, $x \geq 1$.

---

11. Suppose that $y = f(x) = x^2 - 4x + 4$. Then on any interval where the inverse function $y = f^{-1}(x)$ exists, the derivative of $y = f^{-1}(x)$ with respect to x is:

   A. $\dfrac{1}{2x - 4}$

   B. $\dfrac{1}{2y - 4}$, where x and y satisfy the equation $y = x^2 - 4x + 4$.

   C. $\frac{1}{2}x^{-\frac{1}{2}}$

D. $\frac{1}{2}x^{\frac{1}{2}}$

E. Both $\dfrac{1}{2y-4}$, where x and y satisfy the equation $y = x^2 - 4x + 4$, and $\frac{1}{2}x^{-\frac{1}{2}}$.

Answer C. You want to write the inverse function implicitly as $x = f(y)$.

This gives you $x = y^2 - 4y + 4$.

Next, you should try to solve for y, which in this case IS possible.

Notice that $x = y^2 - 4y + 4 = (y-2)^2 \Rightarrow x^{1/2} + 2 = y$.

So, the function $y = f^{-1}(x) = x^{1/2} + 2$ is explicit, and its derivative with respect to x is simply

$$\frac{d}{dx}(f^{-1}(x)) = \frac{1}{2}x^{-1/2}.$$

---

12. Suppose $f(x) = \sin(\pi \cos x)$. On any interval where the inverse function $y = f^{-1}(x)$ exists, the derivative of $f^{-1}(x)$ with respect to x is:

A. $\dfrac{-1}{\cos(\pi \cos x)}$, where x and y are related by the equation $x = \sin(\pi \cos y)$.

B. $\dfrac{-1}{\pi \sin x \cos(\pi \cos x)}$, where x and y are related by the equation $x = \sin(\pi \cos y)$.

C. $\dfrac{-1}{\pi \sin y \cos(\pi \cos y)}$, where x and y are related by the equation $x = \sin(\pi \cos y)$.

D. $\dfrac{-1}{\cos(\pi \cos y)}$, where x and y are related by the equation $x = \sin(\pi \cos y)$.

E. $\dfrac{-1}{\sin y \cos(\pi \cos y)}$, where x and y are related the equation $x = \sin(\pi \cos y)$.

Answer C. First write the inverse function implicitly as $x = f(y)$: $x = \sin(\pi \cos y)$.

Now differentiate implicitly to get

$$1 = \cos(\pi \cos y)\pi(-\sin y)y' = -\pi \sin y \cos(\pi \cos y)y'$$

$$\Rightarrow \frac{d}{dx}(f^{-1}(x)) = y' = \frac{-1}{\pi \sin y \cos(\pi \cos y)},$$

where x and y are related by the equation $x = \sin(\pi \cos y)$.

---

13. Suppose that $f(x)$ is an invertible function (that is, has an inverse function), and that the slope of the tangent line to the curve $y = f(x)$ at the point $(2, -4)$ is -0.2. Then:

A. The slope of the tangent line to the curve $y = f^{-1}(x)$ at the point $(-4, 2)$ is -0.2.

B. The slope of the tangent line to the curve $y = f^{-1}(x)$ at the point $(2, -4)$ is -5.

C. The slope of the tangent line to the curve $y = f^{-1}(x)$ at the point $(2, -4)$ is 5.

D. The slope of the tangent line to the curve $y = f^{-1}(x)$ at the point $(-4, 2)$ is -5.

E. The slope of the tangent line to the curve $y = f^{-1}(x)$ at the point $(-4, 2)$ is 5.

Answer   D.

Recall that the derivative of f at (a, b) is the reciprocal of the derivative of f^{-1} at (b, a).

In other words, the slope of the tangent line to $y = f^{-1}(x)$ at (b, a) is the reciprocal of the slope of the tangent line to $y = f(x)$ at (a, b).

In our problem, $a = 2$ and $b = -4$ so the slope of the tangent line to the curve $y = f^{-1}(x)$ at the point $(-4, 2)$ is $\frac{1}{-0.2} = -5$.

---

14.   Suppose $f(x) = 4x^5 - (1/x^4)$. The slope of the tangent line to the graph of $y = f^{-1}(x)$ at $(3, 1)$ is:

A. $\frac{1}{24}$

B. $\frac{1}{16}$

C. 0.00062

D. 1,620

E. 972

Answer   A. The slope of the tangent line to the graph of $y = f^{-1}(x)$ at $(3, 1)$ is the reciprocal of the slope of the tangent line to the graph of $y = f(x)$ at $(1, 3)$.

You have that $f'(x) = 20x^4 + 4/x^5$, so $f'(1) = 20 + 4 = 24$.

Therefore, the slope of the tangent line to the graph of $y = f^{-1}(x)$ at $(3, 1)$ is $\frac{1}{24}$.

---

15.   Let $f(x) = (x + 3)^3 + 2$. The graph of $y = f^{-1}(x)$ has a vertical tangent at:

A. $(66, 1)$

B. $(2, -3)$

C. $(10, -1)$

D. $(-3, 2)$

E. $(-1, 10)$

Answer   B. (HINT: Think about the symmetry of the 2 graphs.) One way to do this problem is notice that the graph of $y = f^{-1}(x)$ has a vertical tangent at (a, b) if and only if the graph of $y = f(x)$ has a horizontal tangent at (b, a). Setting $f'(x) = 0$ gives you

$$3(x+3)^2 = 0 \Rightarrow (x+3) = 0 \Rightarrow x = -3.$$

So at $(-3, f(-3)) = (-3, 2)$, the graph of $y = f(x)$ has a horizontal tangent.

Thus, the graph of the inverse function $y = f^{-1}(x)$ has a vertical tangent at $(2, -3)$.

Another way to do this is to find $\dfrac{d}{dx}(f^{-1}(x))$ and look for places that its denominator is 0.

---

16.   Suppose that $y = f(x) = \sqrt{2x}$, $x \geq 0$.

Find a value $c > 0$ such that the tangent line to the curve $y = f(x)$ at $x = c$ has the same slope as the tangent line to the curve $y = f^{-1}(x)$ at $x = c$:

A. $c = \dfrac{1}{8}$

B. $c = \dfrac{1}{2}$

C. $c = \left(\dfrac{1}{8}\right)^{\frac{1}{3}}$

D. $c = \left(\dfrac{1}{8}\right)^{\frac{2}{3}}$

E. $c = \left(\dfrac{1}{2}\right)^{\frac{1}{3}}$

Answer   E. Note that the slope of the tangent to the curve $y = f(x)$ at $x = c$ is $f'(c) = (2c)^{-\frac{1}{2}}$.

The slope of the tangent to the curve $y = f^{-1}(x)$ is $\dfrac{d}{dx}(f^{-1}(x))$ evaluated at $x = c$.

To find $\dfrac{d}{dx}(f^{-1}(x))$ write the inverse function as $x = f(y)$: $x = \sqrt{2y} \Rightarrow y = \frac{1}{2}x^2$.

That is, $y = f^{-1}(x) = \frac{1}{2}x^2$, and thus $\dfrac{d}{dx}(f^{-1}(x)) = x$, which evaluated at $x = c$ is just c.

So the slope of the tangent to the curve $y = f(x)$ at $x = c$ is $(2c)^{-\frac{1}{2}}$, while the slope of the tangent to the curve $y = f^{-1}(x)$ at $x = c$ is c. Setting them equal gives you

$$\frac{1}{\sqrt{2c}} = c \Rightarrow 1 = c\sqrt{2c} \Rightarrow 1 = \sqrt{2c^3}$$

$$\Rightarrow \frac{1}{2} = c^3 \Rightarrow c = \left(\frac{1}{2}\right)^{\frac{1}{3}}.$$

17. Which of the following is equal to $\sqrt{1-x^2}$?

I. $\sin(\arccos(x))$

II. $\cos(\arctan(x))$

III. $\cos(\arcsin(x))$

A. I only

B. II only

C. III only

D. I and III only

E. I, II, and III

Answer D.

On I: Think of a right triangle with an angle θ such that $x = \cos\theta$ and $\theta = \arccos x$:

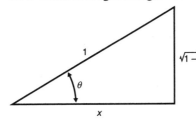

Then $\sin(\arccos(x)) = \sin(\theta) = \sqrt{1-x^2}$.

On II: Think of a right triangle with an angle θ such that $x = \tan\theta$ and $\theta = \arctan x$:

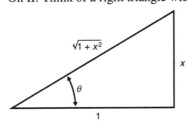

Then $\cos(\arctan(x)) = \cos(\theta) = \dfrac{1}{\sqrt{1+x^2}}$.

On III: Think of a right triangle with an angle θ such that $x = \sin\theta$ and $\theta = \arcsin x$:

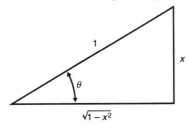

Then $\cos(\arcsin(x)) = \cos(\theta) = \sqrt{1-x^2}$.

18. If $g(t) = \tan^{-1}(t)$, then $g(t)$ is:

A. undefined at $t = 0$

B. defined but not continuous at $t = 0$

C. continuous but not differentiable at $t = 0$

D. continuous and differentiable at $t = 0$

E. none of the above

Answer   D.

$$g(t) = \arctan(t)$$
$$\lim_{t \to 0} g(t) = g(0) = 0$$
$$g'(t) = \frac{1}{t^2 + 1}$$
$$g'(0) = 1$$

19. If $h(x) = \arcsin(x^2 - 2x)$, then:

I. $h(x)$ has a domain of $\{x : -1 \le x \le 1\}$

II. $h'(x) = \dfrac{2x - 2}{\sqrt{1 - x^2}}$

III. $h'(x) = \dfrac{2x - 2}{\sqrt{1 - x^4 + 4x^3 - 4x^2}}$

A. I only

B. II only

C. III only

D. I and II only

E. I and III only

Answer   C.

I is not true. To find the domain of $h(x) = \arcsin(x^2 - 2x)$ you have to find all x values such that $-1 \le x^2 - 2x \le 1$.

$x^2 - 2x$ has a minimum value of -1 at $x = 1$ (from the first derivative test), so you need to solve only the inequality $x^2 - 2x \le 1$, which is equivalent to $x^2 - 2x - 1 \le 0$.

Using the quadratic formula to find the roots of the equation $x^2 - 2x - 1 = 0$ gives $x = 1 \pm \sqrt{2}$.

Checking the value of $x^2 - 2x - 1$ at points smaller than both of the above roots, larger than both of the roots, and between the roots shows that $x^2 - 2x - 1 \leq 0$ if and only if $1 - \sqrt{2} \leq x \leq 1 + \sqrt{2}$. Therefore, the domain is $\{x : 1 - \sqrt{2} \leq x \leq 1 + \sqrt{2}\}$.

II is not true, and III is true because $\dfrac{d}{dx}(\sin^{-1} u) = \dfrac{1}{\sqrt{(1 - u^2)}}\left(\dfrac{du}{dx}\right)$ with $u = x^2 - 2x$, which gives:

$$\frac{d}{dx}(\arcsin(x^2 - 2x))$$
$$= \frac{1}{\sqrt{1 - (x^2 - 2x)^2}} \times (2x - 2) = \frac{2x - 2}{\sqrt{1 - x^4 + 4x^3 - 4x^2}}.$$

---

20. If $5\pi xy = \pi y^2 + 16\arctan(x)$, the tangent line at point $(1, 1)$ has a slope $\approx$

A. 1.000

B. 0.849

C. 0.009

D. -0.818

E. -1.667

Answer D.

$$5\pi xy = \pi y^2 + 16\arctan(x)$$
$$5\pi xy' + 5\pi y = 2\pi yy' + \frac{16}{x^2 + 1} \quad \text{at point } (1, 1)$$
$$5\pi y' + 5\pi = 2\pi y' + 8$$
$$3\pi y' = 8 - 5\pi$$
$$y' = \frac{8}{3\pi} - \frac{5}{3} \approx -.818$$

---

21. $\dfrac{d}{dx}(\arccos(4x)) =$

A. $-\dfrac{4}{\sqrt{(1 - x^2)}}$

B. $-\dfrac{4}{\sqrt{(1 - 16x^2)}}$

C. $-\dfrac{1}{\sqrt{(1 - x^2)}}$

D. $-\dfrac{4}{\sqrt{(1 + 16x^2)}}$

E. $-\dfrac{4}{\sqrt{(1 - 4x^2)}}$

Answer   B.

Using the fact that

$$\frac{d}{dx}(\arccos u) = -\frac{1}{\sqrt{1 - u^2}}\left(\frac{du}{dx}\right),$$

you get

$$\frac{d}{dx}(\arccos(4x)) = -\frac{1}{\sqrt{1 - (4x)^2}}(4) = -\frac{4}{\sqrt{(1 - 16x^2)}}.$$

---

22.   The derivative of $\arctan(x^2 + 3x)$ with respect to x is:

A. $\dfrac{1}{1 + (x^2 + 3x)}$

B. $\dfrac{x^2 + 3x}{1 + x^2}$

C. $\dfrac{2x + 3}{x^4 + 6x^3 + 9x^2 + 1}$

D. $\dfrac{2x + 3}{1 + x^2}$

E. $\dfrac{x^2 + 3x}{x^4 + 6x^3 + 9x^2 + 1}$

Answer   C. Using the fact that $\dfrac{d}{dx}(\arctan u) = \dfrac{1}{1 + u^2}\left(\dfrac{du}{dx}\right)$, you get

$$\frac{d}{dx}(\arctan(x^2 + 3x)) = \frac{1}{1 + (x^2 + 3x)^2}(2x + 3) = \frac{2x + 3}{x^4 + 6x^3 + 9x^2 + 1}.$$

---

23.   Suppose $f(x) = \left(\arcsin \dfrac{1}{x}\right)^2$. Then $f'(x) =$

A. $\dfrac{-2 \arcsin \frac{1}{x}}{x^2\sqrt{1 - (\frac{1}{x})^2}}$

B. $\dfrac{-2x^{-2}}{\sqrt{1 - x^{-2}}}$

C. $\dfrac{-2}{x^2\sqrt{1 - x^2}}$

D. $\dfrac{2\arcsin\frac{1}{x}}{x\sqrt{1-x^{-2}}}$

E. $\dfrac{\frac{2}{x^2}\arcsin\frac{1}{x}}{\sqrt{1-\frac{1}{x^2}}}$

Answer   A.

First you find that

$$f'(x) = \frac{d}{dx}\left(\arcsin\frac{1}{x}\right)^2 = 2\left(\arcsin\frac{1}{x}\right)\frac{d}{dx}\left(\arcsin\frac{1}{x}\right)$$

by the Chain Rule.

Then, using the fact that

$$\frac{d}{dx}(\arcsin u) = \frac{1}{\sqrt{1-u^2}}\left(\frac{du}{dx}\right)$$

you get

$$f'(x) = 2\left(\arcsin\frac{1}{x}\right)\frac{d}{dx}\left(\arcsin\frac{1}{x}\right)$$

$$= 2\left(\arcsin\frac{1}{x}\right)\frac{1}{\sqrt{1-\left(\frac{1}{x}\right)^2}}\left(-\frac{1}{x^2}\right) = \frac{-2\arcsin\frac{1}{x}}{x^2\sqrt{1-\left(\frac{1}{x}\right)^2}}.$$

---

24. $\dfrac{d}{dx}(1+\cos^{-1}(5x))^3 =$

A. $\dfrac{-3(1+\cos^{-1}(5x))^2}{\sqrt{(1-25x^2)}}$

B. $\dfrac{-15(1+\cos^{-1}(5x))^2}{\sqrt{(1-25x^2)}}$

C. $\dfrac{-5(1+\cos^{-1}(5x))^3}{\sqrt{(1-25x^2)}}$

D. $\dfrac{-5(1+\cos^{-1}(5x))^3}{\sqrt{(1-5x^2)}}$

E. $\dfrac{-15(1+\cos^{-1}(5x))^2}{\sqrt{(1+5x^2)}}$

Answer B.

Using the fact that

$$\frac{d}{dx}(\cos^{-1} u) = -\frac{1}{\sqrt{1-u^2}}\left(\frac{du}{dx}\right)$$

and the Chain Rule, you get $\dfrac{d}{dx}(1 + \cos^{-1}(5x))^3$

$$= 3(1 + \cos^{-1}(5x))^2 \frac{d}{dx}(1 + \cos^{-1}(5x))$$

$$= 3(1 + \cos^{-1}(5x))^2\left[\frac{-1}{\sqrt{1-(5x)^2}}(5)\right]$$

$$= \frac{-15(1 + \cos^{-1}(5x))^2}{\sqrt{1-25x^2}}.$$

---

25. $\dfrac{d}{dt}\sqrt{\arcsin(3t-1)} =$

A. $\dfrac{3\sqrt{\arcsin(3t-1)}}{2\sqrt{(-9t^2+6t)}}$

B. $\dfrac{3(\arcsin(3t-1))^{1/2}}{\sqrt{(-9t^2+6t)}}$

C. $\dfrac{1}{\sqrt{(-9t^2+6t)\arcsin(3t-1)}}$

D. $\dfrac{3\sqrt{\arcsin(3t-1)}}{\sqrt{(-9t^2+6t)}}$

E. $\dfrac{3}{2\sqrt{(-9t^2+6t)\arcsin(3t-1)}}$

Answer E.

Using the fact that

$$\frac{d}{dt}(\arcsin u) = \frac{1}{\sqrt{1-u^2}}\left(\frac{du}{dt}\right)$$

and the Chain Rule, you get

$$\frac{d}{dt}\sqrt{\arcsin(3t-1)} = \frac{d}{dt}(\arcsin(3t-1))^{1/2}$$

$$= \frac{1}{2}(\arcsin(3t-1))^{-1/2}\frac{d}{dt}(\arcsin(3t-1))$$

$$= \frac{1}{2}(\arcsin(3t-1))^{-1/2}\left[\frac{1}{\sqrt{1-(3t-1)^2}}(3)\right]$$

$$= \frac{3}{2\sqrt{\left(1-(3t-1)^2\right)\arcsin(3t-1)}}$$

$$= \frac{3}{2\sqrt{(-9t^2+6t)\arcsin(3t-1)}}.$$

Free-Response Questions on Inverse Trigonometric Functions

---

1. Find each of the following:

A. $\dfrac{d}{dx}\left(x \arcsin \sqrt{5x-2}\right) =$

B. Suppose $f(t) = \left(\sin^{-1}\left(\dfrac{1}{t}\right)\right) \sin(2t)$. Then $f'(t) =$

C. Suppose $g(x) = \dfrac{4x^3 - 1}{\arctan(2x)}$. Then $g'(x) =$

D. Find y' if $x = \arccos y + 2y$.

---

2. Let $\theta = f(x) = \arctan x$.

   Use a tangent line approximation $(f(x) \approx f(a) + (x-a)f'(a))$ along with the fact that $\arctan(\sqrt{3}) = \frac{\pi}{3}$ to estimate $\arctan(1.8)$.

3.

A. Consider the curve $y = \arcsin x$ on the interval $[-1, 1]$.

   Find the equation for the tangent line to the curve at $\left(-\frac{\sqrt{3}}{2}, -\frac{\pi}{3}\right)$.

B. Consider the curve $y^2 + y = \arcsin x$.

   Find the equation for the tangent line to the curve at $(.53119, -1.40000)$.

---

4. A searchlight located 600 feet from the nearest point A on a straight road is aimed at a motorcycle traveling at a rate of 45 feet per second on the road.

A. Use inverse trigonometric functions to find the rate at which the searchlight is rotating when the motorcycle is 100 feet from A.

   HINT: Draw a picture, representing your situation as a right triangle. At one corner is the motorcycle moving away from or toward A, at another corner is A, and at the third corner is the searchlight.

B. Suppose instead that the motorcycle is at A at time $t = 0$ and moving along the road (in a positive direction, such that the motorcycle's distance from A is given by the function $d(t) = 20t + t^2$ for $0 \leq t \leq 30$.

   Find an equation (in t) for $\dfrac{d\theta}{dt}$ on the interval $[0,30]$. Then, use your graphing calculator to estimate at what time between $t = 0$ and $t = 30$ the searchlight is rotating the fastest.

Answers

1.

A. $\dfrac{d}{dx}(x \arcsin \sqrt{5x - 2})$

$= \arcsin \sqrt{5x - 2} + x \left[\dfrac{5}{2\sqrt{\left(1 - (\sqrt{5x-2})^2\right)(5x-2)}} \right].$

Remember: Simplification isn't necessary.

$\dfrac{d}{dx}(x \arcsin \sqrt{5x - 2})$

$= \arcsin \sqrt{5x - 2} + x\dfrac{d}{dx}(\arcsin \sqrt{5x - 2})$ by the product rule

$= \arcsin \sqrt{5x - 2} + x \left[\dfrac{1}{\sqrt{\left(1 - (\sqrt{5x - 2})^2\right)}}\left(\dfrac{5}{2}(5x - 2)^{-\frac{1}{2}}\right) \right]$

$= \arcsin \sqrt{5x - 2} + x \left[\dfrac{5}{2\sqrt{\left(1 - (\sqrt{5x - 2})^2\right)(5x - 2)}} \right].$

B. $f'(t) = \dfrac{-\sin(2t)}{t^2\sqrt{1 - \left(\frac{1}{t^2}\right)}} + 2\left(\sin^{-1}\left(\dfrac{1}{t}\right) \right)\cos(2t).$

$f'(t) = \sin(2t)\dfrac{d}{dt}\ \sin^{-1}\left(\dfrac{1}{t}\right)$

$+ \left(\sin^{-1}\left(\dfrac{1}{t}\right) \right)\dfrac{d}{dt}\ \sin(2t)$ by the product rule

$= \sin(2t)\left[\dfrac{1}{\sqrt{1 - \left(\frac{1}{t}\right)^2}}\left(-\dfrac{1}{t^2}\right) \right] + \left(\sin^{-1}\left(\dfrac{1}{t}\right) \right)2\cos(2t)$

$= \dfrac{-\sin(2t)}{t^2\sqrt{1 - \frac{1}{t^2}}} + 2\left(\sin^{-1}\left(\dfrac{1}{t}\right) \right)\cos(2t)$

C. $g'(x) = \dfrac{12x^2 \arctan(2x) - (4x^3 - 1)\left(\frac{2}{1 + 4x^2}\right)}{(\arctan(2x))^2}.$

$g'(x) = \dfrac{d}{dx}\left(\dfrac{4x^3 - 1}{\arctan(2x)} \right)$

$= \dfrac{\arctan(2x)12x^2 - (4x^3 - 1)\dfrac{d}{dx}(\arctan(2x))}{(\arctan(2x))^2}$

by the quotient rule

$$= \frac{12x^2 \arctan(2x) - (4x^3 - 1)\left(\dfrac{2}{1 + 4x^2}\right)}{(\arctan(2x))^2}.$$

D.

$$y' = \frac{1}{2 - (1 - y^2)^{-\frac{1}{2}}}.$$

Differentiating the equation $x = \arccos y + 2y$ implicitly with respect to x, you get

$$1 = -\frac{1}{\sqrt{(1 - y^2)}}y' + 2y'.$$

Then, solving for y' you find

$$1 = \left(2 - (1 - y^2)^{-\frac{1}{2}}\right)y'$$

$$\Rightarrow y' = \frac{1}{2 - (1 - y^2)^{-\frac{1}{2}}}.$$

---

2. $\arctan(1.8) \approx 1.064$.

Using the equation $f(x) \approx f(a) + (x - a)f'(a)$, we get

$$f(1.8) \approx f(\sqrt{3}) + (1.8 - \sqrt{3})f'(\sqrt{3}).$$

The derivative of f is $f'(x) = \dfrac{1}{1 + x^2}$, so $f'(\sqrt{3}) = \dfrac{1}{1 + 3} = \dfrac{1}{4}$.

So $f(1.8) \approx \frac{\pi}{3} + (1.8 - \sqrt{3})f'(\sqrt{3}) = \frac{\pi}{3} + (1.8 - \sqrt{3})(\frac{1}{4}) = 1.064$.

You can confirm that this approximation is close by using your calculator to find its approximation for $\arctan(1.8) \approx 1.0637$.

3.

A. The equation is $y = 2(x + \frac{\sqrt{3}}{2}) - \frac{\pi}{3}$.

$y' = \dfrac{1}{\sqrt{1 - x^2}}$, so the slope of the curve at the particular point $(-\frac{\sqrt{3}}{2} - \frac{\pi}{3})$ is

$$\frac{1}{\sqrt{1 - \left(\dfrac{-\sqrt{3}}{2}\right)^2}} = 2.$$

Using the point slope formula for a line $(y = m(x - x_1) + y_1)$ gives you the equation

$$y = 2\left(x + \frac{\sqrt{3}}{2}\right) - \frac{\pi}{3}.$$

B. The equation is $y = -.65571(x - .53119) - 1.4$.

Implicit differentiation yields $2yy' + y' = \dfrac{1}{\sqrt{1 - x^2}}$, which gives you

$$y' = \frac{1}{(2y + 1)\sqrt{1 - x^2}}.$$

At the point $(.53119, -1.40000)$, you get

$$y' = \frac{1}{(2(-1.4) + 1)\sqrt{1 - .53119^2}} = -.655714095.$$

Then, using the point slope formula $(y = m(x - x_1) + y_1)$ you get

$$y = -.65571(x - .53119) - 1.4.$$

4.

A. The searchlight is rotating at approximately $7.297297297 \times 10^{-2}$ radians per second when the motorcycle is 100 feet from point A.

You could use a drawing here illustrating the situation when the motorcycle is 100 feet from A. You'd have a right triangle oriented like this:

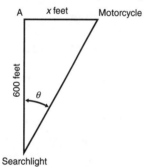

From the drawing above, you have the relationship $\theta = \arctan\left(\dfrac{x}{600}\right)$.

You're interested in $\dfrac{d\theta}{dt}$ when $x = 100$.

Differentiating implicitly with respect to t gives you

$$\frac{d\theta}{dt} = \frac{1}{1 + \left(\dfrac{x}{600}\right)^2}\left(\frac{1}{600}\right)\frac{dx}{dt} = \frac{1}{1 + \left(\dfrac{100}{600}\right)^2}\left(\frac{1}{600}\right)45 = \frac{27}{370} = 0.07297297 \ \frac{\text{radians}}{\text{second}}.$$

B. The correct time is approximately $t = 9.64943192$ seconds.

Here $\theta = \arctan\left(\dfrac{d(t)}{600}\right)$. So $\dfrac{d\theta}{dt} = \dfrac{1}{1 + \left(\dfrac{20t + t^2}{600}\right)^2}\left(\dfrac{1}{600}\right)(20 + 2t)$.

Graphing $\dfrac{d\theta}{dt}$ on a calculator, one finds that the function has a maximum value on the interval $[0, 30]$ at about $t = 9.64943192$. (Don't forget to check the endpoints, too!)

B. Review of Logarithmic and Exponential Functions

►Objective 1   Identify whether a function is algebraic, exponential, or logarithmic (these functions may be given as a graph, a formula, or a table of numbers).

Example 1   Identify whether each of the following could represent an exponential function of the form $y = ab^x$, where $0 < b < 1$.

A.

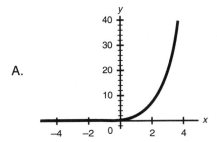

B.

x	-1	0	1	2	3
$f(x)$	2	4	8	16	32

C.

x	-1	0	1	2	3
$f(x)$	8	2	$\frac{1}{2}$	$\frac{1}{8}$	$\frac{1}{32}$

D.

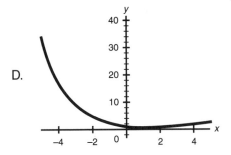

Tips   When you need to compare the growth of logarithmic and exponential functions to other functions, remember that exponential growth functions will always outgrow polynomials and power functions; all three of these will generally outgrow logarithms. (The only exceptions here are the constant polynomials, $p(x) = Constant$, which don't outgrow logarithms.)

Algebraically, you have $\lim\limits_{x \to \infty} \dfrac{a^x}{x^b} = \infty$ and $\lim\limits_{x \to \infty} \dfrac{a^x}{p(x)} = \infty$ for every $a > 1$, $b > 0$, and polynomial $p(x)$ with a positive coefficient for the term with the highest exponent.

$\lim\limits_{x \to \infty} \dfrac{a^x}{\log_c x} = \infty$, $\lim\limits_{x \to \infty} \dfrac{x^b}{\log_c x} = \infty$ and $\lim\limits_{x \to \infty} \dfrac{p(x)}{\log_c x} = \infty$ for every $a > 1, b > 0, c > 0$, and nonconstant polynomial $p(x)$ with a positive coefficient for the term with the highest exponent.

Answers

A.   This is not $y = ab^x$, where $0 < b < 1$. In this picture it would appear that $b > 1$.

B.   This is not $y = ab^x$, where $0 < b < 1$. From this data it would appear that $b > 1$.

C.   This does seem to fit $y = ab^x$, where $0 < b < 1$. It's $y = 2(\frac{1}{4})^x$.

D.   This is not $y = ab^x$, where $0 < b < 1$. Remember that exponential and logarithmic functions are monotonic and thus invertible. Notice on this one how the graph has positive slope around $x = 4$ and is thus not monotonic.

Example 2   Identify whether each of the following could represent a logarithmic function of the form $y = a \, \log_b x$, where $a > 0$.

A.

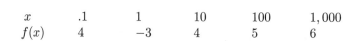

x	.1	1	10	100	$1,000$
$f(x)$	4	-3	4	5	6

B.

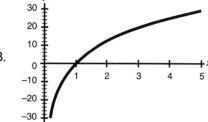

C.

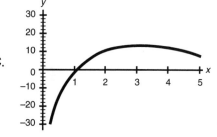

D.

x	.1	1	10	100	$1,000$
$f(x)$	-1	0	1	2	3

Answers

A.   This is not logarithmic. Logarithmic functions are monotonic.

B.   This is logarithmic. It's $y = 20 \log_3 x$.

C.   This is not logarithmic. Logarithmic functions are monotonic. Also realize that

    $\lim\limits_{x \to \infty} a \log_b x = \infty$ if $a > 0$. Notice how the graph goes from increasing to decreasing.

D. This is $y = \log_{10} x$.

►Objective 2   Use the laws of exponents and logarithms to manipulate expressions involving exponential and logarithmic functions.

Example   Which of the following statements are true for x, y, a, b, all greater than zero?

A. $\log_b x + \log_b y = \log_b(x + y)$

B. $e^{x+y} = e^x + e^y$

C. $\log_b x^a = a \log_b x$

D. $(e^x)^b = be^x$

E. $\log_b 0 = 1$

F. $\log_2(a) = \dfrac{\ln a}{\ln 2}$

Answers

A. $\log_b x + \log_b y = \log_b(x + y)$ is false. $\log_b x + \log_b y = \log_b(xy)$ is true.

B. $e^{x+y} = e^x + e^y$ is false. $e^{x+y} = e^x e^y$ is true.

C. $\log_b x^a = a \log_b x$ is true.

D. $(e^x)^b = be^x$ is false. $(e^x)^b = e^{bx}$ is true.

E. $\log_b 0 = 1$ is false; $\log_b x$ has a domain $x > 0$. $\log_b 1 = 0$ is true.

F. $\log_2(a) = \dfrac{\ln a}{\ln 2}$ is true because $\log_a x = \dfrac{\ln x}{\ln a}$.

►Objective 3   Solve problems, using the fact that logarithmic functions and exponential functions are inverses of each other.

Example   Given that the slope of the tangent line to $y = \log_c x$ at point (a, b) is equal to 7, what is the slope of the tangent line to $y = c^x$ at point (b, a)?

Tip   If $g(x)$ is the inverse of $f(x)$, $g'(x) = \dfrac{1}{f'(g(x))}$.

Answer   If $g(x)$ is the inverse of $f(x)$, $g'(x) = \dfrac{1}{f'(g(x))}$, so the slope of $y = c^x$ at $x = b$ is just the reciprocal of the slope of $y = \log_c x$ at $x = a$: $m = \frac{1}{7}$.

►Objective 4   Write equations that model simple exponential growth and decay situations.

Example   A bucket containing salt is leaking in such a way that the amount of salt remaining in the bucket at any time is only $\frac{1}{4}$ as much as there was an hour before; in other words, the quantity of salt in a bucket is $\frac{1}{4}$ times as big every hour. Write an equation that gives the amount of salt in the bucket for any time $t \geq 0$ in hours. (Assume that the bucket has C_o lbs of salt at $t = 0$.)

Answer   The amount of salt needs to be multiplied by $\frac{1}{4}$ each hour (t). We can accomplish this easily by simply raising $\frac{1}{4}$ to the power t.

$$C(t) = C_o \left(\frac{1}{4}\right)^t.$$

If you found another form equivalent to this one, that's fine:

$$C(t) = C_o \left(\frac{1}{4}\right)^t = C_o \frac{1}{4^t} = C_o 4^{-t} = C_o [e(\ln 4)]^{-t} = C_o e^{-t \ln 4}.$$

C.   Derivatives Involving Transcendental Functions

▶Objective 1   Find the derivative of a logarithmic function (with any base).

Example   Find the derivative for each of the following:

     A. $g(x) = 7\pi \ln x$

     B. $y = \log_3 x$

Tip   Remember your change of base formula: $\log_a x = \dfrac{\ln x}{\ln a}$.

Answers

     A. $g(x) = 7\pi \ln x \Rightarrow g'(x) = \dfrac{7\pi}{x}$.

     B. $y = \log_3 x \Rightarrow y = \log_3 x = \dfrac{\ln x}{\ln 3} \Rightarrow \dfrac{dy}{dx} = \dfrac{1}{\ln 3}\left(\dfrac{1}{x}\right) = \dfrac{1}{(\ln 3)x}$.

▶Objective 2   Find the derivative of an exponential function (with any base).

Example   Find the derivative for each of the following:

     A. $h(t) = 4e^t$

     B. $f(x) = 15\pi(4)^x$

Tip Remember that if $y = a^x$, then $y' = (\ln a)a^x$ (because $a^x = e^{(\ln a)x}$).

Answers

A. $h(t) = 4e^t \Rightarrow h'(t) = 4e^t$.

B. $f(x) = 15\pi(4)^x \Rightarrow f'(x) = 15\pi(\ln 4)4^x$.

►Objective 3 Combine these rules (for finding derivatives of logarithmic and exponential functions) with the product, quotient, and Chain Rule to find the derivatives of complicated functions involving logs or exponential functions.

Example Find the derivative for each of the following:

A. $y = \ln(4 - x^3)$

B. $g(t) = \dfrac{e^{\sin t}}{t}$

C. $h(x) = e^x \ln x$

Tip The product, quotient, and Chain Rules apply the same way they do to all functions.

Answers

A. $y = \ln(4 - x^3) \Rightarrow \dfrac{dy}{dx} = \dfrac{1}{(4 - x^3)}(-3x^2) = -\dfrac{3x^2}{4 - x^3}$.

B. $g(t) = \dfrac{e^{\sin t}}{t} \Rightarrow g'(t) = \dfrac{t(\cos t)e^{\sin t} - e^{\sin t}}{t^2} = (e^{\sin t})\dfrac{t \cos t - 1}{t^2}$

C. $h(x) = e^x \ln x \Rightarrow h'(x) = e^x \ln x + \dfrac{e^x}{x}$

►Objective 4 Use the derivative to analyze curves for **all** functions including algebraic, trigonometric, inverse trigonometric, logarithmic, exponential, and combinations of these.

Example (Calculator needed) Given that $g'(x) = (x - 3)\ln x$:

A. Find and classify all relative extrema of $g(x)$. Justify your answers.

B. Find the x values for all inflection points of $g(x)$. Justify your answers.

C. Draw a rough sketch of the graph of $g(x)$ *if $g(x) = 0$ at $x = 3$.*

Tips Don't try to find an explicit expression for $g(x)$.

Use number-line tests to help you justify your answers.

Answers

A. For this one, simply do a first derivative number-line test:

x	$0 < x < 1$	$x = 1$	$1 < x < 3$	$x = 3$	$3 < x$
$g'(x)$	*Positive*	0	*Negative*	0	*Positive*
$g(x)$	Increasing	Rel. *Max*	Decreasing	Rel. *Min*	Increasing

(Notice that $\ln x$ goes from negative to positive at $x = 1$.) Thus, $g(x)$ has a relative maximum at $x = 1$. $g(x)$ has a relative minimum at $x = 3$.

B.

$$g'(x) = (x - 3)\ln x$$
$$g''(x) = \ln x + \frac{x - 3}{x}$$

$g''(x) = \ln x + \frac{(x-3)}{x} = 0$. The solution is $x \approx 1.85455$.

x	$0 < x < 1.85455$	$x = 1.85455$	$1.85455 < x$
$g''(x)$	*Negative*	0	*Positive*
$g(x)$	Concave Down	*Inflection Point*	Concave Up

And $g''(x)$ changes sign at approximately $x \approx 1.85455$.

Thus, $g(x)$ has an inflection point at $x \approx 1.85455$.

C. This should remind you of a cubic, except that the function is only defined for positive x. The graph of $y = g(x)$ should look something like this:

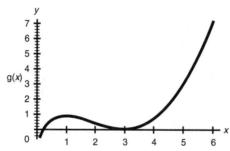

However the exact y values may vary (except at $x = 3$).

▶Objective 5 Use the derivative to optimize situations for **all** functions, including algebraic, trigonometric, inverse trigonometric, logarithmic, exponential, and combinations of these.

Example (**No** calculator)

The rectangle R has vertices of $(a, 0)$, $(3, 0)$, $(3, e^a)$, and (a, e^a), where $0 < a < 3$.

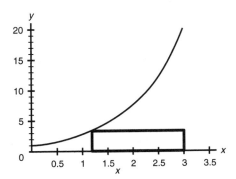

Find the maximum possible value for the area of rectangle R. Justify your answer.

Tip   As with all optimization problem, you need to find a function that gives the quantity being optimized in terms of only one variable.

Answer   The area of a rectangle is $A = bh = (3 - a)e^a$, where the base of the rectangle is $(3 - a)$ and the height is e^a.

$$\frac{dA}{da} = (3 - a)e^a - e^a \text{ from the product rule.}$$

$$\frac{dA}{da} = (3 - a)e^a - e^a = (3 - a - 1)e^a = (2 - a)e^a$$

a	$0 < a < 2$	$a = 2$	$2 < a < 3$
$\dfrac{dA}{da}$	*Positive*	0	*Negative*
Area	Increasing	*Max*	Decreasing

So, the maximum area occurs at $a = 2$.

The maximum area is thus $A = bh = (3 - a)e^a = (3 - 2)e^2 = e^2$.

▶**Objective 6**   Solve problems about rates of change (including rectilinear motion) for **all** functions, including algebraic, trigonometric, inverse trigonometric, logarithmic, exponential, and combinations of these.

Example   Water leaks out of a bucket at a rate given by $r(t) = 4e^{-.2t}$ gallons/minute.

A.   How much water has leaked out after 10 minutes?

B.   Find the time required for five gallons to leak out of the bucket.

Answers

A. $\displaystyle\int_0^{10} 4e^{-.2t}\,dt = 20(1 - e^{-2}) \approx 17.293$ gallons.

B.

$$5 = \int_0^T 4e^{-.2t}dt = \left[-20e^{-.2t} \right]_0^T$$
$$= -20e^{-.2T} - (-20e^{-.2(0)}) = 20 - 20e^{-.2T}, \text{ so}$$
$$-15 = -20e^{-.2T}$$

You could solve this with your calculator now or do a little simplification first:

$$15 = 20e^{-.2T}$$
$$\frac{3}{4} = e^{-.2T}$$
$$\ln \frac{3}{4} = -.2T$$
$$\ln \frac{4}{3} = .2T$$

Thus, $T = 5 \ln \frac{4}{3} \approx 1.43841$ minutes.

►Objective 7   Solve related rates problems for **all** functions, including (algebraic, trigonometric, inverse trigonometric, logarithmic, exponential, and combinations of these.

Example   (Calculator needed)

Line segment $\overline{PQ}$ forms an angle θ with the x-axis as shown below, where point P has the coordinates (x, x^2) and point Q has the coordinates $(1, 0)$. Point P moves along the curve such that $\dfrac{dx}{dt} = 2$ units/second for $0 < x < 1$ and $x = 0$ when $t = 0$.

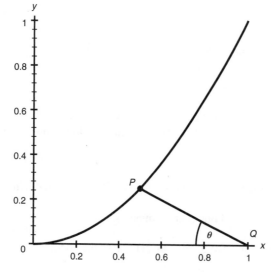

A.   Find a function that gives θ as a function of x.

B.   Find $\dfrac{d\theta}{dt}$ when $x = \frac{1}{2}$.

Tip When doing related rates, remember that you don't need an explicit form for the function.

Answers

A.

$$\tan \theta = \frac{y}{1-x} = \frac{x^2}{1-x}$$

$$\theta = \arctan\left(\frac{x^2}{1-x}\right)$$

(You might also have found this in terms of an arccosine or arcsine.)

B. You could use the Chain Rule on the form $\theta = \arctan\left(\dfrac{x^2}{1-x}\right)$:

$$\frac{d\theta}{dt} = \frac{1}{1+\left(\dfrac{x^2}{1-x}\right)^2} \times \frac{(1-x)2x-(-1)x^2}{(1-x)^2} \times \frac{dx}{dt}.$$

When $x = \frac{1}{2}$ and $\dfrac{dx}{dt} = 2$, we have

$$\frac{d\theta}{dt} = \frac{1}{1+\left(\dfrac{\left(\frac{1}{2}\right)^2}{1-\left(\frac{1}{2}\right)}\right)^2} \times \frac{\left(1-\frac{1}{2}\right)2\left(\frac{1}{2}\right)+\left(\frac{1}{2}\right)^2}{\left(1-\frac{1}{2}\right)^2} \times 2$$

$$\frac{d\theta}{dt} = \frac{1}{1+\left(\dfrac{\frac{1}{4}}{\frac{1}{2}}\right)^2} \times \frac{\frac{1}{2}+\frac{1}{4}}{\frac{1}{4}} \times 2 = \frac{1}{1+\frac{1}{4}} \times \frac{\frac{3}{4}}{\frac{1}{4}} \times 2$$

$$\frac{d\theta}{dt} = \frac{4}{5}(3)2 = \frac{24}{5} \ \frac{\text{radians}}{\text{sec}}.$$

OR

You might have also used the implicit form:

$$\tan \theta = \frac{x^2}{1-x}$$

$$\sec^2(\theta) \times \frac{d\theta}{dt} = \frac{(1-x)2x-(-1)x^2}{(1-x)^2} \times \frac{dx}{dt}$$

$$\theta = \arctan\left(\frac{x^2}{1-x}\right) = \arctan\left(\frac{\left(\frac{1}{2}\right)^2}{1-\frac{1}{2}}\right) = \arctan\left(\frac{1}{2}\right)$$

$$= .463647609$$

$$\sec^2(.463647609) \times \frac{d\theta}{dt} = \frac{\left(1-\frac{1}{2}\right)2\left(\frac{1}{2}\right)+\frac{1}{2}^2}{\left(1-\frac{1}{2}\right)^2} \times 2$$

$$\frac{5}{4} \times \frac{d\theta}{dt} = 3 \times 2$$

$$\frac{d\theta}{dt} = \frac{24}{5} \ \frac{\text{radians}}{\text{sec}}$$

Note that instead of using the calculator to find $\sec^2(\theta)$, you can use a 1-2-$\sqrt{5}$ triangle to show that $\sec^2(\arctan(\frac{1}{2})) = \frac{5}{4}$.

Multiple-Choice Questions on Exponential and Logarithmic Functions

1. Which of the following is **not** true for $f(x) = a^x \quad a > 1$?

A. $f(0) = 1$

B. $f(1) = 0$

C. $\lim\limits_{x \to -\infty} f'(x) = 0$

D. $\lim\limits_{x \to \infty} f(x) = \infty$

E. $\lim\limits_{x \to \infty} f'(x) = \infty$

Answer   B. You know that $f(x) = a^x$ always gives $f(1) = a$, so $f(1) = 0$ is false.

$f(0) = 1$ is true because $a^0 = 1$.

$\lim\limits_{x \to -\infty} f'(x) = 0$ is true because the graph of $f(x) = a^x$ as you go to $-\infty$ gets flatter and flatter.

$\lim\limits_{x \to \infty} f(x) = \infty$ is true because exponential functions where $a > 1$ grow without bound.

$\lim\limits_{x \to \infty} f'(x) = \infty$ is true, since the curve is getting steeper and steeper as you go to the right.

---

2. Which of the following is true for $f(x) = a^x, \quad 0 < a < 1$?

I. $\lim\limits_{x \to \infty} f'(x) = 0$

II. $\lim\limits_{x \to -\infty} f'(x) = \infty$

III. $\lim\limits_{x \to \infty} f(x) = 0$

A. I only

B. II only

C. III only

D. I and III only

E. I, II, and III

Answer   D.

I. $\lim\limits_{x \to \infty} f'(x) = 0$ is **true** because the curve gets flat as you go to the right.

II. $\lim\limits_{x \to -\infty} f'(x) = \infty$ is **false** because the slope is always negative for $f(x) = a^x$, $0 < a < 1$.

III. $\lim\limits_{x\to\infty} f(x) = 0$ is **true**. There is a horizontal asymptote as you go to the right.

---

3. Which of the following could be the graph of $f(x) = a^{-x}, \quad 0 < a < 1$?

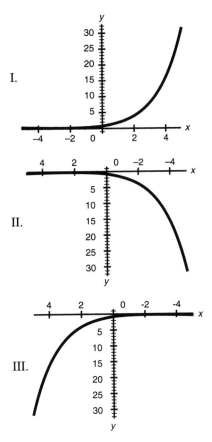

A. I only

B. II only

C. III only

D. I and III only

E. I, II, and III

Answer   D. Notice that $\left(\frac{1}{2}\right)^{-x} = 2^{x}$, so you end up back at the other form of the exponential. You might also simply notice that we reflected the curve across the y-axis when we put a negative in front of the x.

---

4. A modified growth curve has the form $g(x) = C - Aa^{x}$ with $0 < a < 1, C > 0, A > 0$, and $x > 0$. Which of the following statements would be true about $g'(x)$?

I. $\lim\limits_{x\to\infty} g'(x) = 0$

II. $g'(x) > 0$

III. $\lim\limits_{x \to 0+} g'(x) = \infty$

A. I only

B. II only

C. I and II only

D. I and III only

E. I, II, and III

Answer   C.

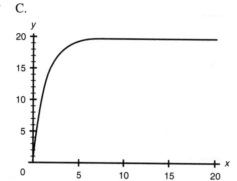

I. $\lim\limits_{x \to \infty} g'(x) = 0$ is true, since the curve flattens out to the right.

II. $g'(x) > 0$ is true, since the curve is always increasing. But

III. $\lim\limits_{x \to 0+} g'(x) = \infty$ is not true.

Notice that $\lim\limits_{x \to 0+} (C - Aa^x) = C - A$, and there isn't a cusp or a vertical tangent line.

---

5. Exponential growth $N(t) = C_0 a^t$ can also be written in the form $N(t) = C_0 e^{kt}$, where k represents the continuous percent growth rate (as a ratio of the investment). Find the value after 7 years of $100 saved in a bank account with 5% interest compounded continuously.

A. $141.91

B. $135.00

C. $105.00

D. $156,000

E. 1.586×10^{17}

Answer   A. $N(t) = C_0 e^{kt} = 100e^{.05t}$, so after 7 years, $N(t) = 100e^{.05(7)} \approx \$141.91$.

6. Consider the function $f(x) = 100 - \dfrac{12,000}{400 + 3,000a^x}$, where $0 < a < 1$.

   Which of the following statements are true?

   I. $\lim\limits_{x \to \infty} f(x) = 70$

   II. $\lim\limits_{x \to -\infty} f(x) = 70$

   III. $\lim\limits_{x \to \infty} f(x) = 100$

   IV. $\lim\limits_{x \to -\infty} f(x) = 100$

   V. $\lim\limits_{x \to \infty} f(x) = 100 - \dfrac{12,000}{400 + 3,000a}$

   A. I and V only

   B. I and IV only

   C. II and III only

   D. III and V only.

   E. I, IV, and V only.

Answer   B. Since $0 < a < 1$, $3,000a^x$ represents exponential decay. So you know that

$$\lim_{x \to \infty} 3,000a^x = 0, \qquad \lim_{x \to -\infty} 3,000a^x = \infty, \qquad \text{and} \qquad \lim_{x \to 0} 3,000a^x = 3,000.$$

Therefore,

$$\lim_{x \to \infty} f(x) = 100 - \frac{12,000}{400 + 0} = 70,$$
$$\lim_{x \to -\infty} f(x) = 100 - 0 = 100, \qquad \text{and}$$
$$\lim_{x \to 0} f(x) = 100 - \frac{12,000}{400 + 3,000} = \frac{1,640}{17}.$$

7. (Exponential decay) An isotope of sodium has a half-life of 20 hours. Suppose an initial sample of this isotope has mass 10 grams. The amount of the isotope (in grams) remaining after t hours is given by:

   A. $10(1.035264924)^t$

   B. $10(0.965936329)^t$

   C. $10e^{-0.965936329t}$

   D. Both $10(1.035264924)^t$ and $10e^{-0.965936329t}$

E. Both $10(0.965936329)^t$ and $10e^{-0.965936329t}$

Answer   B. The amount of isotope remaining after t hours is given by

$$Amount(t) = C_0 a^t.$$

At time $t = 0$, the amount of isotope remaining is 10 grams, so $10 = C_0 a^{(0)} = C_0$. Alternatively, you could have remembered that C_0 here must correspond to the initial amount of the isotope.

So now you have $Amount(t) = 10a^t$.

Since the half-life of the isotope is 20 hours,

$$Amount(20) = 10a^{20} = \frac{1}{2}(10) = 5 \text{ grams}.$$

Solving for a gives you

$$10a^{20} = 5$$
$$\Rightarrow a^{20} = \frac{1}{2}$$
$$\Rightarrow a = \left(\frac{1}{2}\right)^{\frac{1}{20}} = 0.965936329.$$

So the amount (in grams) of isotope remaining after t hours is given by the formula

$$Amount(t) = 10(0.965936329)^t.$$

If you like, you can rewrite this as

$$10(.965936329)^t = 10e^{(\ln 0.965936329)^t}$$
$$10e^{-0.03465735903t}.$$

---

8.   The *decibel level* of a sound is given by the equation $D = 10\log_{10}(\frac{I}{I_0})$, where I is the sound's intensity and I_0 corresponds to the intensity of the weakest sound that can be detected by the human ear. A sound of 140 decibels produces pain in the average human ear. Approximately how many times greater than I_0 must the intensity I of a sound be to reach this decibel level?

A. $22,026$ times greater

B.   14 times greater

C.   1.14612 times greater

D.   2.63906 times greater

E. 10^{14} times greater

Answer   E. You want to solve the equation $140 = 10\log_{10}(\frac{I}{I_0})$ for the ratio $(\frac{I}{I_0})$. You get

$$10\log_{10}\left(\frac{I}{I_0}\right)$$

$$\Rightarrow 14 = \log_{10}\left(\frac{I}{I_0}\right)$$

$$\Rightarrow 10^{14} = \left(\frac{I}{I_0}\right).$$

So I needs to be 10^{14} greater than I_0 in intensity for the sound to produce pain in the average human!

---

9.   Which of these correctly describes $\ln(4x)$?

A. $\displaystyle\int_{1}^{4x} \frac{1}{t}\,dt$

B. $\displaystyle\int_{1}^{x} \frac{1}{4t}\,dt$

C. $\displaystyle\int_{1}^{4x} \frac{1}{4t}\,dt$

D. $4\displaystyle\int_{1}^{x} \frac{1}{t}\,dt$

E.   None of these.

Answer   A. This is an area function, and the variable goes in the upper limit of the integral.

---

10.   You can tell that $y = e^{2x}$ is always positive for which of the following reasons?

I. It's an exponential function with a base > 0.

II. It's a square.

III. Its derivative is always positive.

A.   I only

B.   II only

C.   III only

D.   I and II

E.   I and III

Answer   D. All three statements are true, but only I and II can tell you that this function is positive.

$y = e^{2x}$ is positive because it's an exponential function with base $e > 0$. You can also tell that it's positive because $y = e^{2x} = (e^x)^2$ is a square. The fact that its derivative is positive means that $y = e^{2x}$ is increasing, but it doesn't imply that $y = e^{2x}$ is itself positive.

---

11.   Find the derivative of $y = 2^{17 \log_2(35x)}$.

A. $\dfrac{dy}{dx} = 2^{17 \log_2(35x)} 35 \ln 2$

B. $\dfrac{dy}{dx} = 595(35x)^{16}$

C. $\dfrac{dy}{dx} = 2^{17 \log_2(35x)}$

D. $\dfrac{dy}{dx} = (35)2^{17 \log_2(35x)}$

E. $\dfrac{dy}{dx} = (35)2^{16 \log_2(35x)}$

Answer   B. Note there's a nice shortcut because you have 2 to the $\log_2$ of something, and these functions are inverses.

So $y = 2^{17 \log_2(35x)} = (35x)^{17}$, and the derivative is $\dfrac{dy}{dx} = 17(35x)^{16}35 = 595(35x)^{16}$.

---

12.   Which of these is the correct derivative of $y = 4^x$?

A. $\dfrac{dy}{dx} = \dfrac{e^{4x}}{4}$

B. $\dfrac{dy}{dx} = \dfrac{4^x}{\ln 4}$

C. $\dfrac{dy}{dx} = \dfrac{4^x}{2 \ln 2}$

D. $\dfrac{dy}{dx} = \dfrac{e^4}{e^x}$

E.   None of these.

Answer   E. The factor of $\ln 4 = 2 \ln 2$ is multiplied by the derivative, not divided.

You can remember how to find the derivative of $y = a^x$ where $a > 0$ without having to memorize this derivative rule:

Start with $y = 4^x = (e^{\ln 4})^x = e^{x \ln 4}$. Then the derivative is: $\dfrac{dy}{dx} = (\ln 4)4^x$.

---

13.   If $f(x) = 2e^{3x}$, then $f'(x) =$

A. $2e^{3x}$

B. $3e^{3x}$

C. $6e^{3x}$

D. $6e^{3x-1}$

E. $6xe^{3x-1}$

Answer   C. Using the Chain Rule and the fact that $\dfrac{d}{dx}(e^x) = e^x$, you get

$$f'(x) = \frac{d}{dx}(2e^{3x}) = 2\frac{d}{dx}(e^{3x})$$

$$= 2(e^{3x})\frac{d}{dx}(3x) \text{ (by the Chain Rule)}$$

$$= 2(e^{3x})(3) = 6e^{3x}.$$

---

14. $\dfrac{d}{dx}\ln(12x+4) =$

A. $\dfrac{1}{12x+4}$

B. $\dfrac{1}{12x}$

C. $\dfrac{1}{x}$

D. $\dfrac{12}{12x+4}$

E.   None of the above.

Answer   D. Using the Chain Rule and the fact that $\dfrac{d}{dx}\ln(x) = \dfrac{1}{x}$, you have

$$\frac{d}{dx}\ln(12x+4) = \frac{1}{12x+4} \times \frac{d}{dx}(12x+4)$$

$$= \frac{1}{12x+4}(12).$$

$$= \frac{12}{12x+4}.$$

---

15. $\dfrac{d}{dx}\log_5(\sqrt{x}-1) =$

A. $\dfrac{1}{2\sqrt{x}(\sqrt{x}-1)}$

B. $\dfrac{\ln 5}{2\sqrt{x}(\sqrt{x}-1)}$

C. $\dfrac{(\ln 5)\sqrt{x}}{2(\sqrt{x}-1)}$

D. $\dfrac{\sqrt{x}}{2(\ln 5)(\sqrt{x}-1)}$

E. None of the above

Answer E. Using the Chain Rule and the fact that $\dfrac{d}{dx}\log_5 x = \dfrac{1}{x(\ln 5)}$, you get

$$\frac{d}{dx}\log_5(\sqrt{x}-1) = \frac{1}{(\ln 5)(\sqrt{x}-1)}\frac{d}{dx}(\sqrt{x}-1)$$

$$= \frac{1}{(\ln 5)(\sqrt{x}-1)}\left(\frac{1}{2\sqrt{x}}\right)$$

$$= \frac{1}{2(\ln 5)\sqrt{x}(\sqrt{x}-1)}.$$

---

16. If $f(x) = 5^{3x} + 2x^2$, then $f'(x) =$

A. $3 \times 5^{3x} + 4x$

B. $5^{3x}\log_5(3x) + 4x$

C. $3(\ln 5)5^{3x} + 4x$

D. $5^{3x}\ln 5 + 4x$

E. $5^{3x}\log_5(3x)$

Answer C. Using the Chain Rule and the fact that $\dfrac{d}{dx}(5^x) = (\ln 5)5^x$ you get

$$f'(x) = \frac{d}{dx}(5^{3x} + 2x^2)$$

$$= \frac{d}{dx}(5^{3x}) + \frac{d}{dx}(2x^2)$$

$$= (\ln 5)5^{3x}\frac{d}{dx}(3x) + 4x$$

$$= (\ln 5)5^{3x}(3) + 4x$$

$$= 3(\ln 5)5^{3x} + 4x.$$

---

17. If $f(x) = \ln(\cos x)$, then $f'(x) =$

A. $\dfrac{-1}{\sin x \cos x}$

B. $\dfrac{-\sin x}{\cos x}$

C. $\dfrac{\sin x}{\cos x}$

D. $\dfrac{1}{\cos x}$

E. None of the above

Answer B. Using the Chain Rule and the fact that $\dfrac{d}{dx}\ln x = \dfrac{1}{x}$, you get

$$\begin{aligned}
f'(x) &= \frac{d}{dx}\ln(\cos x)\\
&= \frac{1}{\cos x}\frac{d}{dx}(\cos x)\\
&= \frac{1}{\cos x}(-\sin x)\\
&\frac{-\sin x}{\cos x}.
\end{aligned}$$

---

18. $\dfrac{d}{dx}((\ln x)2^x) =$

A. $(\ln 2)\dfrac{2^x}{x} + (\ln x)2^x$

B. $\dfrac{(\ln 2)2^x}{x}$

C. $2^x + (\ln 2)(\ln x)2^x$

D. $\dfrac{2^x}{x} + (\ln 2)(\ln x)2^x$

E. $\dfrac{2^x}{x} + (\ln 2)2^x + (\ln x)2^x$

Answer D. Using the product rule along with the Chain Rule to differentiate $2^x = e^{(\ln 2)x}$, you get

$$\begin{aligned}
\frac{d}{dx}(\ln x)2^x &= 2^x\frac{d}{dx}(\ln x) + (\ln x)\frac{d}{dx}2^x\\
&= 2^x\left(\frac{1}{x}\right) + \ln x(\ln 2)2^x\\
&= \frac{2^x}{x} + (\ln 2)(\ln x)2^x.
\end{aligned}$$

---

19. If $f(x) = 3e^{-2x^2}$, then $f'(x) =$

A. $3e^{-2x^2}$

B. $-12x^2 e^{-2x^2}$

C. $-12xe^{-2x^2}$

D. $-6x^2 e^{-2x^2}$

E. None of the above

Answer   C. By the chain rule you have

$$f'(x) = \frac{d}{dx}(3e^{-2x^2}) = 3e^{-2x^2} \times \frac{d}{dx}(-2x^2) = 3e^{-2x^2}(-4x) = -12xe^{-2x^2}.$$

---

20. $\dfrac{d}{dx}(\log_2(3x)) =$

A. $\dfrac{1}{3x}$

B. $\dfrac{1}{x}$

C. $\dfrac{3}{x(\ln 2)}$

D. $\dfrac{1}{x(\ln 2)}$

E. None of the above

Answer   D. By the chain rule you have

$$\frac{d}{dx}\log_2(3x) = \frac{1}{(\ln 2)(3x)} \times \frac{d}{dx}(3x)$$
$$= \frac{3}{(\ln 2)(3x)}$$
$$= \frac{1}{x(\ln 2)}.$$

---

21.   If $f(x) = \ln[(4x^4 + 3)^3]$, then $f'(x) =$

A. $3(\dfrac{16x^3}{4x^4 + 3})^2$

B. $3\ln[(4x^4 + 3)]^2(\dfrac{16x^3}{4x^4 + 3})$

C. $\dfrac{48}{4x}$

D. $\dfrac{48x^3}{(4x^4 + 3)}$

E. $\dfrac{48x^3}{(4x^4 + 3)^3}$

Answer D. First, rewrite $f(x) = \ln[(4x^4 + 3)^3] = 3\ln(4x^4 + 3)$ to simplify things. Then

$$f'(x) = 3\frac{1}{(4x^4 + 3)}(16x^3) = \frac{48x^3}{(4x^4 + 3)}.$$

---

22. If $f(x) = [\ln(3x^2 - 36)]^2$, then $f'(x) =$

A. $\dfrac{24x}{3x^2 - 36}$

B. $\dfrac{2\ln(3x^2 - 36)}{3x^2 - 36}$

C. $\dfrac{6x}{3x^2 - 36}$

D. $\dfrac{12x\ln(3x^2 - 36)}{3x^2 - 36}$

E. $\dfrac{12x}{3x^2 - 36}$

Answer D.

$$\begin{aligned}
f'(x) &= \frac{d}{dx}[\ln(3x^2 - 36)]^2 \\
&= 2[\ln(3x^2 - 36)]^1 \times \frac{d}{dx}[\ln(3x^2 - 36] \text{ by the Chain Rule} \\
&= 2\ln(3x^2 - 36) \times \frac{1}{3x^2 - 36} \times \frac{d}{dx}(3x^2 - 36), \text{ again by the Chain Rule} \\
&= \frac{2\ln(3x^2 - 36)}{3x^2 - 36} \times 6x \\
&= \frac{12x\ln(3x^2 - 36)}{3x^2 - 36}.
\end{aligned}$$

---

23. $\dfrac{d}{dx}\sqrt{1 + e^{3x}} =$

A. $\dfrac{3}{2\sqrt{1 + e^{3x}}}e^{3x}$

B. $\dfrac{e^{3x}}{(1 + e^{3x})^{-\frac{1}{2}}}$

C. $\frac{3}{2}(1 + e^{3x})^{-\frac{1}{2}}$

D. $3(1 + e^{3x})^{-\frac{1}{2}}$

E. $\frac{e^{3x}}{2}(1 + e^{3x})^{-\frac{1}{2}}$

Answer   A. Using the Chain Rule, you get

$$\frac{d}{dx}\sqrt{1 + e^{3x}} = \frac{1}{2}(1 + e^{3x})^{-\frac{1}{2}}\frac{d}{dx}(1 + e^{3x})$$
$$= \frac{1}{2}(1 + e^{3x})^{-\frac{1}{2}}(3e^{3x})$$
$$= \frac{3}{2\sqrt{1 + e^{3x}}}e^{3x}.$$

---

24.   If $f(x) = e^{1/x} + \dfrac{1}{e^x}$, then $f'(x) =$

A. $\dfrac{e^{\frac{1}{x}}}{x^2} - \dfrac{x}{e^x}$

B. $\dfrac{1}{x^2}e^{\frac{1}{x}} + \dfrac{1}{e^x}$

C. $\dfrac{-e^{\frac{1}{x}}}{x^2} - \dfrac{1}{e^x}$

D. $\dfrac{-e^{\frac{1}{x}}}{x^2} - \dfrac{x}{e^x}$

E. $-x^{-2}e^{\frac{1}{x}} + \dfrac{1}{e^x}$

Answer   C. First, rewrite $f(x)$ as $f(x) = e^{(x^{-1})} + e^{-x}$ (it's always easier to differentiate when fractions have been turned into negative exponents). Then,

$$f'(x) = e^{(x^{-1})} \times (-x^{-2}) + e^{-x} \times (-1) = \frac{-e^{\frac{1}{x}}}{x^2} - \frac{1}{e^x}.$$

---

25. $\dfrac{d}{dx}(3^{\cos^2 x}) =$

A. $2(\ln 3)(3\cos^2 x)(\cos x)(\sin x)$

B. $-2(\ln 3)(3^{\cos^2 x})(3\cos^2 x)(\cos x)(\sin x)$

C. $-2(\ln 3)(3^{\cos^2 x})(\cos x)(\sin x)$

D. $-2(\ln 3)(3^{\cos^2 x})(\sin x)$

E. $2(\ln 3)(3^{\cos^2 x}) \cos x$

Answer C. You have

$$
\begin{aligned}
\frac{d}{dx} 3^{\cos^2 x} &= (\ln 3) 3^{\cos^2 x} \times \frac{d}{dx}(\cos^2 x) \\
&= (\ln 3) 3^{\cos^2 x} (2 \cos x)(-\sin x) \\
&= -2(\ln 3)(3^{\cos^2 x})(\cos x)(\sin x).
\end{aligned}
$$

---

26. $\dfrac{d}{dx}(3^{\ln(3x)}) =$

A. $\dfrac{3^{\ln(3x)} \ln 3}{x}$

B. $e^{\ln^2 3} x^{\ln 3} \ln 3$

C. $e^{\ln 3} x^{\ln 3} \ln 3$

D. $\dfrac{(3)3^{\ln(3x)} \ln(3x)}{x}$

E. $\dfrac{3^{\ln(3x)} \ln(3x)}{x}$

Answer A. Using the Chain Rule repeatedly, you get

$$
\begin{aligned}
\frac{d}{dx} 3^{\ln(3x)} &= (\ln 3) 3^{\ln(3x)} \frac{d}{dx} \ln(3x) \\
&= (\ln 3) 3^{\ln(3x)} \frac{1}{3x}(3) \\
&= \frac{3^{\ln(3x)} \ln 3}{x}.
\end{aligned}
$$

---

27. If $y = \ln(\dfrac{x}{\tan x})$, then $y' =$

A. $\dfrac{\tan x - \sec^2 x}{\tan^2 x}$

B. $-2 \tan x - \sec^2 x$

C. $\dfrac{x(\tan x - \sec^2 x)}{\tan^3 x}$

D. $2 \tan x \sec^2 x$

E. $\dfrac{\tan x - x \sec^2 x}{x \tan x}$

Answer   E. Using the Chain Rule repeatedly, you get

$$y' = \frac{d}{dx}\ln(\frac{x}{\tan x}) = \frac{1}{(\frac{x}{\tan x})} \times \frac{d}{dx}\left(\frac{x}{\tan x}\right)$$

$$= \frac{\tan x}{x} \times \frac{(\tan x) - x\sec^2 x}{\tan^2 x}$$

$$= \frac{\tan x - x\sec^2 x}{x\tan x}.$$

Free-Response Questions on Logarithmic and Exponential Functions

Note: Don't simplify these unless you absolutely have to. Simplification is a great place to make errors, and it makes it difficult to tell if your answer is correct.

---

1. For each of the following, use implicit differentiation to find y'. If possible, express y' explicitly as a function of x.

A. $y = (x+1)^{2x}$

B. $12^y + \log_5(xy) - 4x = -3$

---

2. Consider the curve $y = \log_4 x - \frac{1}{5}x$.

    Find the coordinates of the global maximum on $(0, \infty)$, and show that the graph of the curve is concave down for all $x > 0$. (Only analytic methods are valid here).

---

3. One of the most accurate formulas for predicting the height of a preschool child is called the Jenss model. If x represents the child's age in years (between $\frac{1}{4}$ and 6), the height $h(x)$ (in centimeters) of the child can be approximated by $h(x) = 79.041 + 6.39x - e^{3.261-0.993x}$.

A. Using this model, predict the height and the rate of growth when a child reaches 2 years of age.

B. Using this model, predict at what age a child's rate of growth is largest, and at what age its rate of growth is smallest. Remember, the model is valid only on the domain $\frac{1}{4} \le x \le 6$. Justify your answer.

---

4. The table below gives data from an experiment in which a bacteria culture is placed in a limited nutrient environment (t is the number of hours and $P(t)$ is the number of bacteria at time t). The culture started with 2,000 bacteria, and each hour the size of the bacteria population was determined and recorded.

t (hrs)	0	1	2	3	4	5	6
$P(t)$	2,000	3,390	5,802	9,835	16,720	28,411	48,330

A. Using just the data provided, estimate the rate of population growth of the bacteria at $t = 3$ hours. Explain your method.

B. Using two points of data from the table above, determine constants C_o and b_t which can be used to model the function $P(t)$ as an exponential function $P(t) = C_o(b^t)$. Your exponential function doesn't have to satisfy all the data points in the table exactly, just two of them. (Hint: To make it easier for yourself, one point you should use is at $t = 0$.)

C. Using the exponential model for $P(t)$ that you derived in part b, estimate the rate of growth of the bacteria population at $t = 3$ hours with $P'(3)$. Compare this result to your result in part A.

Answers

1.

A.

$$y' = \left(2\ln(x+1) + \frac{2x}{x+1}\right)(x+1)^{2x}.$$

The first step is to take the natural log of both sides, $\ln y = 2x\ln(x+1)$. Then, differentiating implicitly gives you

$$\frac{y'}{y} = 2\ln(x+1) + \frac{2x}{x+1}$$

$$\Rightarrow y' = \left(2\ln(x+1) + \frac{2x}{x+1}\right)y = \left(2\ln(x+1) + \frac{2x}{x+1}\right)(x+1)^{2x}$$

B.

$$y' = \frac{4 - \frac{1}{x(\ln 5)}}{\left[(\ln 12)12^y + \frac{1}{y(\ln 5)}\right]}.$$

Differentiating implicitly gives you

$$\Rightarrow (\ln 12)12^y y' + \frac{1}{(\ln 5)(xy)}(xy' + y) - 4 = 0$$

$$\Rightarrow \left[(\ln 12)12^y + \frac{1}{y(\ln 5)}\right]y' = 4 - \frac{1}{x(\ln 5)} \Rightarrow y' = \frac{4 - \frac{1}{x(\ln 5)}}{\left[(\ln 12)12^y + \frac{1}{y(\ln 5)}\right]}.$$

---

2. The global maximum over the domain $(0, \infty)$ is $y = 0.203999713$ and occurs at $x = 3.606737603$.

First, use the first derivative to find the critical point(s): $y' = \frac{1}{(\ln 4)x} - \frac{1}{5}$.

Setting this equal to 0 gives you

$$\frac{1}{(\ln 4)x} - \frac{1}{5} = 0,$$

whose solution is $x = \frac{5}{\ln 4} \approx 3.606737602$.

The y value at $x = \frac{5}{\ln 4}$ is $\log_4\left(\frac{5}{\ln 4}\right) - \frac{1}{5}\left(\frac{5}{\ln 4}\right) \approx 0.203999713$.

The second derivative is

$$y'' = -\frac{1}{(\ln 4)x^2},$$

which is always negative, and the graph is concave down for all $x > 0$.

Therefore, the critical point $(3.606737603, 0.203999713)$ is the global maximum.

3.

A. At age 2 years, the child's height is predicted to be 88.24229859 centimeters, while the child's rate of growth is predicted to be 9.9436505 centimeters per year.

The rate of growth (in centimeters per year) is given by

$$h'(x) = 6.39 + 0.993e^{3.261 - 0.993x}.$$

At age 2 (years), the child's height is predicted to be

$$h(2) = 79.041 + 6.39(2) - e^{3.261 - 0.993(2)} = 88.24229859 \text{ centimeters,}$$

and the child's rate of growth is predicted to be

$$h'(2) = 6.39 + 0.993e^{3.261 - 0.993(2)} = 9.9436505$$

centimeters per year.

B. The maximum growth rate occurs at $\frac{1}{4}$ years old, while the minimum growth rate occurs at 6 years old. First, we find the critical points:

$$h'(x) = 6.39 + 0.993e^{3.261 - 0.993x}. \text{ Setting } h'(x) = 0,$$
$$6.39 + 0.993e^{3.261 - 0.993x} = 0$$
$$\Rightarrow 0.993e^{3.261 - 0.993x} = -6.39$$
$$\Rightarrow e^{3.261 - 0.993x} = \frac{-6.39}{0.993} = -6.435045317.$$

But there are no solutions to this equations, since the exponential is always positive! Thus, there are no critical points, and the extrema values must occur at the endpoints of the intervals.

At $x = \frac{1}{4}$, the rate of growth is

$$h'\left(\frac{1}{4}\right) = 6.39 + 0.993e^{3.261 - 0.993(.25)} \approx 26.59086419 \text{ cm per year,}$$

while at $x = 6$, the rate of growth is

$$h'(6) = 6.39 + 0.993e^{3.261 - 0.993(6)} \approx 6.45693558 \text{ cm per year.}$$

Note that this matches our intuitive idea that a child's rate of growth tends to decrease as the child gets older and bigger.

4.

A. The estimate is attained by approximating the derivative $P'(t)$ with a difference quotient. Alternatively this can be seen as approximating the instantaneous rate of change with an average rate of change over the smallest possible interval provided.

Possible quotients are

$$rate \approx \frac{P(4) - P(3)}{4 - 3} = \frac{16,720 - 9,835}{1} = 6,885 \text{ bacteria per hour}$$

$$rate \approx \frac{P(3) - P(2)}{3 - 2} = \frac{9,835 - 5,802}{1} = 4,033 \text{ bacteria per hour}$$

$$rate \approx \frac{P(4) - P(2)}{4 - 2} = \frac{16,720 - 5,802}{2} = 5,459 \text{ bacteria per hour}$$

Your explanation should contain something about approximating the derivative $P'(t)$ with a difference quotient or approximating the instantaneous rate of change with an average rate of change over a small interval.

B. There are many possible answers. They should be close to $2,000(1.7^t)$ ($C_0 = 2,000$, $b = 1.7$). If you follow the hint, taking $t = 0$ and $P(t) = 2,000$ gives $C_0 = 2,000$ automatically. Then, taking any other point will enable you to solve for b:

$$t = 1 : 3,390 = 2,000(b^1) \Rightarrow b = \frac{3,390}{2,000} = \frac{339}{200} = 1.695$$

$$t = 2 : 5,802 = 2,000(b^2) \Rightarrow b = \left(\frac{5,802}{2,000}\right)^{\frac{1}{2}} = 1.703$$

$$t = 3 : 9,835 = 2,000(b^3) \Rightarrow b = \left(\frac{9,835}{2,000}\right)^{\frac{1}{3}} = 1.701$$

$$t = 4 : 16,720 = 2,000(b^4) \Rightarrow b = \left(\frac{16,720}{2,000}\right)^{\frac{1}{4}} = 1.700$$

$$t = 5 : 28,411 = 2,000(b^5) \Rightarrow b = \left(\frac{28,411}{2,000}\right)^{\frac{1}{5}} = 1.700$$

$$t = 6 : 48,330 = 2,000(b^6) \Rightarrow b = \left(\frac{48,330}{2,000}\right)^{\frac{1}{6}} = 1.700$$

C. Growth rate at $t = 3$ hours is approximately $5,214$ (your answer may vary but should be close). You need to find $P'(t)$ correctly and then plug in $t = 3$. You should get something close to

$$P'(t) = 2,000(\ln 1.7)1.7^t$$

and

$$P'(3) = 2,000(\ln 1.7)1.7^3 \approx 5,214 \text{ bacteria per hour.}$$

Your answer will depend on what you got in part B.

D. Integrals Involving Transcendental Functions

▶Objective 1   Find antiderivatives involving transcendental functions.

Example   Solve each of the following:

A. $\displaystyle\int \frac{4}{x+7}\,dx$

B. $\displaystyle\int \frac{5}{\sqrt{1-x^2}}\,dx$

C. $\displaystyle\int \frac{3}{\sqrt{1-x}}\,dx$

Tip   Make sure you can find antiderivatives for logarithms and exponentials of any base, as well as antiderivatives associated with the inverse trigonometric functions.

You must become so familiar with these that you can tell when a particular integral will lead to a transcendental function (or not!).

Answers

A. $\displaystyle\int \frac{4}{x+7}\,dx = 4\ln|x+7| + C$

B. $\displaystyle\int \frac{5}{\sqrt{1-x^2}}\,dx = 5\arcsin x + C$

C. $\displaystyle\int \frac{3}{\sqrt{1-x}}\,dx = \int 3(1-x)^{-\frac{1}{2}}\,dx = -6\sqrt{(1-x)} + C.$

▶Objective 2   Use substitution (if necessary) to find more complicated antiderivatives and definite integrals involving transcendental functions.

Example   (**no** calculator) Find each of the following:

A. $\displaystyle\int \frac{x}{1+x^4}\,dx$

B. $\displaystyle\int \frac{x^3}{1+x^4}\,dx$

C. $\displaystyle\int_0^1 te^{t^2}\,dt$

D. $\displaystyle\int_0^1 \frac{1}{\sqrt{4-x^2}}\,dx$

Tip   Use substitution only when you can't find the antiderivative without substitution.

Answers

A. $\displaystyle\int \frac{x}{1+x^4}\, dx = \frac{1}{2}\arctan x^2 + C.$

B. $\displaystyle\int \frac{x^3}{1+x^4}\, dx = \frac{1}{4}\ln(1+x^4) + C.$

   Notice that there are no absolute values here; we know $1 + x^4$ is always positive.

C. $\displaystyle\int_0^1 te^{t^2}\, dt = \left[\frac{1}{2}e^{t^2}\right]_0^1 = \frac{1}{2}e - \frac{1}{2}.$ (Don't forget to plug in 0.)

D. $\displaystyle\int_0^1 \frac{1}{\sqrt{4-x^2}}\, dx = \left[\arcsin\frac{1}{2}x\right]_0^1 = \frac{\pi}{6}.$

▶**Objective 3**   Solve problems related to area for **all** functions covered in this class (algebraic, trigonometric, inverse trigonometric, logarithmic, exponential, and combinations of these).

Example   Write a function that gives the area $A(X)$ under the curve of $y = e^x$ on the domain $0 \le x \le X$.

Answer

$$A(X) = \int_0^X e^x\, dx = e^X - e^0 = e^X - 1.$$

▶**Objective 4**   Solve problems related to average values for **all** functions covered in this class (algebraic, trigonometric, inverse trigonometric, logarithmic, exponential, and combinations of these).

Example   The average value of $g(x) = \dfrac{2}{\sqrt{1-x^2}} + C$ on the domain $-\frac{1}{2} \le x \le \frac{1}{2}$ is equal to π.

Find the value of C. (Justify your answer.)

Tip   Recall, the average value of a function f over the interval $[a, b]$ is given by

$$Avg = \frac{1}{b-a}\int_a^b f(x)\, dx.$$

Answer

$$Avg = \frac{1}{b-a}\int_a^b f(x)\, dx = \frac{1}{1}\int_{-\frac{1}{2}}^{\frac{1}{2}}\left(\frac{2}{\sqrt{1-x^2}} + C\right) dx$$

$$= \left[2\arcsin(x) + Cx\right]_{-\frac{1}{2}}^{\frac{1}{2}} = \left(\frac{2\pi}{6} + \frac{C}{2}\right) - \left(-\frac{2\pi}{6} - \frac{C}{2}\right)$$

$$= \frac{2\pi}{3} + C.$$

So, we need $\frac{2\pi}{3} + C = \pi \Rightarrow C = \frac{\pi}{3}.$

▶Objective 5   Solve problems related to volume for **all** functions covered in this class (algebraic, trigonometric, inverse trigonometric, logarithmic, exponential, and combinations of these).

Example   (calculator needed) The region R is bounded by $y = e^x$ and the line $y = 2x + 1$.

Find the volume of a solid formed by revolving the region R about the x-axis.

Tip   When you use your calculator to find approximate intersection points, also use the calculator
to evaluate the definite integral.

Answer $y = e^x$ and the line $y = 2x + 1$ intersect when $e^x - 2x - 1 = 0$. Solutions are: $x = 0$,
1.256 4312

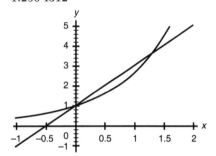

The area of any cross section is given by

$$A(x) = \pi R^2 - \pi r^2 = \pi(2x + 1)^2 - \pi(e^x)^2,$$

and the volume is thus

$$V = \int_0^{1.2564312} \left(\pi(2x + 1)^2 - \pi(e^x)^2\right) dx \approx 4.360963103.$$

Example 2   The region Q in the xy plane is bounded by the curve $y = \dfrac{1}{\sqrt{x^2 + 1}}$, the x-axis, $x = 1$,
and $x = k$, where $k > 1$.

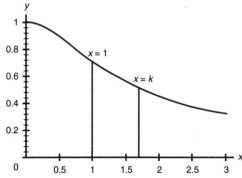

A solid with Q as a base has cross sections perpendicular to the x-axis that are equilateral
triangles. The solid has a volume of $\frac{\pi\sqrt{3}}{48}$. What is the value of k?

Answer The area of an equilateral triangle is given by $A = \frac{1}{2}bh = \frac{\sqrt{3}}{4}b^2$. Since $b = \dfrac{1}{\sqrt{x^2+1}}$,

$$A(x) = \frac{\sqrt{3}}{4}\left(\frac{1}{\sqrt{x^2+1}}\right)^2 = \frac{\sqrt{3}}{4}\left(\frac{1}{x^2+1}\right).$$

The volume of the solid is thus given by

$$V = \int_a^b A(x)\,dx = \int_1^k \frac{\sqrt{3}}{4}\left(\frac{1}{x^2+1}\right)dx = \left[\frac{\sqrt{3}}{4}\arctan x\right]_1^k$$

$$= \frac{\sqrt{3}}{4}\arctan(k) - \frac{\sqrt{3}}{4}\arctan(1).$$

Setting this equal to the volume:

$$\frac{\sqrt{3}}{4}\arctan(k) - \frac{\sqrt{3}}{4}\left(\frac{\pi}{4}\right) = \frac{\pi\sqrt{3}}{48}$$

$$\arctan(k) - \frac{\pi}{4} = \frac{\pi}{12}$$

$$\arctan(k) = \frac{\pi}{12} + \frac{\pi}{4} = \frac{\pi}{3}$$

$$k = \tan\left(\frac{\pi}{3}\right)$$

$$k = \sqrt{3} \approx 1.732.$$

▶**Objective 6** Solve problems related to motion for **all** functions covered in this class (algebraic, trigonometric, inverse trigonometric, logarithmic, exponential, and combinations of these).

Example The acceleration of an object is given by $a(t) = \dfrac{1}{\sqrt{1 - \left(\frac{t}{4}\right)^2}}$ in m/sec^2 for $0 \le t < 4$, and the object has a velocity of 2 m/sec when $t = 0$.

A. Find an expression for $v(t)$ for $0 \le t < 4$.

B. Estimate the net change in position of the object during the third second.

Tips For part B, think of $0 \le t \le 1$ as the first second.

Use your calculator when you can't find an antiderivative.

Answers

A.

$$a(t) = \frac{1}{\sqrt{1 - \left(\frac{t}{4}\right)^2}}$$

$$v(t) = 4\arcsin\left(\frac{t}{4}\right) + C$$

but $v(0) = 2$, so $2 = 4\arcsin(0) + C$

$$C = 2.$$

Therefore

$$v(t) = 4\arcsin\left(\frac{t}{4}\right) + 2\left(\frac{m}{s}\right), \text{ where } 0 \le t < 4.$$

B. Change in position

$$= \int_a^b v(t)\,dt = \int_2^3 v(t)\,dt = \int_2^3 \left(4\arcsin\left(\frac{t}{4}\right) + 2\right)\,dt \approx 4.715 \text{ meters}$$

▶**Objective 7** Use the definite integral to accumulate various quantities for **all** functions including algebraic, trigonometric, inverse trigonometric, logarithmic, exponential, and combinations of these.

Example Money goes into a bank account at a continuous rate given by $r(t) = Ce^{kt}$ in dollars/day.

A. If at time $t = 0$, the rate was .02 dollars/day and at time $t = 100$ the rate was .50 dollars/day, then what is the rate equation? (that is, solve for C and k)

B. How much money (to the nearest cent) accumulated in the account during the time interval $0 \le t \le 365$?

Answers

A. At $t = 0$ the rate was .02 dollars/day.

$$r(0) = Ce^{k(0)}$$
$$C = .02$$
$$r(100) = .02e^{k(100)} = .50$$
$$100k = \ln\left(\frac{.5}{.02}\right) \approx 3.218875825$$
$$k \approx 0.032188758$$

So, $r(t) \approx .02\, e^{(0.032188758)t}$.

B. Money accumulated $= \displaystyle\int_0^{365} .02\, e^{(0.032188758)t}\,dt \approx \$78,668.95.$

▶**Objective 8** Use the Fundamental Theorem of Calculus with transcendental functions.

Example Let $h(x) = \displaystyle\int_1^{3x+1} \frac{1}{t}\,dt.$

A. Find $h'(x)$.

B. Find $h(x)$.

C. State the domain of $h(x)$.

Answers   From the Fundamental Theorems of Calculus:

A. $h'(x) = \frac{3}{3x+1}$

B. $h(x) = \int_1^{3x+1} \frac{1}{t}\, dt = \left[\ln t\right]_1^{3x+1} = \ln(3x+1) - \ln 1 = \ln(3x+1).$

C. We need $3x + 1 > 0$, so the domain is $\{x | x > -\frac{1}{3}\}$.

Multiple-Choice Questions Involving Transcendental Functions

---

1. Which of the following curves has a rate of change at any x that is twice the value of the function at x?

A. $y = 2e^x$

B. $y = (e^x)^2$

C. $y = e^{x^2}$

D. $y = 2e^x$

E. $y = 2^x$

Answer   B. $y = (e^x)^2 = e^{2x} \implies y' = 2e^{2x}$, which is equal to $2y$.

---

2. Consider the graph of the function h given by $h(x) = e^{-x^2}$ for $0 \leq x$. There is a point of inflection at $x =$

A. There are no inflection points.

B. 0

C. $-\sqrt{2}$

D. $\sqrt{2}$

E. $\frac{\sqrt{2}}{2}$

Answer   E.

$$h(x) = e^{-x^2}$$
$$h'(x) = -2xe^{-x^2}$$
$$h''(x) = -2e^{-x^2} + 4x^2e^{-x^2} = 2e^{-x^2}(2x^2 - 1)$$
$$h''(x) = 0 \text{ when } 2x^2 - 1 = 0$$

Thus, $x = \frac{\sqrt{2}}{2}$, since the negative root is not in the domain.

	$0 \leq x \leq \frac{\sqrt{2}}{2}$	$x = \frac{\sqrt{2}}{2}$	$\frac{\sqrt{2}}{2} < x$
$h''(x)$	*Negative*	0	*Positive*
$h(x)$	Concave Down	*Inflection Point*	Concave Up

---

3.   The line tangent to the curve $y = \log_a x$ at $x = 1$ is:

A. $y = 1$

B. $y = x - 1$

C. $y = \frac{1}{\ln a}(x - 1)$

D. $y = \frac{1}{\ln a}(x - 1) + 1$

E. $\frac{1}{\ln a}x$

Answer   C.

$$\frac{dy}{dx} = \frac{d}{dx}(\log_a x) = \frac{d}{dx}\left(\frac{\ln x}{\ln a}\right) = \frac{1}{x \ln a}.$$

Therefore, at $x = 1$, $y = \log_a(1) = 0$, and the tangent line to $y = \log_a x$ has a slope $m = \frac{1}{\ln a}$. By the point slope form:

$$y = \frac{1}{\ln a}(x - 1) + 0 = \frac{1}{\ln a}(x - 1).$$

---

4.   The slope of the line normal to the curve $y = a^x (a \neq 1)$ at $x = 0$ is:

A. $\frac{-1}{\ln a}$

B. $\ln a$

C.   0

D. -1

E.   1

Answer A. $y' = (\ln a)a^x$, so at $x = 0$ the slope of the tangent is $\ln a$.

The slope of the normal line is $\frac{-1}{\ln a}$.

---

5. Consider the graph of the function h given by $h(x) = e^{-x^2}$ for $0 \le x$.

Let $A(x)$ be the area of the shaded rectangle shown in the figure below.

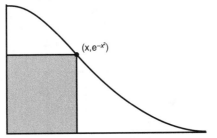

Which of the following statements are true?

I. The slope of the graph of h is a minimum at $x = \frac{\sqrt{2}}{2}$.

II. The graph of h has a point of inflection at $x = \frac{\sqrt{2}}{2}$.

III. $A(x)$ is a maximum at $x = \frac{\sqrt{2}}{2}$.

A. I only

B. II only

C. I and II only

D. I, II, and III

E. III only

Answer D.

We found in question 2 that h has a point of inflection at $x = \frac{\sqrt{2}}{2}$. Also, h'' changes sign from negative to positive at $x = \frac{\sqrt{2}}{2}$, so h' is a minimum there. Thus, I and II are both true. To see that III is also true, note that

$$A(x) = bh = x(e^{-x^2})$$
$$A'(x) = e^{-x^2} - 2x^2 e^{-x^2} = -e^{-x^2}(2x^2 - 1)$$
$$A'(x) = 0 \quad \text{gives} \quad -e^{-x^2}(2x^2 - 1) = 0$$

Since e^{-x^2} is never 0, the only value satisfying this equation is $x = \frac{\sqrt{2}}{2}$ (the negative root is not in the domain). Now we analyze the sign of $A'(x)$:

	$0 < x < \frac{\sqrt{2}}{2}$	$x = \frac{\sqrt{2}}{2}$	$\frac{\sqrt{2}}{2} < x$
$A'(x)$	Positive	0	Negative
$A(x)$	Increasing	Relative *Max*	Decreasing

Thus $A(x)$ is a maximum at $x = \frac{\sqrt{2}}{2}$.

---

6. The Statue of Liberty is about 150 ft tall from the base of her feet to the top of her torch. (She has a nose 4.5 feet long!) The statue also stands on a pedestal that's about 150 ft tall. How far away (x) in feet from the base of the pedestal should an observer be to maximize the angle of view (α)?

A. 300

B. 150

C. 0

D. 212.132

E. 425.264

Answer   D.

First you need α as a function of x:

$$\tan(\alpha_{\text{Pedestal}}) = \frac{150}{x}$$

$$\tan(\alpha_{\text{Both}}) = \frac{300}{x}$$

$$\alpha_{\text{statue}} = \alpha_{\text{Both}} - \alpha_{\text{Pedestal}}$$

$$= \arctan\left(\frac{300}{x}\right) - \arctan\left(\frac{150}{x}\right),$$

where $0 < x$. The derivative is

$$\alpha'(x) = \frac{1}{1 + (\frac{300}{x})^2}\left(\frac{-300}{x^2}\right) - \frac{1}{1 + (\frac{150}{x})^2}\left(\frac{-150}{x^2}\right)$$

$$= \frac{-300}{x^2 + (300)^2} + \frac{150}{x^2 + (150)^2}.$$

and this is equal to zero when

$$\frac{150}{x^2 + (150)^2} = \frac{300}{x^2 + (300)^2}$$

$$x^2 + (300)^2 = 2x^2 + 2(150)^2$$

$$x^2 = (300)^2 - 2(150)^2 = 45,000$$

$$x = 150\sqrt{2} \approx 212.132 \text{ feet}$$

x	$0 < x < 150\sqrt{2}$	$150\sqrt{2}$	$150\sqrt{2} < x$
$\alpha'(x)$	*Positive*	0	*Negative*
$\alpha(x)$	Increasing	Relative *Max*	Decreasing

---

7. $\displaystyle\lim_{x \to 1} \frac{e^{2x} - e^2}{x - 1} =$

(Hint: Interpret this limit as an instantaneous rate of change.)

A. 0

B. 1

C. e^2

D. $2e^2$

E. Does not exist

Answer D. This is just the definition of derivative, so the answer is the slope of $y = e^{2x}$ at $x = 1$:

$$y' = 2e^{2x}, \quad \text{which is} \quad 2e^2 \quad \text{at } x = 1.$$

---

8. A storage tank has dimensions given by revolving the curve $y = \frac{1}{10}(e^x - 1)$ from $x = 0$ to $x = 4$ about the y-axis, where x and y are measured in feet, as shown.

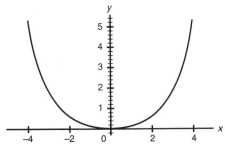

Water is flowing into the tank at a constant rate of 1.000 ft^3 per minute. If h represents the depth of the water in the tank, how fast is the depth of the water increasing when the water is 2 feet deep ($h = 2$)?

(Hint: Find the volume in the tank as a function of h, where the function $V(h)$ includes a definite integral that you do not evaluate.)

A. $0.3285 \frac{\text{ft}}{\text{min}}$

B. $0.4469 \frac{\text{ft}}{\text{min}}$

C. $1.000 \frac{\text{ft}}{\text{sec}}$

D. $.03434 \frac{\text{ft}}{\text{min}}$

E. $3.0445 \frac{\text{ft}}{\text{min}}$

Answer   D. First realize that $y = \frac{1}{10}(e^x - 1)$ is the same as $x = \ln(10y + 1)$.

Thus, the volume in the tank as a function of h is $V(h) = \pi \int_0^h (\ln(10y + 1))^2 \, dy$.

Using the chain rule and the First Fundamental Theorem of Calculus gives

$$\frac{dV}{dt} = \frac{dV}{dh} \times \frac{dh}{dt} = \pi(\ln(10h + 1))^2 \frac{dh}{dt}.$$

(Notice that you don't actually find the antiderivative here.) So, at the instant

$$\frac{dV}{dt} = 1.000,$$

$$1 = \pi(\ln 21)^2 \frac{dh}{dt} \Rightarrow \frac{1}{\pi(\ln 21)^2} = \frac{dh}{dt} \Rightarrow \frac{dh}{dt}$$

$$= 0.03434 \frac{\text{ft}}{\text{min}}.$$

---

9. The curve $y = x + \log_3(x^2 + 5)$ has points of inflection at $x =$

A. $\frac{\pm\sqrt{10}}{\ln 3}$

B. $\pm(\ln 3)\sqrt{5}$

C. $\pm\frac{2}{3}\sqrt{10}$

D. $\pm\sqrt{5}$

E. There are no points of inflection.

Answer   D.

To find the inflection points, you need to set the second derivative to 0. We have

$$y' = 1 + \frac{2x}{(\ln 3)(x^2 + 5)},$$

and the second derivative is

$$y'' = \frac{2(x^2 + 5) - 2x(2x)}{(\ln 3)(x^2 + 5)^2} = \frac{-2x^2 + 10}{(\ln 3)(x^2 + 5)^2}.$$

The second derivative is 0 precisely whenever the numerator in the above expression is 0. Solving this gives you

$$-2x^2 + 10 = 0 \Rightarrow x^2 = 5 \Rightarrow x = \pm\sqrt{5}.$$

Next you need to check to see if the concavity changes at these two x values.

$y'' = \dfrac{-2x^2 + 10}{(\ln 3)(x^2 + 5)^2}$ is negative to the left of $x = -\sqrt{5}$ and positive just to the right of $x = -\sqrt{5}$, so the curve has a point of inflection there.

$y'' = \dfrac{-2x^2 + 10}{(\ln 3)(x^2 + 5)^2}$ is positive just to the left of $x = \sqrt{5}$ and negative to the right of $x = \sqrt{5}$ so the curve has a point of inflection there.

---

10. Consider the function $f(x) = 20x^2 e^{-3x}$ on the domain $[0, \infty)$. On its domain, the curve $y = f(x)$

A. attains its maximum value at $x = \frac{3}{4}$ and does have a minimum value.

B. attains its maximum value at $x = \frac{2}{3}$ and does not have a minimum value.

C. attains its maximum value at $x = \frac{3}{4}$ and attains its minimum value at $x = 0$.

D. attains its maximum value at $x = \frac{2}{3}$ and attains its minimum value at $x = 0$.

E. attains its maximum value at $x = \frac{\sqrt{3}}{3}$ and does not have a minimum value.

Answer   D. The function $f(x)$ on the domain $[0, \infty)$ is always non-negative, and in fact is 0 only when $x = 0$. So the $f(x)$ has a local (and global) minimum of 0 at $x = 0$. To find the maximum of $f(x)$ take the derivative and set it to 0:

$$f'(x) = 40xe^{-3x} + 20x^2(-3)e^{-3x} \quad \text{(by the product rule)}$$
$$= 40xe^{-3x} - 60x^2 e^{-3x}$$
$$= xe^{-3x}(40 - 60x).$$

Then $xe^{-3x}(40 - 60x) = 0 \Rightarrow$ either $xe^{-3x} = 0$ or $(40 - 60x) = 0$. In the first case, $xe^{-3x} = 0$ only when $x = 0$, and in the other case $(40 - 60x) = 0$ only when $x = \frac{2}{3}$.

You know already that $x = 0$ is a minimum. To check that $x = \frac{2}{3}$ is a maximum, use either the second derivative test, or the fact that $f'(x) = xe^{-3x}(40 - 60x)$ is positive just to the left of $x = \frac{2}{3}$ and negative to the right. This also suffices to show that the function attains its global maximum $x = \frac{2}{3}$.

---

11. Consider the function $f(x) = 20x^2 e^{-3x}$ on the domain $[0, \infty)$.

The curve has points of inflection at $x =$

A. -0.2440 and 0.9107

B. 0.2357 and 1.1785

C. 0.1953 and 1.1381

D. 0.2357 only

E. 1.1381 only

Answer C. The first derivative $f'(x)$ is given in the feedback to the problem above.
Using the product rule, the second derivative is

$$
\begin{aligned}
f''(x) &= \left[\frac{d}{dx}(xe^{-3x})\right](40 - 60x) + (xe^{-3x})\frac{d}{dx}(40 - 60x) \\
&= (e^{-3x} - 3xe^{-3x})(40 - 60x) + (xe^{-3x})(-60) \\
&= e^{-3x}(1 - 3x)(40 - 60x) + e^{-3x}(-60x) \\
&= e^{-3x}(180x^2 - 120x - 60x + 40) + e^{-3x}(-60x) \\
&= e^{-3x}(180x^2 - 240x + 40)
\end{aligned}
$$

Since e^{-3x} is always positive, the second derivative is 0 when $180x^2 - 240x + 40 = 0$. Use your calculator to find the roots: .1953 and 1.1381. You can check that $180x^2 - 240x + 40$ is positive for $x < .1953$, negative for x between .1953 and 1.1381, and positive again for $x > 1.1381$. Therefore, $f''(x)$ follows that same pattern (remember that e^{-3x} is always positive), and thus both x values are inflection points.

---

12. $3,000$ young trout are introuduced into a large fish pond. The number of trout still alive after t years is modeled by the formula $N(t) = 3,000(.9)^t$. What is the rate of population **decrease** when the number of trout in the pound reaches $2,000$?

A. Approximately 520 trout per year

B. Approximately 211 trout per year

C. Approximately $2,000$ trout per year

D. Approximately 494 trout per year

E. None of the above

Answer B. First, find at what time t the population reaches $2,000$. You get

$$N(t) = 2,000 \Rightarrow 3,000(.9)^t = 2,000$$

$$\Rightarrow (.9)^t = \frac{2}{3}$$

$$\Rightarrow t\ln(.9) = \ln\left(\frac{2}{3}\right)$$

$$\Rightarrow t = \frac{\ln\left(\frac{2}{3}\right)}{\ln(.9)} \approx 3.848359184 \text{ years.}$$

Then, find $\dfrac{dN}{dt}$ at $t = \dfrac{\ln(\frac{2}{3})}{\ln(.9)}$. You get $\dfrac{dN}{dt} = 3,000(\ln .9)(.9)^t$, and setting $t = \dfrac{\ln(\frac{2}{3})}{\ln(.9)}$ gives you approximately -210.7210313 fish per year. This means that the rate of **decrease** in the fish population is approximately 211 trout per year at the time of concern.

13. A 25 ft ladder is leaning against a vertical wall.

At what rate (with respect to time) is the angle θ between the ground and the ladder changing, if the top of the ladder is sliding down the wall at the rate of r inches per second, at the moment that the top of the ladder is h feet from the ground?

(Look for an equation in terms of h and θ.)

A. $\dfrac{1}{25\sqrt{1 - \left(\frac{h}{25}\right)^2}}$

B. $\dfrac{r}{300\sqrt{1 - \left(\frac{h}{25}\right)^2}}$

C. $\dfrac{1}{25\sqrt{1 + \left(\frac{h}{25}\right)^2}}$

D. $\dfrac{r}{25\sqrt{1 + \left(\frac{h}{25}\right)^2}}$

E. $\dfrac{r}{12\sqrt{1 - \left(\frac{h}{25}\right)^2}}$

Answer B. The angle θ is related to h by the equation

$$\sin\theta = \frac{h}{25} \Rightarrow \theta = \arcsin\left(\frac{h}{25}\right).$$

You're looking for $\dfrac{d\theta}{dt}$. Using the fact that $\dfrac{dh}{dt}$ is the rate of change of the height in feet per second, which is equal to $\dfrac{r}{12}$, we have:

$$\frac{d\theta}{dt} = \frac{1}{\sqrt{1 - \left(\frac{h}{25}\right)^2}} \frac{1}{25} \frac{dh}{dt}$$

$$= \frac{1}{25\sqrt{1 - \left(\frac{h}{25}\right)^2}}\left(\frac{r}{12}\right)$$

$$= \frac{r}{300\sqrt{1 - \left(\frac{h}{25}\right)^2}}.$$

14. Consider the graph of the function g given by $g(x) = \frac{\ln(x^2+1)}{x}$ for $0 < x$. Let $A(x)$ be the area of the shaded rectangle shown in the figure below, where x is the coordinate of the lower right corner of the rectangle, with $x > 0$.

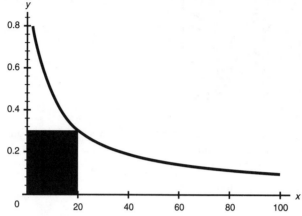

The x value $(x > 0)$ that maximizes $A(x)$ is:

A. 1.9802913

B. 19.7540002

C. 17.6330301

D. 25.9392111

E. None of the above—there is no maximum.

Answer   E. The area of the rectangle is given by

$$A(x) = x \left(\frac{\ln(x^2 + 1)}{x} \right)$$

(length times height). So $A(x) = \ln(x^2 + 1)$. Searching for critical points, you have

$$A'(x) = \frac{1}{x^2 + 1}(2x) = \frac{2x}{x^2 + 1}.$$

The derivative is thus positive for all $x > 0$, which indicates that $A(x)$ has no maximum on the interval $x > 0$. Note that there are no endpoints to check.

15. The position of a particle moving in a straight line is given by $s(t) = e^{-t}\cos(5t)$ for $t > 0$, where t is in seconds. If the particle changes direction at time T seconds, then T must satisfy the equation:

A. $\cos(5T) = 0$

B. $5T = \arctan(-\frac{1}{5})$

Apex Learning

C. $5e^{-t}\sin(5t) = 0$

D. $\tan(5T) = -\frac{1}{5}$

E. $\cot(5T) = 5$

Answer   D. The particle changes direction only at places where the velocity is 0 (but not necessarily at **all** places where the velocity is 0). The velocity is given by

$$v(t) = s'(t) = -e^{-t}\cos(5t) + e^{-t}(-5\sin(5t)) = -e^{-t}(\cos(5t) + 5\sin(5t)).$$

Setting the velocity equal to 0 gives you

$$-e^{-t}(\cos(5t) + 5\sin(5t)) = 0 \Rightarrow \cos(5t) + 5\sin(5t) = 0 \quad (\text{since } -e^{-t} \text{ is never 0})$$
$$\Rightarrow 5\sin(5t) = -\cos(5t)$$
$$\Rightarrow \frac{\sin(5t)}{\cos(5t)} = -\frac{1}{5}$$
$$\Rightarrow \tan(5t) = -\frac{1}{5}.$$

Notice that you can't go further to say that this means

$$5t = \arctan\left(-\frac{1}{5}\right),$$

because of the restricted range of arctan.

---

16.   The position of a particle moving in a straight line is given by $s(t) = e^{-t}\cos(5t)$ for $t > 0$, where t is in seconds. If the particle's speed is a relative maximum at time T, then T must satisfy the equation:

A. $\tan(5T) = 2.4$

B. $\tan(5T) = 4.8$

C. $\cos(5T) = 0$

D. $e^{-T} = \dfrac{5\sin(5T) - 25\cos(5T))}{(\cos(5T) + 5\sin(5T))}$

E. $e^{-T} = \dfrac{(\cos(5T) + 5\sin(5T))}{-5\sin(5T) + 25\cos(5T))}$

Answer   A. The particle's velocity attains a relative maximum or minimum only at times when the acceleration is 0 (but not necessarily at all these times). A particle's **speed** is at a relative maximum when the absolute value of the velocity attains a relative maximum, and therefore only at times where the acceleration is 0. The velocity is given by

$$v(t) = s'(t) = -e^{-t}(\cos(5t) + 5\sin(5t)).$$

The acceleration is given by

$$a(t) = v'(t) = e^{-t}(\cos(5t) + 5\sin(5t))$$
$$- e^{-t}(-5\sin(5t) + 25\cos(5t))$$
$$= e^{-t}(\cos(5t) + 5\sin(5t)) + e^{-t}(5\sin(5t) - 25\cos(5t))$$
$$= e^{-t}(\cos(5t) + 5\sin(5t) + 5\sin(5t) - 25\cos(5t))$$
$$= e^{-t}(10\sin(5t) - 24\cos(5t)).$$

Since e^{-t} is always positive, the acceleration is 0 only when

$$10\sin(5t) - 24\cos(5t) = 0$$
$$\Rightarrow 10\sin(5t) = 24\cos(5t)$$
$$\Rightarrow \frac{\sin(5t)}{\cos(5t)} = \frac{24}{10}$$
$$\Rightarrow \tan(5t) = 2.4.$$

Multiple-Choice Questions on Integrating Transcendental Functions

1. $\int \ln x \, dx =$

A. $\frac{1}{x}$

B. $\frac{1}{x} + C$

C. $|\ln(\frac{1}{x})| + C$

D. $|\frac{1}{x}| + C$

E. $x \ln x - x + C$

Answer E. You can check this by differentiating each choice.

2. $\int_0^4 \frac{x}{1+x^2} \, dx =$

A. $\arctan 4$

B. $\frac{1}{2} \ln 17$

C. $\ln 17$

D. $\frac{1}{2} \arctan 4$

E. $2 \arctan 4$

Answer   B. You can let $u = 1 + x^2$. Then $du = 2x\,dx$, so $x\,dx = \frac{1}{2}\,du$.

When $x = 0$, $u = 1$, and when $x = 4$, $u = 17$. The integral becomes:

$$\int_0^4 \frac{x}{1+x^2}\,dx = \frac{1}{2}\int_1^{17}\frac{1}{u}\,du = \frac{1}{2}\ln|u|\Big|_1^{17} = \frac{1}{2}\ln 17.$$

---

3.   Use the substitution $u = \sqrt{x}$ to find an exact answer for $\displaystyle\int_1^4 \frac{1}{\sqrt{x}(1+x)}\,dx =$

A. $2\arctan 2 - \frac{\pi}{2}$

B. 0.6435

C. 5

D. 5.104

E. $2\arctan 4 - \frac{\pi}{2}$

Answer   A. Let $u = \sqrt{x}$. Then $du = \dfrac{1}{2\sqrt{x}}\,dx$, when $x = 1$, $u = 1$, and when $x = 4$, $u = 2$. What about the $1 + x$? If $u = \sqrt{x}$, then $x = u^2$.

So the integral becomes

$$\int_1^4 \frac{1}{\sqrt{x}(1+x)}\,dx = 2\int_1^2 \frac{1}{1+u^2}\,du$$
$$= 2(\arctan u)\Big|_1^2 = 2\arctan 2 - \frac{\pi}{2}.$$

---

4.   Find an exact answer for the area of the region between the curves $y = 2^x$ and $y = 3^x$, between $x = 0$ and $x = 2$.

A. $\frac{9}{\ln 2} - \frac{4}{\ln 3}$

B. $\frac{9}{\ln 3} - \frac{4}{\ln 2}$

C. $\frac{8}{\ln 3} - \frac{3}{\ln 2}$

D. 2

E. 2.954

Answer   C. The area here is

$$\int_0^2 (3^x - 2^x)\,dx = \frac{3^x}{\ln 3} - \frac{2^x}{\ln 2}\Big|_0^2 = \frac{8}{\ln 3} - \frac{3}{\ln 2}.$$

5. The area occupied by a mold on the surface of some old yogurt in the back of the refrigerator is increasing at a rate of $0.15e^{\frac{1}{2}t}$ square centimeters per day at time t days. How much more area is covered after 4 days than was covered after 2 days?

A. 1.917 square centimeters

B. 0.515 square centimeters

C. $.3(e^4 - e^2)$ square centimeters

D. 1.401 square centimeters

E. $.3(e^2 + e)$ square centimeters

Answer   D. The change in the area is represented by the integral $\int_{2}^{4} 0.15e^{\frac{1}{2}t}dt \approx 1.401$, so there is 1.401 square centimeters more covered after 4 days than after 2 days.

6. $\int 5^{-2x} \, dx =$

A. $\frac{-2}{\ln 5} \times 5^{-2x} + C$

B. $\frac{-1}{2\ln 5} \times 5^{-2x} + C$

C. $\frac{-1}{2\ln 5} \times 5^{x} + C$

D. $\frac{-2}{\ln 5} \times 5^{x} + C$

E. $\frac{-1}{2\ln 2} \times 5^{-2x} + C$

Answer   B. By substituting with $u = -2x$, you get $du = -2\,dx \Rightarrow dx = \dfrac{du}{-2}$, and thus

$$\int 5^{-2x}\,dx = \int \frac{5^u}{-2}\,du = \frac{-1}{2\ln 5}5^u + C = \frac{-1}{2\ln 5}5^{-2x} + C.$$

7. $\int x^2 e^{(3x^3)} \, dx =$

A. $e^{(3x^3)} + C$

B. $\frac{-1}{2\ln 5}5^{-2x} + C$

C. $\frac{x}{9}e^{(3x^3)} + C$

D. $\frac{x}{3}e^{(3x^3)} + C$

E. $\frac{1}{9}e^{(3x^3)} + C$

Answer   E. $u = 3x^3$ looks like a good substitution. Then $du = 9x^2\,dx$, and

$$\int x^2 e^{(3x^3)}\,dx = \int \frac{1}{9}e^{(u)}\,du = \frac{1}{9}e^u + C = \frac{1}{9}e^{(3x^3)} + C.$$

---

8. $\displaystyle\int \frac{1}{x^2 + 25}\,dx =$

(Hint: Factor a 25 out of the denominator.)

A. $\frac{1}{25}\arctan x + C$

B. $\frac{1}{25}\arctan \frac{x}{5} + C$

C. $\frac{1}{5}\arctan \frac{x}{5} + C$

D. $\frac{\ln(x^2 + 25)}{2x} + C$

E. $\ln(x^2 + 25) + C$

Answer   C. This looks like an inverse trigonometric function, in particular $\arctan u$.

The trick is to rewrite the integrand to look more like $\dfrac{1}{1 + u^2}$.

You get

$$\int \frac{1}{x^2 + 25}\,dx = \int \frac{1}{25(\frac{x^2}{25} + 1)}\,dx = \frac{1}{25}\int \frac{1}{(\frac{x}{5})^2 + 1}\,dx.$$

Now, taking $u = \frac{x}{5}$ gives you

$$du = \frac{1}{5}\,dx \Rightarrow dx = 5\,du,$$

and thus,

$$\frac{1}{25}\int \frac{1}{(\frac{x}{5})^2 + 1}\,dx = \frac{1}{25}\int \frac{5}{(u)^2 + 1}\,du$$
$$= \frac{5}{25}\arctan u + C = \frac{1}{5}\arctan \frac{x}{5} + C.$$

---

9. $\displaystyle\int \frac{x}{\sqrt{1 - x^4}}\,dx =$

(Hint: try the substitution $u = x^2$.)

A. $x\arcsin(x^2) + C$

B. $2\arcsin(x^2) + C$

C. $\frac{1}{2}\arcsin(x^2) + C$

D. $x \arcsin(x) + C$

E. $\frac{1}{2} \arcsin(u) + C$

Answer   C. This looks a bit like $\arcsin u$. Let's rewrite the integral to look more like $\int \frac{du}{\sqrt{1-u^2}}\, du$.

The x^4 needs to be a u^2, so take $u = x^2$. Then you get $du = 2x\, dx$, and

$$\int \frac{x}{\sqrt{1-x^4}}\, dx = \int \frac{1}{2} \frac{1}{\sqrt{1-u^2}}\, du$$
$$= \frac{1}{2} \arcsin u + C = \frac{1}{2} \arcsin(x^2) + C.$$

---

10. $\displaystyle \int x\left(23^{-(x^2)}\right) dx =$

A. $-\frac{1}{2}\left(23^{-(x^2)}\right) + C$

B. $-\frac{2}{\ln 23}\left(23^{-(x^2)}\right) + C$

C. $-\frac{1}{2\ln 23}\left(23^{-(x^2)}\right) + C$

D. $-x(\ln 23)\left(23^{-(x^2)}\right) + C$

E. $-\frac{\ln 23}{2}\left(23^{-(x^2)}\right) + C$

Answer   C. The substitution to take is $u = -x^2$, which gives you $x\, dx = -\frac{1}{2}\, du$. Then

$$\int x23^{-(x^2)}\, dx = \frac{-1}{2} \int 23^u\, du$$
$$= \frac{-1}{2\ln 23} 23^u + C = -\frac{1}{2\ln 23} 23^{-(x^2)} + C.$$

---

11. $\displaystyle \int \frac{\sin \frac{\theta}{2}}{\cos \frac{\theta}{2} + 4}\, d\theta =$

A. $-2\ln(\cos \frac{\theta}{2} + 4) + C$

B. $2\ln|\sin \frac{\theta}{2} + 4| + C$

C. $-\frac{1}{2}\ln|\cos \frac{\theta}{2} + 4| + C$

D. $-\frac{1}{2}\ln|\sin \frac{\theta}{2} + 4| + C$

E. None of these

Answer A. Making the substitution $u = \cos\frac{\theta}{2}$, you get $du = -\frac{1}{2}\sin\frac{\theta}{2}\,d\theta$. Then

$$\int \frac{\sin\frac{\theta}{2}}{\cos\frac{\theta}{2} + 4}\,d\theta = \int \frac{-2}{u + 4}\,du = -2\int \frac{1}{u + 4}\,du$$

$$= -2\ln|u + 4| + C = -2\ln\left|\cos\frac{\theta}{2} + 4\right| + C.$$

Notice that you can eliminate the absolute value sign, since $\cos\frac{\theta}{2} + 4$ is always positive. Thus the answer is $-2\ln(\cos\frac{\theta}{2} + 4) + C$.

---

12. $\displaystyle\int \frac{\sin x}{\cos^2 x + 1}\,dx =$

(Hint: Try the substitution $u = \cos x$.)

A. $\arctan(\cos x) + C$

B. $\frac{1}{2}\ln(\cos^2 x + 1) + C$

C. $-\frac{1}{2}\arctan(\cos x) + C$

D. $-\frac{1}{2}\ln(\cos^2 x + 1) + C$

E. $-\arctan(\cos x) + C$

Answer E. Taking $u = \cos x$ gives you $du = -\sin x\,dx$, and thus,

$$\int \frac{\sin x}{(\cos x)^2 + 1}\,dx = \int \frac{-1}{(u)^2 + 1}\,du = -\int \frac{1}{u^2 + 1}\,du$$

$$= -\arctan u + C = -\arctan(\cos x) + C.$$

---

13. $\displaystyle\int \frac{5^{\tan x}}{\cos^2 x}\,dx =$

(Hint: Try either $u = 5^{\tan x}$ or $u = \tan x$.)

A. $5^{\tan x} + C$

B. $\frac{1}{2\ln 5}(5^{\tan x})^2 + C$

C. $\frac{\ln 5}{2}(5^{\tan x})^2 + C$

D. $\left(\frac{1}{\ln 5}\right)5^{\tan x} + C$

E. $(\ln 5)5^{\tan x} + C$

Answer D. Setting $u = 5^{\tan x}$, you have $du = (\ln 5)5^{\tan x}\sec^2 x\,dx = \dfrac{(\ln 5)5^{\tan x}}{\cos^2 x}\,dx$. Thus,

$$\int \frac{5^{\tan x}}{\cos^2 x}\,dx = \frac{1}{\ln 5}\int du = \frac{1}{\ln 5}u + C = \frac{1}{\ln 5}5^{\tan x} + C.$$

Alternatively, if you use $u = \tan x$, you get $du = \sec^2 x\,dx$, and thus,

$$\int \frac{5^{\tan x}}{\cos^2 x}\,dx = \int 5^u\,du = \frac{1}{\ln 5}5^u + C = \frac{1}{\ln 5}5^{\tan x} + C.$$

14. $\displaystyle \int \frac{2^x}{\sqrt{2^x+1}}\, dx =$

A. $\displaystyle \frac{\ln\sqrt{2^x+1}}{\ln 2} + C$

B. $2\sqrt{2^x+1} + C$

C. $\ln 2\sqrt{\ln(2^x+1)} + C$

D. $\displaystyle \frac{2}{\ln 2}\sqrt{2^x+1} + C$

E. $(\ln 2)\sqrt{2^x+1} + C$

Answer D. Setting $u = 2^x$ gives you $du = (\ln 2)2^x\, dx$, and thus,

$$\int \frac{2^x}{\sqrt{2^x+1}}\, dx = \frac{1}{\ln 2}\int \frac{1}{\sqrt{u+1}}\, du$$
$$= \frac{1}{\ln 2}2\sqrt{u+1} + C = \frac{2}{\ln 2}\sqrt{2^x+1} + C.$$

15. $\displaystyle \int \frac{x-2}{x^2-4x+12}\, dx =$

A. $\ln|x-6| + C$

B. $2\ln(x-2)^2 + C$

C. $2\ln|x^2-4x+12| + C$

D. $\frac{1}{2}\ln|x^2-4x+12| + C$

E. Both $\ln|x-6| + C$ and $2\ln|x^2-4x+12| + C$

Answer D. Setting $u = x^2 - 4x + 12$ gives you $du = (2x-4)\, dx = 2(x-2)\, dx$. Thus,

$$\int \frac{x-2}{x^2-4x+12}\, dx = \frac{1}{2}\int \frac{1}{u}\, dx = \frac{1}{2}\ln|x^2-4x+12| + C.$$

Actually, you can drop the absolute values if you want, since $x^2 - 4x + 12$ is always positive!

16. $\displaystyle \int \cot(4t)\, dt =$

(Hint: Rewrite the integrand in terms of sines and cosines.)

A. $\frac{1}{4}\ln|\sin(4t)| + C$

B. $-\frac{1}{4}\csc^2(4t) + C$

C. $4\ln|\sin(4t)| + C$

D. $-\frac{1}{4}\sec(4t)| + C$

E. None of these

Answer A. First you rewrite the integral in terms of sines and cosines. You have

$$\int \cot(4t)dt = \int \frac{\cos(4t)}{\sin(4t)}dt.$$

Substituting $u = \sin(4t)$, you get $du = 4\cos(4t)dt$, and thus,

$$\int \frac{\cos(4t)}{\sin(4t)}dt = \int \frac{1}{4u}\,du = \frac{1}{4}\ln|u| + C = \frac{1}{4}\ln|\sin(4t)| + C.$$

---

17. The region bounded by $y = e^x$, $y = e$, and the y-axis is revolved around the y-axis. The volume of the resulting solid is:

A. $\frac{1}{2}\pi e^{2e} - \frac{1}{2}\pi e^2$

B. $\frac{1}{2}e^{2e} - \frac{1}{2}e^2$

C. $\frac{1}{2}\pi e^{2e}$

D. 0.718

E. 2.257

Answer E.

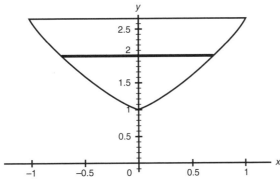

Your slices should be horizontal to give the volume of any slice as $dV = \pi(\ln(y))^2\,dy$.

Using your calculator,

$$V = \int_1^e A(y)dy = \int_1^e \pi(\ln(y))^2 dy \approx 2.257.$$

18. A region R is bounded by the x-axis, y-axis, $x = \sqrt{3}$, and $y = \dfrac{1}{\sqrt{x^2 + 1}}$.

Suppose a solid has region R as a base and is such that every cross section of the solid perpendicular to the x-axis is an equilateral triangle. The plane perpendicular to the x-axis at $x = k$ cuts this volume in half. If the volume of the whole solid is $\dfrac{\pi\sqrt{3}}{12}$, find k.

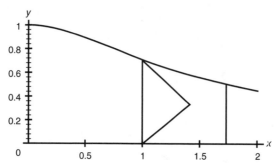

(Hint: The area of an equilateral triangle is $A = \frac{\sqrt{3}}{4}b^2$, where b is the length of one side.)

A. $\dfrac{\sqrt{3}}{2}$

B. $\dfrac{\sqrt{3}}{3}$

C. $\dfrac{\sqrt{3}}{4}$

D. 0.453

E. 0.906

Answer   B. You're using the area of an equilateral triangle, $A = \frac{\sqrt{3}}{4}b^2$, where b is the length of a side of the triangle. Then you know the volume of any slice to be

$$dV = A(x)\,dx = \frac{\sqrt{3}}{4}\left(\frac{1}{\sqrt{x^2 + 1}}\right)^2 dx = \frac{\sqrt{3}}{4}\left(\frac{1}{x^2 + 1}\right)dx$$

so $V = \displaystyle\int_0^{\sqrt{3}} \frac{\sqrt{3}}{4}\left(\frac{1}{x^2 + 1}\right)dx = \frac{\sqrt{3}}{4}\arctan(\sqrt{3}) = \frac{1}{12}\pi\sqrt{3}$, as given in the problem.

To find k you need half of the volume:

$$\frac{\pi\sqrt{3}}{24} = \int_0^{k} \frac{\sqrt{3}}{4}\left(\frac{1}{x^2 + 1}\right)dx = \frac{\sqrt{3}}{4}\arctan(k)$$

$$\implies \arctan(k) = \frac{\pi}{6} \qquad \implies k = \tan\left(\frac{\pi}{6}\right) = \frac{1}{3}\sqrt{3}.$$

19. A particle has a velocity on the x-axis given by $v(t) = 2t^2 - 2^t$ for $t \geq 0$.

How many times will the particle cross its original position $(t = 0)$?   (Use a calculator.)

A. None

B. One

C. Two

D. Three

E. More than three

Answer   C. If $v(t) = 2t^2 - 2^t$, then $x(t) = \frac{2}{3}t^3 - \frac{1}{\ln 2}2^t + C$.

Setting $x(t) = x(0)$, you find two roots when $t > 0$, using the calculator.

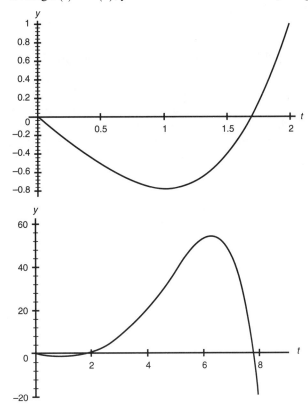

20. Oil is leaking from a tanker at the rate of $2,000e^{-0.2t}$ gallons per hour, where t is measured in hours. How much oil has leaked out of the tanker after 10 hours?   (Use a calculator.)

A. 54 gallons

B. 271 gallons

C. 865 gallons

D. 8,647 gallons

E. 14,778 gallons

Answer D. The total oil will just be the sum of all $R(t)\, dt$ from $0 \le t \le 10$:

$$\text{Total} = \int_0^{10} R(t)\, dt = \int_0^{10} 2,000e^{-0.2t}\, dt \approx 8,646.647 \approx 8,647 \text{ gallons.}$$

Note that you could have found

$$\int_0^{10} 2,000e^{-0.2t}\, dt = -10,000e^{-.2t}\Big|_0^{10}$$

$$= -10,000e^{-2} + 10,000 = 10,000(1 - e^{-2}) \approx 8,647,$$

but you would still have wanted to use your calculator to find the approximate value.

More Practice with Integrating Transcendental Functions

Note: Remember, don't simplify these unless you absolutely have to. Simplification is a great place to make errors, and it makes it difficult to tell if your answer is correct. Also, remember that there is always more than one way to the correct answer. You might make a different, still completely correct, substitution from the one you see here. If you get the same answer we did, you're probably right.

---

1. Evaluate the definite integrals below exactly, by substituting and changing the limits of integration.

A. $\displaystyle\int_1^2 xe^{2x^2}\, dx$

B. $\displaystyle\int_{-1}^0 \frac{8x^3}{x^4 + 2}\, dx$

C. $\displaystyle\int_0^{1/2} \frac{1}{1 + 4x^2}\, dx$

D. $\displaystyle\int_{-4}^{-1} \frac{e^x}{\sqrt{1 - e^{2x}}}\, dx$

E. $\displaystyle\int_{-1}^2 4^{-3x+1}\, dx$

---

2. Evaluate the indefinite integrals below. You'll have to decide for yourself whether to use substitution, inverse trig functions, or attempt to be clever!

A. $\displaystyle\int x\tan(x^2)\, dx$ (Hint: First rewrite the integrand in terms of sines and cosines.)

B. $\displaystyle\int \frac{1}{\sqrt{9 - 4x^2}}\, dx$

C. $\displaystyle\int \frac{2^x}{1+2^x}\,dx$

D. $\displaystyle\int \frac{1}{x(1-\ln x)}\,dx$

E. $\displaystyle\int \frac{3^x}{1+9^x}\,dx$ (Hint: $9^x = 3^{2x}$)

Answers

_____

1.

A.

$$\int_1^2 xe^{2x^2}\,dx = \frac{1}{4}e^8 - \frac{1}{4}e^2$$

Setting $u = 2x^2$ gives you $du = 4x\,dx$, $u(1) = 2$, $u(2) = 8$, and thus

$$\int_1^2 xe^{2x^2}\,dx = \frac{1}{4}\int_2^8 e^u\,du$$

$$= \frac{1}{4}\Big[e^u\Big]_2^8$$

$$= \frac{1}{4}e^8 - \frac{1}{4}e^2.$$

B.

$$\int_{-1}^0 \frac{8x^3}{x^4+2}\,dx = 2\ln 2 - 2\ln 3$$

Setting $u = x^4 + 2$, you get $du = 4x^3\,dx$, $u(-1) = 3$, $u(0) = 2$, and thus

$$\int_{-1}^0 \frac{8x^3}{x^4+2}\,dx = 2\int_3^2 \frac{1}{u}\,du = 2\Big[\ln u\Big]_3^2 = 2\ln 2 - 2\ln 3.$$

You can also make the substitution $u = x^4$, in which case the integral becomes

$$2\int_1^0 \frac{1}{u+2}\,du = 2\Big[\ln(u+2)\Big]_1^0 = 2\ln 2 - 2\ln 3.$$

C.

$$\int_0^{\frac{1}{2}} \frac{1}{1+4x^2}\,dx = \frac{1}{8}\pi.$$

By setting $u = 2x$, you get $du = 2\,dx$, $u(\frac{1}{2}) = 1$, $u(0) = 0$.

$$\int_0^{\frac{1}{2}} \frac{1}{1+4x^2}\,dx = \int_0^1 \frac{1}{2}\frac{1}{1+u^2}\,du = \frac{1}{2}\int_0^1 \frac{1}{1+u^2}\,du$$

$$= \frac{1}{2}\Big[\arctan u\Big]_0^1 = \frac{1}{2}\Big[\arctan 1 - \arctan 0\Big] = \frac{1}{2}\Big[\frac{\pi}{4}\Big] = \frac{1}{8}\pi.$$

D.

$$\int_{-4}^{-1} \frac{e^x}{\sqrt{1-e^{2x}}}\, dx = \arcsin(e^{-1}) - \arcsin(e^{-4}).$$

By setting $u = e^x$, you get $du = e^x\, dx$, $u(-1) = e^{-1}$, $u(-4) = e^{-4}$, and thus

$$\int_{-4}^{-1} \frac{e^x}{\sqrt{1-e^{2x}}}\, dx = \int_{e^{-4}}^{e^{-1}} \frac{1}{\sqrt{1-u^2}}\, du$$

$$= \left[\, \arcsin u \,\right]_{e^{-4}}^{e^{-1}} = \arcsin(e^{-1}) - \arcsin(e^{-4}).$$

E.

$$\int_{-1}^{2} 4^{-3x+1}\, dx = -\frac{4^{-5}}{3\ln 4} + \frac{4^4}{3\ln 4}.$$

Setting $u = -3x+1$, you get $du = -3\, dx$, $u(-1) = 4$, $u(2) = -5$, and thus

$$\int_{-1}^{2} 4^{-3x+1}\, dx = \frac{-1}{3}\int_{4}^{-5} 4^u\, du = -\frac{1}{3\ln 4}\left[4^u \right]_{4}^{-5} = -\frac{4^{-5}}{3\ln 4} + \frac{4^4}{3\ln 4}.$$

2.

A.

$$\int x\tan(x^2)\, dx = -\frac{1}{2}\ln|\cos(x^2)| + C.$$

Rewriting the integral as

$$\int x\frac{\sin(x^2)}{\cos(x^2)}\, dx,$$

and setting $u = \cos(x^2)$, you get $du = -2x\sin(x^2)\, dx$, and thus

$$\int x\frac{\sin(x^2)}{\cos(x^2)}\, dx = -\frac{1}{2}\int \frac{1}{u}\, du = -\frac{1}{2}\ln|u| + C = -\frac{1}{2}\ln|\cos(x^2)| + C.$$

B.

$$\int \frac{1}{\sqrt{9-4x^2}}\, dx = \frac{1}{2}\arcsin(\tfrac{2}{3}x) + C.$$

To make $\dfrac{1}{\sqrt{9-4x^2}}$ look more like the derivative of the inverse sine function, we must factor 9 out of the radical in the denominator:

$$\int \frac{1}{\sqrt{9-4x^2}}\, dx = \int \frac{1}{\sqrt{9\left(1-\frac{4}{9}x^2\right)}}\, dx = \int \frac{1}{3\sqrt{\left(1-\frac{4}{9}x^2\right)}}\, dx$$

$$= \frac{1}{3}\int \frac{1}{\sqrt{\left(1-\frac{4}{9}x^2\right)}}\, dx = \frac{1}{3}\int \frac{1}{\sqrt{\left(1-\left(\frac{2}{3}x\right)^2\right)}}\, dx.$$

Then setting $u = \frac{2}{3}x$, you get $du = \frac{2}{3}\,dx$, and thus

$$\frac{1}{3} \int \frac{1}{\sqrt{\left(1 - \left(\frac{2}{3}x\right)^2\right)}}\,dx = \frac{1}{3} \int \frac{3}{2} \frac{1}{\sqrt{(1 - u^2)}}\,du = \frac{1}{2} \int \frac{1}{\sqrt{(1 - u^2)}}\,du$$

$$= \frac{1}{2}\arcsin u + C = \frac{1}{2}\arcsin\left(\frac{2}{3}x\right) + C.$$

C.

$$\int \frac{2^x}{1 + 2^x}\,dx = \frac{\ln(1 + 2^x)}{\ln 2} + C.$$

Setting $u = 1 + 2^x$, you get $du = (\ln 2)2^x\,dx$, and thus

$$\int \frac{2^x}{1 + 2^x}\,dx = \int \frac{1}{(\ln 2)u}\,du = \frac{1}{\ln 2} \int \frac{1}{u}\,du = \frac{\ln u}{\ln 2} + C$$

$$= \frac{\ln(1 + 2^x)}{\ln 2} + C.$$

(Note that the absolute value isn't needed, since $1 + 2^x > 0$ for all x.)

D.

$$\int \frac{1}{x(1 - \ln x)}\,dx = -\ln|1 - \ln x| + C.$$

Taking $u = 1 - \ln x$, you get $du = -\frac{1}{x}\,dx$, and thus

$$\int \frac{1}{x(1 - \ln x)}\,dx = \int -\frac{1}{u}\,du = -\ln|u| + C = -\ln|1 - \ln x| + C.$$

E.

$$\int \frac{3^x}{1 + 9^x}\,dx = \frac{\arctan 3^x}{\ln 3} + C.$$

Using the hint, you can rewrite the integral as

$$\int \frac{3^x}{1 + 9^x}\,dx = \int \frac{3^x}{1 + 3^{2x}}\,dx = \int \frac{3^x}{1 + (3^x)^2}\,dx.$$

This looks like a function that's related to the inverse tangent. So setting $u = 3^x$, you get $du = (\ln 3)3^x\,dx$, and thus

$$\int \frac{3^x}{1 + (3^x)^2}\,dx = \int \frac{1}{\ln 3} \frac{1}{1 + u^2}\,du = \frac{1}{\ln 3} \int \frac{1}{1 + u^2}\,du$$

$$= \frac{1}{\ln 3}\arctan u + C = \frac{1}{\ln 3}\arctan(3^x) + C.$$

Chapter 10   *Differential Equations*

This chapter reviews objectives related to differential equations—equations in terms of a function and one or more of its derivatives. Some of the most important applications of calculus arise from solving differential equations. You already have solved some differential equations: when you find an antiderivative $y = F(x)$ for a function $f(x)$, you have just solved the differential equation $\dfrac{dy}{dx} = f(x)$. Now we will extend this idea a bit to the class of *separable* differential equations and examine how we can solve them graphically and algebraically.

A. Differential Equations and Slope Fields

▶Objective 1   Identify the order of a differential equation.

Example   Identify the order of each of the following differential equations.

A. $\dfrac{dy}{dt} = \dfrac{d^2y}{dt^2} + 4y - x$

B. $y' = 2x$

C. $\dfrac{dy}{dt} = \dfrac{y^2}{t^3}$

Tip   The order of a differential equation is related to the highest order derivative in the equation.

Answer

A. $\dfrac{dy}{dt} = \dfrac{d^2y}{dt^2} + 4y - x$ Second Order

B. $y' = 2x$ First Order

C. $\dfrac{dy}{dt} = \dfrac{y^2}{t^3}$ First Order

►Objective 2   Identify slope fields associated with given differential equations.

Example   Which of the following slope fields is consistent with the differential equation $\dfrac{dy}{dx} = y^2$?

I.

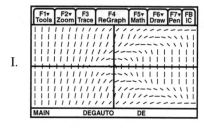

II.

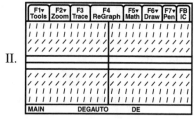

III.

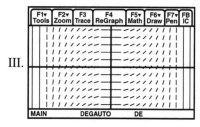

Tip   Slope fields provide a good way to visualize what's going on with first-order differential equations, so it's worth your time to think about slope fields.

Answers

I.

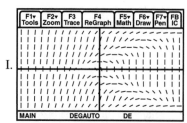

Looking at the first quadrant, it can be seen that the slope at any point in the x, y plane depends on both the y and the x value.

Apex Learning

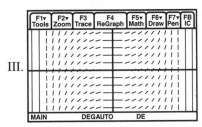

II.

Yes! This slope field is from $\dfrac{dy}{dx} = y^2$. Notice that the slope is always greater than or equal to zero, and that it's independent of the x value. When $y = 0$, the slope is zero, and as y goes to infinity or to negative infinity (irrespective of x) the slope gets steeper.

III.

This slope field is from $\dfrac{dy}{dx} = x^2$. Notice that on this one the slope at any point doesn't depend on y. As x varies, the slope varies, but for a certain x value the slope doesn't change as we change y.

▶Objective 3   Identify differential equations associated with given slope fields.

Example   Which of the following differential equations is consistent with the following slope field?

I. $\dfrac{dy}{dx} = \dfrac{x}{y^2}$

II. $\dfrac{dy}{dx} = \dfrac{x}{y}$

III. $\dfrac{dy}{dx} = \dfrac{x^2}{y}$

Tip   Notice that each of these is separable.

Answer   I. Yes! $\dfrac{dy}{dx} = \dfrac{x}{y^2}$ has a slope field of

II. $\dfrac{dy}{dx} = \dfrac{x}{y}$ has a slope field of

III. $\dfrac{dy}{dx} = \dfrac{x^2}{y}$ has a slope field of

▶**Objective 4**   Separate the variables in first-order differential equations.

Example   Separate but do not solve each of the following differential equations. (If the differential equation isn't separable, then simply state that fact.)

A. $\dfrac{dy}{dx} = \dfrac{x}{y^2}$

B. $\dfrac{dy}{dx} = x - \pi$

C. $y' = \dfrac{x}{y^2 - 5}$

D. $\dfrac{dy}{dx} = \dfrac{x}{y^2} - 5$

Tip   You must work the equation algebraically until it has the form $g(y)\,dy = f(x)\,dx$. Remember that $y' = \dfrac{dy}{dx}$.

Apex Learning

Answers

A. $\dfrac{dy}{dx} = \dfrac{x}{y^2} \Rightarrow y^2\,dy = x\,dx.$

B. $\dfrac{dy}{dx} = x - \pi \Rightarrow dy = (x - \pi)\,dx.$

C. $y' = \dfrac{dy}{dx} = \dfrac{x}{y^2 - 5} \Rightarrow (y^2 - 5)\,dy = x\,dx.$

D. $\dfrac{dy}{dx} = \dfrac{x}{y^2} - 5$: This one isn't separable.

▶**Objective 5**   Solve first-order separable differential equations.

Example 1

A. Find general solutions to each of the following differential equations:

   (i) $\dfrac{dy}{dx} = \dfrac{x + 4}{y}$ (ii) $y' = \dfrac{x^2}{y - 1}$

B. Assuming that each of these curves pass through the point $(0, 1)$, find specific solutions to both differential equations.

Tips   Use separation of variables. Don't forget your constant of integration.

Answers

A.   For (i):

$$\frac{dy}{dx} = \frac{x + 4}{y}$$
$$y\,dy = (x + 4)dx$$
$$\int y\,dy = \int (x + 4)dx$$
$$\frac{1}{2}y^2 = \frac{1}{2}x^2 + 4x + C$$

For (ii):

$$y' = \frac{dy}{dx} = \frac{x^2}{y - 1}$$
$$(y - 1)dy = x^2\,dx$$
$$\int (y - 1)dy = \int x^2\,dx$$
$$\frac{1}{2}y^2 - y = \frac{1}{3}x^3 + C$$

B. For (i):

$$\frac{dy}{dx} = \frac{x+4}{y}$$

$$\frac{1}{2}y^2 = \frac{1}{2}x^2 + 4x + C.$$

Plugging in $x = 0, y = 1$,

$$C = \frac{1}{2}.$$

so that

$$\frac{1}{2}y^2 = \frac{1}{2}x^2 + 4x + \frac{1}{2},$$

For (ii):

$$\frac{dy}{dx} = \frac{x^2}{y-1}$$

$$\frac{1}{2}y^2 - y = \frac{1}{3}x^3 + C.$$

Plugging in $x = 0, y = 1$,

$$C_2 = -\frac{1}{2},$$

so that

$$\frac{1}{2}y^2 - y = \frac{1}{3}x^3 - \frac{1}{2}.$$

Example 2 Given that $\frac{dy}{dx} = \frac{x+4}{y}$ represents the slope of a function $g(x)$ such that $g(0) = 1$, find an expression to explicitly represent the function $g(x)$.

Tip Remember that function graphs must pass the vertical line test.

Answer You do exactly what you did in the first example to begin:

$$\frac{dy}{dx} = \frac{x+4}{y}$$

$$\frac{1}{2}y^2 = \frac{1}{2}x^2 + 4x + C$$

$$\frac{1}{2}y^2 = \frac{1}{2}x^2 + 4x + \frac{1}{2}.$$

But now you need an explicit expression for $g(x)$, so you isolate y:

$$y = \pm\sqrt{x^2 + 8x + 1}.$$

Since the point $(0, 1)$ lies on your function, you know that you must use the positive root:

$$g(x) = \sqrt{x^2 + 8x + 1}$$

►**Objective 6**   Translate differential equations from words into math.

Example   Write a differential equation equivalent to each of these:

A.   The rate of change of y with respect to x is directly proportional to y and inversely proportional to x.

B.   A chemical decomposition proceeds at a rate equal to the square of the chemical present.

Tip   Find the implied derivative in the statement. Note that the negative is needed, since the chemical is decomposing.

Answers

A. $\dfrac{dy}{dx} = k\dfrac{y}{x}$

B. $\dfrac{dy}{dt} = -y^2$ where y is the amount of chemical present.

►**Objective 7**   Translate differential equations from math into words.

Example   Write a sentence to describe the differential equation $\dfrac{dy}{dx} = k\frac{x}{y^2}$.

Answer   The rate of change of y with respect to x is both directly proportional to x and inversely proportional to the square of y.

►**Objective 8**   Solve differential equations given verbally.

Example 1   The slope of a curve is equal to the value of x at any point on the curve. If the curve passes through the point $(2, 9)$, what is the equation of the curve?

Answer

$$\frac{dy}{dx} = x$$
$$dy = x\,dx$$
$$\int dy = \int x\,dx$$
$$y = \frac{1}{2}x^2 + c.$$

Since $y = 9$ when $x = 2$,

$$9 = \frac{1}{2}(2^2) + c$$
$$c = 7$$
$$y = \frac{1}{2}x^2 + 7.$$

Example 2 A chemical decomposition proceeds at a rate (in grams per hour) that is equal to the square of the chemical present, and there are 14 grams of the chemical present at $t = 0$.

Write a function that gives the amount of chemical as a function of t.

Tip Don't forget your constant of integration.

Answer

$$\frac{dy}{dt} = -y^2$$

$$\frac{dy}{y^2} = -dt$$

$$\int \frac{dy}{y^2} = -\int dt$$

$$-\frac{1}{y} = -t + C$$

$$y = \frac{1}{t + C}$$

$y = 14$ when $t = 0$

$$14 = \frac{1}{C}$$

$$C = \frac{1}{14}$$

$$y = f(t) = \frac{1}{t + \frac{1}{14}} = \frac{14}{14t + 1}$$

Multiple-Choice Questions on Differential Equations and Slope Fields

1. The order of the differential equation $y''' = ay'' + by' - \dfrac{y}{x}$ is:

A. first

B. second

C. third

D. negative

E. undefined

Answer C.

2. Which of the following differential equations is consistent with the following slope field:

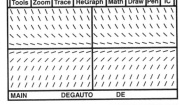

A. $\dfrac{dy}{dx} = y$

B. $\dfrac{dy}{dx} = -y$

C. $\dfrac{dy}{dx} = -x$

D. $\dfrac{dy}{dx} = x$

E. $\dfrac{dy}{dx} = x - y$

Answer   B. Here's what each of the slope fields looks like:

$\dfrac{dy}{dx} = y$:

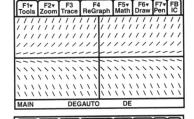

$\dfrac{dy}{dx} = -y$:

$\dfrac{dy}{dx} = -x$:

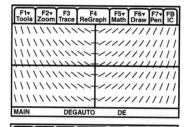

$$\frac{dy}{dx} = x:$$

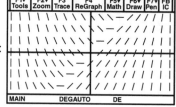

$$\frac{dy}{dx} = x - y:$$

---

3. Which of these slope fields describes the differential equation $\frac{dy}{dx} = xy$?

A.

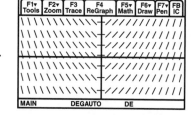

B.

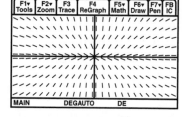

C.

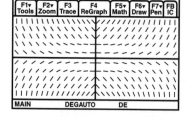

D.

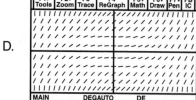

Apex Learning

E.

Answer   C.

The differential equation $\dfrac{dy}{dx} = xy$ indicates negative slopes in the 2nd and 4th quadrants and positive slopes in the 1st and 3rd quadrants. Also notice that $\dfrac{dy}{dx} = xy$ indicates zero slopes at points on both the x- and the y-axis.

---

4.   Which of the following differential equations does **not** represent a typical family of conic sections?

A. $\dfrac{dy}{dx} = \dfrac{x}{y}$

B. $\dfrac{dy}{dx} = 2\dfrac{x}{y}$

C. $\dfrac{dy}{dx} = -\dfrac{2x}{y}$

D. $\dfrac{dy}{dx} = x$

E. $\dfrac{dy}{dx} = \dfrac{y}{x}$

Answer   E. The incorrect option is: $\dfrac{dy}{dx} = \dfrac{y}{x}$. This is a family of lines! When you solve $\dfrac{dy}{dx} = \dfrac{y}{x}$ you get:

$$\frac{1}{y}\,dy = \frac{1}{x}\,dx$$

$$\int \frac{1}{y}\,dy = \int \frac{1}{x}\,dx$$

$$\ln y = \ln x + C$$

$$e^{\ln y} = e^{C_1}e^{\ln x}$$

$$y = Cx,$$

which makes perfect sense when you see the slope field and some arbitrary solutions:

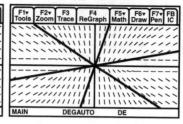

5. Which type of conic section is represented by the differential equation $\dfrac{dy}{dx} = -\dfrac{x+1}{y}$?

    Hint: Look at the slope field **and** try to solve the differential equation.

    A. Circles

    B. Parabolas

    C. Hyperbolas

    D. Ellipses

    E. None of these

Answer   A. You may have just used the slope field:

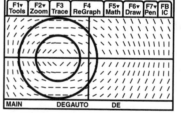

The solutions is given by:

$$\frac{dy}{dx} = -\frac{x+1}{y}$$
$$y\,dy = -(x+1)dx$$
$$\int y\,dy = \int -(x+1)dx$$
$$\frac{1}{2}y^2 = -\frac{1}{2}x^2 - x + C_1,$$

which is the family of circles centered on $P(-1,0)$, since you can change the constant to allow for the factoring of the right side:

$$y^2 = -x^2 - 2x - 1 + C_2$$
$$y^2 = -(x+1)^2 + C_2$$
$$y^2 + (x+1)^2 = r^2.$$

6. Which of the following equations follow from the differential equation $\dfrac{dy}{dx} = -4y\dfrac{x}{x-1}$?

A. $\dfrac{1}{y}\,dy = \dfrac{-4x}{x-1}\,dx$

B. $\displaystyle\int \dfrac{1}{y}\,dy = \int \dfrac{4x}{1-x}\,dx$

C. $\dfrac{1}{4y}\,dy = \dfrac{-x}{x-1}\,dx$

D. All three of these

E. This differential equation is not separable.

Answer   D. They'll all lead to the solution of the differential equation.

---

7. Find the solution to the differential equation $\dfrac{dy}{dx} = \dfrac{2}{x}$ for $x > 0$, given that $y = 3$ when $x = 1$.

A. $y = \ln x + 3$

B. $y = 2\ln x + 3$

C. $y^2 = 2x$

D. $y^2 = 2x + 7$

E. $y^2 = x + 8$

Answer   B.

$$\frac{dy}{dx} = \frac{2}{x}$$
$$dy = \frac{2}{x}\,dx$$
$$\int dy = \int \frac{2}{x}\,dx$$
$$y = 2\ln|x| + C$$
$$y = 2\ln x + C \quad (\text{since } x > 0)$$

Since the curve passes through point $(1,3)$, we have $3 = 2\ln 1 + C$, so

$$C = 3$$
$$y = 2\ln x + 3.$$

8. Which of the following differential equations are separable?

I. $y' = x^2$

II. $\dfrac{dy}{dx} = \dfrac{2}{x} + y$

III. $\dfrac{dy}{dx} = \dfrac{2+y}{x}$

A. I only

B. III only

C. I and II only

D. I and III only

E. I, II, and III

Answer   D.

I. $y' = x^2$ is $\dfrac{dy}{dx} = x^2$, which separates into $dy = x^2\, dx$.

II. $\dfrac{dy}{dx} = \dfrac{2}{x} + y$ is not separable.

III. $\dfrac{dy}{dx} = \dfrac{2+y}{x}$ separates into $\dfrac{1}{2+y}\, dy = \dfrac{1}{x}\, dx$.

9. Which of the following is **not** a differential equation?

A. $\dfrac{d^2y}{dt^2} = t^2 + \dfrac{y}{4}$

B. $4 + y = 2x + \dfrac{dx}{dy}$

C. $(x - x^3)\, dx = 4 - y^2$

D. $y' = 0$

E. $t^2 = s + \left(\dfrac{ds}{dt}\right)^2$

Answer   C. It is not a differential equation, and actually makes very little sense. What's that dx doing there all by itself? Maybe if there was a dy around, you could argue that the equation must at least be related to a differential equation.

A differential equation must contain derivatives, and the equation $(x - x^3)dx = 4 - y^2$ does not. Notice that the equation $(x - x^3)\dfrac{dy}{dx} = 4 - y^2$ *would* be a differential equation in y and x.

10. A *separable differential equation* is a first-order differential equation that can be algebraically manipulated to look like:

A. $f(x)\,dx + f(y)\,dy$

B. $f(y)\,dy = g(x)\,dx$

C. $f(x)\,dx = f(y)\,dy$

D. $g(y)\,dx = f(x)\,dx$

E. None of the above.

Answer   B. A *separable differential equation* is a differential equation that can be algebraically manipulated to look like $f(y)\,dy = g(x)\,dx$ or for that matter $f(x)\,dx = g(y)\,dy$.

The important thing is not the letters representing the functions, but the fact that you isolate all the x's in one function on one side and all the y's in another function on the other side.

11. Which of the following could be the slope field for the differential equation $y' = x^2 y$?

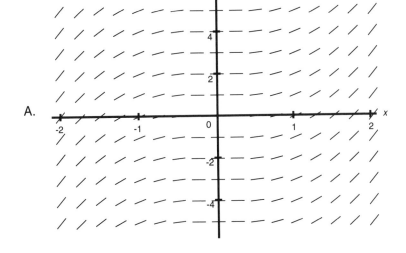

A.

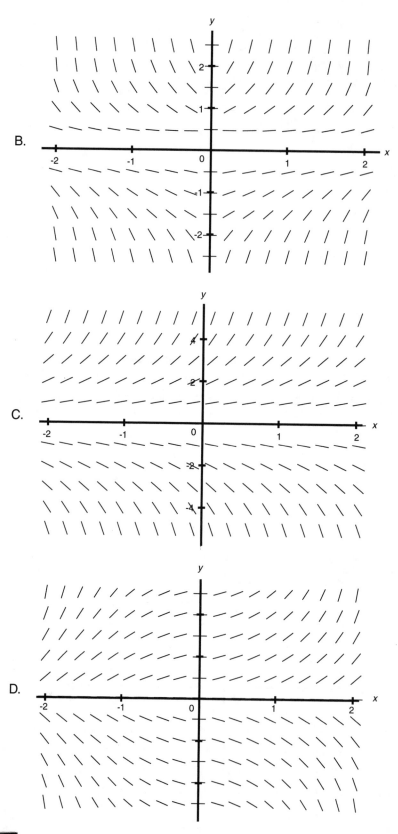

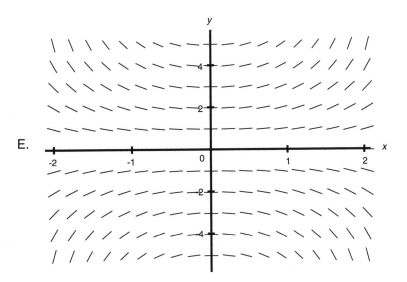

E.

Answer   D. The equation indicates that the slope of the line segments should be positive when y is positive, negative when y is negative, and 0 at $x = 0$ and $y = 0$. Graph D is the only choice that satisfies all these conditions.

---

12.   The following graph could be the slope field plot for which differential equation below?

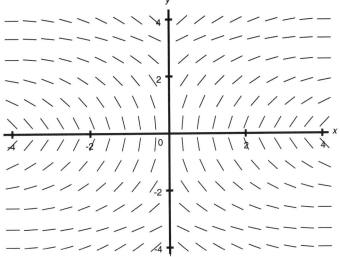

(You should try to do this without your calculator, by examining the slope field in each of the four quadrants as well as on or near the axes.)

A. $y' = \dfrac{x}{y}$

B. $y' = \dfrac{1}{x^2 y}$

C. $y' = \dfrac{1}{yx}$

D. $y' = \dfrac{y}{x^2}$

E. $y' = \dfrac{x^2}{y}$

Answer C. The most important thing to notice about the slope field plot is that the line segments appear to become vertical on the x- and y-axes.

So the equation $y' = f(x, y)$ should have a fraction in it with both x and y on the bottom, so that the derivative goes to positive or negative infinity as either x goes to 0 or y goes to 0.

$y' = \dfrac{1}{x^2 y}$ indicates positive slope whenever y is positive, and the slope field plot above doesn't satisfy this requirement.

So, the answer must be $y' = \dfrac{1}{yx}$ by elimination. You can check to verify that this equation does gives the correct positive and negative slopes in each of the four quadrants.

---

13. Which of the following differential equations is/are **not** separable?

I. $\dfrac{dy}{dt} = \dfrac{1+t}{yt}$

II. $\dfrac{d^2 y}{dx^2} = 3xy + \dfrac{dy}{dx}$

III. $\dfrac{dy}{dx} = y^2 + yx$

A. I only

B. II only

C. III only

D. II and III

E. I and III

Answer D.

Equation I is separable, since

$$\frac{dy}{dt} = \frac{1+t}{yt} \Rightarrow y\,dy = \frac{1+t}{t}\,dt.$$

Equation II is not separable, because of the $\dfrac{d^2 y}{dx^2}$ term. There is no way to separate the variables here.

Equation III is not separable. Algebraic manipulation gives you

$$\frac{dy}{dx} = y(y + x),$$

and there is just no way to get this into the form $f(x)\,dx = g(y)\,dy$.

14. Separating the variables in the equation $ye^{-x}\dfrac{dy}{dx} = x$ gives you the equation:

A. $ye^{-x}\,dy = x\,dx$

B. $y\,dy = xe^{x}\,dx$

C. $y\,dy = xe^{-x}\,dx$

D. $y\,dy = xe^{-1/x}\,dx$

E. None of these.

Answer   B. You can first rewrite the equation as $y\dfrac{1}{e^x}\dfrac{dy}{dx} = x$.

Then multiplying both sides of this equation by $e^x\,dx$ gives you $y\,dy = xe^x\,dx$.

---

15. Suppose you want to find an equation for a curve whose slope at any point (x, y) is $\dfrac{x}{6y^2}$.

Using separation of variables solve the differential equation $\dfrac{dy}{dx} = \dfrac{x}{6y^2}$.

A. $6y^3 = x^2 + C$

B. $3y^3 = 2x^2 + C$

C. $\frac{1}{3}y^3 = \frac{1}{2}x^2 + C$

D. $2y^3 = x^2 + C$

E. $2y^3 = \frac{1}{2}x^2 + C$

Answer   E. Separating the equation $\dfrac{dy}{dx} = \dfrac{x}{6y^2}$ gives you $6y^2\,dy = x\,dx$.

Then integrating both sides, you get

$$\int 6y^2\,dy = \int x\,dx$$

$$\Rightarrow 2y^3 + C_1 = \frac{1}{2}x^2 + C_2$$

$$\Rightarrow 2y^3 = \frac{1}{2}x^2 + C,$$

which you get by combining both constants on the right into a single constant C.

---

16. Suppose you want to find an equation for a curve that passes through the point $(1, 2)$ and whose slope at any point (x, y) is $\dfrac{2x^3}{y}$.

The solution to the differential equation $\dfrac{dy}{dx} = \dfrac{2x^3}{y}$ with initial condition $y(1) = 2$ is:

A. $2y^2 = x^4 + 7$

B. $y^2 = 2x^4 + 2$

C. $y^2 = x^4 + 3$

D. $\frac{1}{2}y^2 = \frac{1}{2}x^4 + 4$

E. $\frac{1}{2}y^2 = \frac{1}{2}x^4 + 3$

Answer C. The differential equation is $\dfrac{dy}{dx} = \dfrac{2x^3}{y}$ is separable to $y\,dy = 2x^3\,dx$.

Integrating both sides gives you

$$\int y\,dy = \int 2x^3\,dx$$
$$\Rightarrow \frac{1}{2}y^2 + C_1 = \frac{1}{2}x^4 + C_2$$
$$\Rightarrow y^2 + 2C_1 = x^4 + 2C_2$$
$$\Rightarrow y^2 = x^4 + C,$$

which you get by combining both constants on the right side and renaming it C.

The initial condition indicates the equation $y^2 = x^4 + C$ must satisfy $y = 2$ when $x = 1$.
Plugging those values in gives you

$$2^2 = 1^4 + C \Rightarrow 4 - 1 = C \Rightarrow C = 3.$$

Therefore, the final solution to the differential equation with the initial condition given is the curve $y^2 = x^4 + 3$.

---

17. Solve the differential equation $\dfrac{dy}{dx} = \dfrac{1+x}{xy}$ with initial condition $y(1) = -2$.

    (Hint: Isolate $\dfrac{1+x}{x}$ on one side of the equation, break it into the sum of two fractions.)

    A. $\frac{1}{2}y^2 = \ln|x| + x + 1$

    B. $\frac{1}{2}y^2 = \frac{1}{2}(x+1)^2$

    C. $\frac{1}{2}y^2 = \frac{1}{2}(x+1)^2 + 3$

    D. $\frac{1}{2}y^2 = \ln|x| + x - 3$

    E. This one is separable, but it's impossible to solve.

Answer   A. The differential equation $\frac{dy}{dx} = \frac{1+x}{xy}$ separates into $y\,dy = \frac{1+x}{x}\,dx$.

Integrating both sides gives you

$$\int y\,dy = \int \frac{1+x}{x}\,dx$$

$$\frac{1}{2}y^2 + C_1 = \int \frac{1}{x}\,dx + \int \frac{x}{x}\,dx$$

$$\frac{1}{2}y^2 + C_1 = \int \frac{1}{x}\,dx + \int 1\,dx$$

$$\frac{1}{2}y^2 + C_1 = \ln|x| + C_2 + x + C_3$$

$$\frac{1}{2}y^2 = \ln|x| + x + C$$

(combining the constants together).

Now, this equation must satisfy $y = -2$ when $x = 1$. Plugging in those values gives you

$$\frac{1}{2}(-2)^2 = \ln|1| + 1 + C$$

$$\Rightarrow 2 = 0 + 1 + C$$

$$\Rightarrow C = 1.$$

Therefore, the solution to this differential equation with the initial condition given is

$$\frac{1}{2}y^2 = \ln|x| + x + 1.$$

B. Solving Separable Differential Equations

Exponential Growth and Decay and Related Applications

The model for exponential growth and decay can be written as

$$y = C_0 a^t \qquad \text{or} \qquad y = Ae^{kt}$$

where the constants C_0 and A both represent the *initial* quantity (at $t = 0$). The parameters a and k control the rate of growth or decay as summarized below:

	Decay	Growth
$y = C_0 a^t$	$0 < a < 1$	$1 < a$
$y = Ae^{kt}$	$k < 0$	$0 < k$

This model arises as a solution to a very important differential equation.

►Objective 1   Solve the differential equation $\dfrac{dy}{dt} = ky$.

Answer $\dfrac{dy}{dt} = ky$ is separable so its general solution can be obtained:

$$\frac{dy}{y} = k\,dt$$

$$\int \frac{dy}{y} = \int k\,dt$$

$$\ln|y| = kt + C$$

$$e^{\ln|y|} = e^{(kt+C)}$$

$$|y| = e^c e^{kt}$$

$$|y| = Ae^{kt}$$

where

$$A = e^c > 0.$$

When y is positive, $|y| = y$, which gives $y = Ae^{kt}$.

When y is negative, $|y| = -y$, which gives $y = -Ae^{kt}$.

We can write a general solution by allowing A to take on positive **and** negative values:

$$y = Ae^{kt} \text{ is a solution to } \frac{dy}{dt} = ky.$$

►Objective 2   Model situations using the solution to $\dfrac{dy}{dt} = ky$.

Example   The rate at which the population of a group of organisms grows is directly proportional to the number of organisms in the population. If the population at time zero is $3,500$ and the population after one year is $5,250$, what will the population be after 3 years?

Tip $y = Ae^{kt}$ is a solution to $\dfrac{dy}{dt} = ky$.

Answer

$$y = Ae^{kt}$$

$$A = 3,500$$

$$y = 3,500e^{kt}$$

$$5,250 = 3,500e^{k(1)}$$

$$\frac{5,250}{3,500} = e^{k(1)}$$

$$k = \ln\frac{5,250}{3,500} = \ln\frac{3}{2}$$

$$y = 3,500e^{(\ln\frac{3}{2})t}$$

At time $t = 3$

$$y = 3,500e^{(\ln\frac{3}{2})3} = 11,812.5.$$

Alternate Solution Recognize that the solution to $\dfrac{dy}{dt} = ky$ is an exponential model.

Then note that the growth factor $a = \frac{5,250}{3,500} = \frac{3}{2}$, so $y = 3,500(\frac{3}{2})^t$. At time $t = 3$

$$y = 3,500\left(\frac{3}{2}\right)^3 = \frac{23,625}{2} = 11,812.5.$$

▶Objective 3 Solve separable differential equations that are similar in form to $\dfrac{dy}{dt} = ky$.

Example Solve the separable differential equation $\dfrac{dy}{dt} = 4y - 8$, given that $y = 4$ when $t = 0$.

Tip This is a separable differential equation.

Answer

$$\frac{dy}{dt} = 4y - 8 = 4(y - 2)$$
$$\frac{dy}{y - 2} = 4dt$$
$$\int \frac{dy}{y - 2} = \int 4dt$$
$$\ln|y - 2| = 4t + C$$
$$|y - 2| = e^{(4t+C)} = Ae^{4t}$$

where $A > 0$. If we allow A to take on negative values, then we can remove the absolute values:

$$y - 2 = Ae^{4t}$$
$$y = Ae^{4t} + 2.$$

Since $y = 4$ when $t = 0$,

$$4 = A + 2$$
$$A = 2$$

Thus $y = 2e^{4t} + 2$.

▶Objective 4 Set up differential equations to model situations.

Example A sailboat accelerates $\left(\dfrac{dv}{dt}\right)$ at a rate that's proportional to the difference between the velocity of the wind (25 mph) from the velocity of the boat when the wind is blowing in a direction that is directly opposite to the direction that the boat is moving in. Write a differential equation to model this situation.

Tip   Use a parameter (k) to represent the proportionality constant.

Answer

$$\frac{dv}{dt} = k(v - 25)$$

▶Objective 5   Solve separable differential equations that model situations.

Example   A 100 gallon tank of salt water has 50 lbs of salt initially dissolved in the tank. The tank is being stirred so that the concentration of the salt water is constant throughout the tank. Fresh water is pouring into the tank at $\frac{1}{4}$ gallons per minute; salt water is also pouring out of the tank at $\frac{1}{4}$ gallons per minute.

A.   Write a differential equation that describes the situation (involving $\frac{ds}{dt}$, which is the change in the amount of salt (lbs) as a function of time in minutes).

B.   Find a solution to the differential equation that gives the amount of salt s (lbs) as a function of time in minutes.

C.   How much salt is in the tank at the end of one day (24 hours)?

Tip   Note that $\frac{ds}{dt}$ is related to the amount of salt s. The units can help you set up the rate here.

Answers

A.   No salt is entering the tank, so we only need to determine how much salt is leaving the tank. The concentration of salt in the water is s pounds per 100 gallons, and it's leaving at $\frac{1}{4}$ gallon per minute, so the salt is leaving at $\frac{1}{4} \times \frac{s}{100} = \frac{1}{400}s$ pounds per minute. The salt is decreasing, so the rate is negative:

$$\frac{ds}{dt} = -\frac{1}{400}s.$$

B.   This is of the form $\frac{dy}{dx} = ky$, so that

$$s = Ae^{-\frac{1}{400}t}.$$

Since $s = 50$ when $t = 0$,

$$s = 50e^{-\frac{1}{400}t}.$$

C.   One day is 24 hours, so $24 \times 60 = 1,440$ minutes. $s = 50e^{-\frac{1,440}{400}} \approx 1.366$ lbs of salt remain in the tank.

Here are some other applications of differential equations:

Newton's Law of Cooling: The temperature $T(t)$ of a body in a surrounding medium of constant temperature T_s at any time t is given by $T(t) - T_s = Ae^{kt}$, or $T(t) = Ae^{kt} + T_s$, where A is the initial difference in the temperatures and k is a constant that depends on the properties of the body and the surrounding medium.

Falling Bodies in Air: The air resistance depends on the properties of the object and its speed. The effect of air resistance on the velocity of an object is modeled well with this simple differential equation $\frac{dv}{dt} = -kv$, where k is a constant. Put in the information about gravity (the constant g), and you have the differential equation for a falling body, including air resistance: $\frac{dv}{dt} = g - kv$.

Logistic Growth of a Population in a Bounded Environment: In such a population, the relative growth rate (that is, the proportion of new animals to the existing population) is proportional to the amount of elbow room available in the habitat. In other words, if P is the population at time t, M is the carrying capacity, and k is the constant of proportionality,

$$\frac{\frac{dP}{dt}}{P} = k\left(1 - \frac{P}{M}\right), \quad \text{or} \quad \frac{dP}{dt} = \frac{k}{M}P(M - P)$$

This is a separable differential equation that requires some special algebraic techniques (partial fractions) to obtain the solution:

$$P = \frac{M}{1 + Ae^{-kt}}$$

Note: The logistic model is included in the BC Calculus course description, but is not an AB Calculus topic.

Example 1   The differential equation $25 - \frac{dP}{dt} = \frac{1}{4}P$ models the amount of rabbits in a particular rabbit population for time $0 \leq t \leq 5$ measured in years.

   A.   Find a general solution of this differential equation.

   B.   Given that the population of rabbits is initially 300, find a specific solution to the differential equation.

   C.   Estimate the maximum number of rabbits present in the population during the third year. (No partial rabbits please!)

Tip $0 \leq t \leq 1$ is the first year.

Answers

A.

$$25 - \frac{dP}{dt} = \frac{1}{4}P$$
$$\frac{dP}{dt} = 25 - \frac{1}{4}P = \frac{1}{4}(100 - P)$$
$$\frac{dP}{100 - P} = \frac{1}{4}dt$$
$$\int \frac{dP}{100 - P} = \int \frac{1}{4}dt$$
$$-\ln|100 - P| = \frac{1}{4}t + C$$
$$|100 - P| = e^{-(\frac{1}{4}t+C)} = Ae^{-\frac{1}{4}t}(A > 0)$$
$$100 - P = Ae^{-\frac{1}{4}t}(A \neq 0)$$

The general solution is, thus,

$$P = 100 - Ae^{-\frac{1}{4}t}$$

B.

$$P = 300 \text{ when } t = 0$$
$$300 = 100 - A$$
$$A = -200$$
$$P = 100 + 200e^{-\frac{1}{4}t}$$

C.

$$P = 100 + 200e^{-\frac{1}{4}t} \implies \frac{dP}{dt} = -50e^{-\frac{1}{4}t} \text{ is always negative, so that } P \text{ is always}$$
decreasing. Thus the maximum value will occur at the beginning of the third year, or when $t = 2$.

$$P = 100 + 200e^{-\frac{1}{4}t}$$
$$P(2) = 100 + 200e^{-\frac{1}{4}(2)} = 100 + 200e^{-\frac{1}{2}} \approx 221 \text{ rabbits.}$$

Example 2 Find the maximum value of the function $g(x)$ if $g(1) = 1$ and the tangent line to the curve $y = g(x)$ at any point on the curve (x, y) has a slope given by $-\frac{x}{y}$.

Tip Write a differential equation to describe the situation.

Answer   The slope at any point on the function is given by

$$\frac{dy}{dx} = -\frac{x}{y}.$$

$$\int y\,dy = -\int x\,dx$$

$$\frac{1}{2}y^2 = -\frac{1}{2}x^2 + C.$$

But since $x = 1$ when $y = 1$,

$$\frac{1}{2}1^2 = -\frac{1}{2}1^2 + C$$
$$C = 1$$
$$\frac{1}{2}y^2 = -\frac{1}{2}x^2 + 1$$
$$x^2 + y^2 = 2$$
$$y = \pm\sqrt{2 - x^2}.$$

But since the function passes through the point $(1, 1)$, we know that the function is given by $g(x) = \sqrt{2 - x^2}$ and the first derivative is given by

$$g'(x) = -\frac{x}{\sqrt{(2 - x^2)}}$$

(or you could use the original expression for $\frac{dy}{dx} = -\frac{x}{y}$, noticing that $y \geq 0$).

x	$x < 0$	$x = 0$	$0 < x$
$g'(x)$	*Positive*	0	*Negative*
$g(x)$	Increasing	Relative *Max*	Decreasing

So, the maximum value of $g(x)$ is given by $g(0) = \sqrt{2 - 0^2} = \sqrt{2}$.

Multiple-Choice Questions on Solving Separable Differential Equations

1.   Alice has a savings account where she's saving for college. She deposits $50 a month into the account, which earns 4% interest. Which of these differential equations would model this situation? ($A(t)$ is the amount in the account at any time t, where t is measured in years. For simplicity's sake, assume that all the deposits and interest are being applied continuously, so the growth of the account is smooth.)

A. $A(t) = 600t + .04A$

B. $\frac{dA}{dt} = 50 + .04A$

C. $\dfrac{dA}{dt} = 50 + .04t$

D. $\dfrac{dA}{dt} = 600 + .04A$

E. $\dfrac{dA}{dt} = 600 + .04t$

Answer D. The amount in the account is increasing by $600 per year (12 deposits) and by the 4% interest per year. So, $\dfrac{dA}{dt} = 600 + .04A$.

---

2. The velocity of a falling body is affected by gravity, of course. In problems like this we often say "neglect air resistance." However, suppose we want to include air resistance in our model for the velocity of a falling body. Which of these differential equations might describe the velocity of a falling body, with air resistance? (The acceleration due to gravity is about 9.8 meters per second squared, and we're taking upward to be the positive direction.)

A. $\dfrac{dv}{dt} = -9.8 - 0.1v$

B. $\dfrac{dv}{dt} = -9.8 - 0.1t$

C. $\dfrac{dv}{dt} = -9.8 + 0.1v$

D. $\dfrac{dv}{dt} = 9.8 - 0.1v$

E. None of these

Answer A. The -9.8 (from gravity) points down, since we're taking "up" to be the positive direction. Moveover, no matter the direction of motion of the object, the air resistance acts to slow the object down. So, we must put a minus sign out front of the v to get the term $-0.1v$. If the velocity is positive, this term is negative which decreases the speed. If the velocity is negative, this term is positive so it acts to increase the velocity. However, since speed is the absolute value of the velocity, in this case the speed decreases as well.

---

3. $y\dfrac{dy}{dx} = x \sin y$ is a separable differential equation.

Which of the following equations would you get after separating the variables?

A. $\dfrac{y}{\sin y} = x$

B. $\dfrac{y}{\sin y}\, dy = \dfrac{x^2}{2} + C$

C. $\dfrac{y}{\sin y}\, dy = x\, dx$

D. $\dfrac{\sin y}{y}\dfrac{1}{dy} = \dfrac{1}{x}\dfrac{1}{dx}$

E. None of these

Answer C. Use algebra to get all the y's on one side and all the x's on the other, making sure your dx and dy both end up "upstairs":

$$y\frac{dy}{dx} = x\sin y$$

$$\frac{y}{\sin y}\,dy = x\,dx$$

---

4. Solve the differential equation $\dfrac{dy}{dx} = (x+1)(y^2+1)$, given that $y=1$ when $x=0$.

A. $\arctan y = \dfrac{x^2}{2} + x + C$

B. $\arctan y = \dfrac{x^2}{2} + x + \dfrac{\pi}{4}$

C. $\arctan y = \ln|x+1| + \dfrac{\pi}{4}$

D. $\arctan y = x^2 + x + C$

E. This has no solution.

Answer B. First, separate the variables and integrate:

$$\frac{dy}{dx} = (x+1)(y^2+1)$$

$$\frac{1}{1+y^2}\,dy = (x+1)\,dx$$

$$\int \frac{1}{1+y^2}\,dy = \int (x+1)\,dx$$

$$\arctan y = \frac{x^2}{2} + x + C$$

Then use the initial conditions to solve for C:

$$\arctan 1 = \frac{0^2}{2} + 0 + C,$$

or $\dfrac{\pi}{4} = C$. So the solution is

$$\arctan y = \frac{x^2}{2} + x + \frac{\pi}{4}$$

5. Separate the variables in this differential equation $\dfrac{dy}{dx} = x + y$.

A. $\dfrac{1}{y} dy = x\,dx$

B. $y\,dy = x\,dx$

C. $y\,dy = \dfrac{1}{x}\,dx$

D. $\dfrac{1}{y}\,dy = \dfrac{1}{x}\,dx$

E. This isn't separable.

Answer E. There isn't any way to get this in the form $f(y)\,dy = g(x)\,dx$. Separating the variables doesn't work to solve this differential equation.

---

6. Solve the differential equation $\dfrac{dy}{dx} = \dfrac{y^2 - 1}{y}$.

A. $\frac{1}{2}\ln|y^2 - 1| = x + C$

B. $\ln|y^2 - 1| = x + C$

C. $\arctan y = x + C$

D. $y \arctan y = x + C$

E. This isn't separable.

Answer A. Separate the variables and integrate:

$$\frac{dy}{dx} = \frac{y^2 - 1}{y}$$

$$\frac{y}{y^2 - 1}\,dy = dx$$

$$\int \frac{y}{y^2 - 1}\,dy = \int dx$$

$$\frac{1}{2}\ln|y^2 - 1| = x + C$$

---

7. Suppose that Saleem has money in a savings account that pays him interest of 5% per year on the account balance. Now, Saleem takes out $50 a month for various frivolities, and his mother secretly deposits $240 every 6 months into his savings account. Assuming that interest is paid and money is deposited and withdrawn from the account in a continuous fashion, the balance $B = B(t)$ (in dollars) remaining in Saleem's savings account at time t (in months) is best modeled by the differential equation:

A. $\dfrac{dB}{dt} = \dfrac{0.05B}{12} - 10$

B. $\dfrac{dB}{dt} = 0.05B - 10$

C. $\dfrac{dB}{dt} = \dfrac{0.05B}{12} - 190$

D. $\dfrac{dB}{dt} = \dfrac{0.05B}{12} - 50$

E. $\dfrac{dB}{dt} = 0.05B - 120$

Answer A. First, convert all the rates into the proper units, which are dollars per month (since t is in months). Thus, Saleem is withdrawing funds at the rate of \$50 a month, and his mother is depositing funds at the rate of $\dfrac{\$240}{6 \text{ months}} = \$40$ a month.

Furthermore, the bank is depositing money into the account at the rate of $0.05B(t)$ dollars a **year** (corresponding to an increase of 5% on the current balance $B(t)$), which converts to $\frac{0.05B}{12}$ dollars a month. Combining all these rates, remembering to subtract those rates that represent withdrawals, you get

$$\frac{dB}{dt} = \frac{0.05B}{12} + 40 - 50 = \frac{0.05B}{12} - 10.$$

---

8. Suppose the number of bacteria in a culture increases by 50% every hour if left on its own. Assuming that biologists decide to remove approximately one thousand bacteria from the culture every ten minutes, which of the following equations best models the population $P = P(t)$ of the bacteria culture, where t is in hours?

A. $\dfrac{dP}{dt} = .5P - 1,000$

B. $\dfrac{dP}{dt} = .5P - 6,000$

C. $\dfrac{dP}{dt} = 1.5P - 6,000$

D. $\dfrac{dP}{dt} = 1.5P - 1,000$

E. $\dfrac{dP}{dt} = -.5P - 1,000$

Answer B. $\dfrac{dP}{dt}$ represents the growth rate of the culture in bacteria per hour.

The population of the culture **increases** on its own at the rate of $.5P$, which corresponds to an increase of 50% of the current population. The biologists are **removing** bacteria at the rate of $6,000$ bacteria per hour, which corresponds to their removing $1,000$ every ten minutes. Then the rate of growth $\dfrac{dP}{dt}$ should be equal to rate of increase minus rate of decrease, giving you

$$\frac{dP}{dt} = .5P - 6,000.$$

When setting up these differential equations, it's often wise to consider all those things that correspond to adding to the quantity involved (rate increases) and all those things that correspond to taking away from the quantity involved (rate decreases), and then combine them as

Overall Rate = Rate of Increase − Rate of Decrease.

---

9. According to Newton's law of cooling, the rate at which an object's temperature changes is directly proportional to the difference in temperature between the object and the surrounding medium. If $T(t)$ represents the temperature of the object ($°C$) at time t (in hours), and T_s represents the constant temperature of the surrounding medium, then the differential equation best describing the rate of change in the temperature of the object is:

A. $\dfrac{dT}{dt} = (kT_s - T)$ for some positive constant of proportionality k.

B. $\dfrac{dT}{dt} = T_s - T$.

C. $\dfrac{dT}{dt} = k(T - T_s)$ for some positive constant of proportionality k.

D. $\dfrac{dT}{dt} = k(T_s - T)$ for some positive constant of proportionality k.

E. $\dfrac{dT}{dt} = \dfrac{k}{(T_s - T)}$ for some positive constant of proportionality k.

Answer   D. $\dfrac{dT}{dt}$ represents the rate of change in the temperature (in $°C$ per hour). This rate of change is directly proportional to the difference in temperature between the object and the surrounding medium, so you should have either that $\dfrac{dT}{dt} = k(T - T_s)$ or $\dfrac{dT}{dt} = k(T_s - T)$ with k being a positive constant of proportionality that likely depends on the object.

To decide which one to use, notice that for $\dfrac{dT}{dt} = k(T - T_s)$ the temperature is increasing when the temperature of the object (T) is greater than the temperature of the surrounding medium (T_s), which doesn't make good common sense.

On the other hand, the equation $\dfrac{dT}{dt} = k(T_s - T)$ has the temperature increasing when the temperature of the object (T) is less than the temperature of the surrounding medium (T_s), which does make good common sense.

Therefore, the differential equation best describing the rate of change in the temperature of the object is

$$\frac{dT}{dt} = k(T_s - T)$$

for some positive constant of proportionality k.

---

10. Suppose a certain country's population has constant relative birth and death rates of 97 births per thousand people per year and 47 deaths per thousand people per year respectively. Assume also that approximately 30,000 people emigrate (leave) from the country every year. Which of the following equations best models the population $P = P(t)$ of the country, where t is in years?

A. $\frac{dP}{dt} = 50P + 30,000$

B. $\frac{dP}{dt} = .05P - 30,000$

C. $\frac{dP}{dt} = .05P + 30,000$

D. $\frac{dP}{dt} = .5P - 30,000$

E. $\frac{dP}{dt} = .5P - 30$

Answer B. The birth rate is 97 births per thousand people per year, so $97(\frac{P}{1,000})$ represents the birth rate in people per year.

Similarly, the death rate of 47 deaths per thousand people per year corresponds to a death rate of $47(\frac{P}{1000})$ people per year. The rate of emigration is simply 30,000 people per year.

Therefore, the population increases $97(\frac{P}{1,000})$ people per year due to the birth rate, and decreases at rate by $47(\frac{P}{1,000}) + 30,000$ people per year due to rate of death and emigration.

Algebraically, this is

$$\frac{dP}{dt} = 97\left(\frac{P}{1,000}\right) - \left[47\left(\frac{P}{1,000}\right) + 30,000\right]$$

$$= 97\left(\frac{P}{1,000}\right) - 47\left(\frac{P}{1,000}\right) - 30,000$$

$$= 50\left(\frac{P}{1,000}\right) - 3,0000$$

$$\frac{dP}{dt} = .05P - 30,000.$$

11. *Learning curves* are studied by psychologists interested in the theory of learning. A learning curve is the graph of a function $P_L(t)$ that represents the performance level of someone who has trained at a skill for t hours. Thus, $\dfrac{dP_L}{dt}$ represents the rate at which the performance level improves. By convention, the derivative $P_L(t)$ is taken to be a positive function.

If M (a positive constant) is the maximum performance level of which the learner is capable, then which differential equations could be a reasonable model for learning (or more precisely, performance level)? Use your common sense applied to the practical meaning behind each equation to determine which of the following are reasonable.

I. $\dfrac{dP_L}{dt} = k(M - P_L)$ for some positive constant k.

II. $\dfrac{dP_L}{dt} = k(P_L)$ for some positive constant k.

III. $\dfrac{dP_L}{dt} = k(M - P_L)^{1/2}$ for some positive constant k.

IV. $\dfrac{dP_L}{dt} = \dfrac{k}{(M - P_L)}$ for some positive constant k.

A. I only

B. I and II only

C. III only

D. I and III only

E. IV only

Answer   D. The problem tells us that the P_L graph has a horizontal asymptote at $P_L = M$.

This tells us two things: 1) $\lim\limits_{t\to\infty} P_L = M$, and 2) $\lim\limits_{t\to\infty} \dfrac{dP_L}{dt} = 0$.

In I and III, as $M - P_L \to 0$, so does $\dfrac{dP_L}{dt}$, so they match our information.

In II, $\lim\limits_{t\to\infty} \dfrac{dP_L}{dt} = kM$, which doesn't fit.

In IV, as $M - P_L \to 0$, $\dfrac{dP_L}{dt} \to \infty$, which doesn't fit.

12. Using the separation of variables technique, solve the following differential equation with the given initial condition $y' = -4y + 36$ and $y(2) = 10$.

A. $\ln|y - 9| = -4x + 8$

B. $\ln|y - 9| = -4x - 8$

C. $\ln|y| = -4x + \ln 10 + 8$

D. $\ln |y + 9| = -4x + \ln 19 + 8$

E. $\ln |y + 9| = -4x - 8 + \ln 19$

Answer   A. Rewrite the equation as $\dfrac{dy}{dx} = -4(y-9)$ and separate variables to get $\dfrac{1}{y - 9} dy = -4 \, dx$. Integrating both sides gives you

$$\ln |y - 9| + C_1 = -4x + C_2$$
$$\Rightarrow \ln |y - 9| = -4x + C$$

(by combining both constants and replacing them with C).

Then, the initial condition $y(2) = 10$ gives you the equation

$$\ln |10 - 9| = -4(2) + C$$
$$\Rightarrow \ln 1 = -8 + C$$
$$\Rightarrow 0 + 8 = C.$$

So, the solution to the differential equation with the given initial condition is

$$\ln |y - 9| = -4x + 8.$$

---

13.   Use separation of variables to solve the differential equation $y' = e^y \sin x$ with initial condition $y(-\pi) = 0$.

A. $e^{-y} = \cos x + 2$

B. $e^y = \cos x + 2$

C. $e^{-y} = -\sin x + 2$

D. $e^{-y} = -\cos x + 2$

E. $e^{-y} = \cos x$

Answer   A. Separating variables, you get

$$\frac{dy}{dx} = e^y \sin x$$
$$\Rightarrow e^{-y} dy = \sin x dx$$

Then, integrating both sides, gives you

$$\int e^{-y} \, dy = \int \sin x \, dx$$
$$\Rightarrow -e^{-y} + C_1 = -\cos x + C_2$$
$$\Rightarrow e^{-y} = \cos x + C$$

(Notice that the two constants have been combined as one—this always seems to happen!)

Then, the initial condition $y(-\pi) = 0$ says that the equation $e^{-y} = \cos x + C$ must be satisfied when $x = -\pi$ and $y = 0$. Plugging in these values, you get

$$e^{(-0)} = \cos(-\pi) + C$$
$$\Rightarrow 1 = -1 + C$$
$$\Rightarrow C = 2.$$

Therefore, the solution to the differential equation with the given initial conditions is the equation

$$e^{-y} = \cos x + 2.$$

---

14. Solve the following differential equation with initial conditions:
$$y'' = e^{-2t} + 10e^{4t}; \qquad y(0) = 1, \qquad y'(0) = 0.$$

A. $y = \frac{5}{8}e^{4t} + \frac{1}{4}e^{-2t} - 3t + \frac{1}{8}$

B. $y = e^{-2t} + e^{4t} - 1$

C. $y = \frac{1}{4}e^{-2t} + \frac{5}{8}e^{4t} - 2t + \frac{1}{8}$

D. $y = -\frac{1}{2}e^{-2t} + \frac{5}{2}e^{4t} - 2$

E. $y = 4e^{-2t} + 16e^{4t} - 16t + 19$

Answer   C. You don't need separation of variables here.

You can just antidifferentiate to get $y' = -\frac{1}{2}e^{-2t} + \frac{5}{2}e^{4t} + C$.

Then, use the condition $y'(0) = 0$.

Solve for the constant C:

$$0 = -\frac{1}{2}e^{-2(0)} + \frac{5}{2}e^{4(0)} + C$$
$$\Rightarrow 0 = -\frac{1}{2} + \frac{5}{2} + C = 2 + C$$
$$\Rightarrow C = -2.$$

So you have $y' = -\frac{1}{2}e^{-2t} + \frac{5}{2}e^{4t} - 2.$

Then, antidifferentiating again gives you

$$y = \frac{1}{4}e^{-2t} + \frac{5}{8}e^{4t} - 2t + C.$$

Using the initial condition $y(0) = 1$ to solve for the constant C:

$$1 = \frac{1}{4}e^{-2(0)} + \frac{5}{8}e^{4(0)} - 2(0) + C$$
$$\Rightarrow 1 = \frac{1}{4} + \frac{5}{8} + C = \frac{7}{8} + C$$
$$\Rightarrow C = \frac{1}{8}.$$

So, you have the solution to the differential equation with the given initial conditions:

$$y = \frac{1}{4}e^{-2t} + \frac{5}{8}e^{4t} - 2t + \frac{1}{8}$$

15. Use separation of variables to solve the following differential equation with initial condition:

$$\frac{dy}{dx} = \frac{yx + 5x}{x^2 + 1} \quad \text{and} \quad y(3) = 5.$$

A. $y^2 = \ln(x^2 + 1) + 25 - \ln 10$

B. $\ln|y + 5| = \ln(x^2 + 1)$

C. $\ln|y + 5| = \arctan 3 + \ln 10 - \arctan 3$

D. $\ln|y + 5| = \frac{1}{2}\ln(x^2 + 1) + \frac{1}{2}\ln(10)$

E. $y = \ln(x^2 + 1) + 50 - \ln 10$

Answer   D. Separating the variables gives you

$$\frac{dy}{dx} = \frac{yx + 5x}{x^2 + 1} = \frac{(y + 5)x}{x^2 + 1}$$

$$\Rightarrow \frac{1}{y + 5}dy = \frac{x}{x^2 + 1}dx$$

Then, integrating gives you

$$\int \frac{1}{y + 5}dy = \int \frac{x}{x^2 + 1}dx$$

$$\Rightarrow \ln|y + 5| = \frac{1}{2}\ln(x^2 + 1) + C.$$

(From now on, we'll just combine the two constants immediately as C.)

Next, using the initial condition $y(3) = 5$ to solve for C:

$$\ln|5 + 5| = \frac{1}{2}\ln(3^2 + 1) + C$$

$$\Rightarrow \ln 10 = \frac{1}{2}\ln(10) + C$$

$$\Rightarrow C = \frac{1}{2}\ln(10).$$

Therefore, the solution to the differential equation with the given initial conditions is

$$\ln|y + 5| = \frac{1}{2}\ln(x^2 + 1) + \frac{1}{2}\ln(10)$$

Note that using laws of logarithms, you can write this as

$$\ln|y + 5| = \frac{1}{2}\ln(10x^2 + 10),$$

or even

$$\ln|y + 5| = \ln\sqrt{10x^2 + 10}.$$

16. Use the separation of variables to solve the following differential equation with initial condition:

$$4x\sqrt{1-t^2}\frac{dx}{dt} - 1 = 0 \qquad \text{and} \qquad x(0) = -2.$$

A. $2x^2 = \arcsin t - 8$

B. $2x^2 = \arcsin t + 8$

C. $2x^2 = \arccos t + 8 - \frac{1}{2}\pi$

D. $2x^2 = \arccos t + 8$

E. $2x^2 = -\arcsin t - 8$

Answer   B. Separating the variables gives you

$$4x\sqrt{1-t^2}\frac{dx}{dt} - 1 = 0$$

$$\Rightarrow 4x\sqrt{1-t^2}\frac{dx}{dt} = 1$$

$$\Rightarrow 4x\,dx = \frac{1}{\sqrt{1-t^2}}dt.$$

Then integrating both sides yields

$$\int 4x\,dx = \int \frac{1}{\sqrt{1-t^2}}dt$$

$$2x^2 = \arcsin t + C.$$

To solve for C, use the initial condition $x(0) = -2$. You get

$$2(-2)^2 = \arcsin 0 = C$$

$$\Rightarrow 8 = 0 + C$$

$$\Rightarrow C = 8.$$

Therefore, the solution to the differential equation with the given initial conditions is

$$2x^2 = \arcsin t + 8.$$

17. Use separation of variables to solve the following differential equation with initial conditions:

$$\frac{dy}{dx} = e^{2x+3y} \qquad \text{and} \qquad y(0) = 1.$$

(Hint: Use a property of exponentials to rewrite the differential equation.)

A. $-\frac{1}{3}e^{-3y} = \frac{1}{2}e^{2x} - \frac{1}{3} - \frac{1}{2}e^2$

B. $\frac{1}{3}e^{-3y} = \frac{1}{2}e^{2x} - \frac{1}{2}e^2 + \frac{1}{3}$

C. $-\frac{1}{3}e^{-3y} = \frac{1}{2}e^{2x} - \frac{1}{2} - \frac{1}{3}e^{-3}$

D. $\frac{1}{3}e^{-3y} = \frac{1}{2}e^{2x} - \frac{1}{2} + \frac{1}{3}e^{-3}$

E. $-\frac{1}{3}e^{-3y} = \frac{1}{2}e^{2x} - \frac{5}{6}$

Answer C. You can write e^{2x+3y} as $e^{2x}e^{3y}$. Then separating variables gives you

$$\frac{dy}{dx} = e^{2x}e^{3y}$$
$$\Rightarrow e^{-3y}\,dy = e^{2x}\,dx.$$

Integrating both sides, you get

$$\int e^{-3y}\,dy = \int e^{2x}\,dx$$
$$\Rightarrow -\frac{1}{3}e^{-3y} = \frac{1}{2}e^{2x} + C.$$

Then, use the initial condition $y(0) = 1$. You have

$$-\frac{1}{3}e^{-3y} = \frac{1}{2}e^{2x} + C$$
$$\Rightarrow -\frac{1}{3}e^{-3(1)} = \frac{1}{2}e^{2(0)} + C$$
$$\Rightarrow -\frac{1}{3}e^{-3} = \frac{1}{2} + C$$
$$\Rightarrow C = -\frac{1}{2} - \frac{1}{3}e^{-3}.$$

Therefore, the solution to the differential equation with the given initial conditions is

$$-\frac{1}{3}e^{-3y} = \frac{1}{2}e^{2x} - \frac{1}{2} - \frac{1}{3}e^{-3}.$$

---

18. In the generalized solution $y = Ae^{kx}$, the k represents:

I. Horizontal stretching/compressing of the solution

II. A proportionality constant between $\dfrac{dy}{dx}$ and y

III. The slope of the tangent line to $y = Ae^{kx}$ when $y = 1$

A. II only

B. I and II only

C. III only

D. I, II, and III

E.   None of these

Answer   D.

I is true, since large values of k shrink the functions horizontally, and small values of k stretch the function horizontally. Also note that negative values of k reflect the function about the y-axis (they make a graph that's symmetric across the y-axis).

II is true because $\dfrac{dy}{dx} = ky$.

III is true because $\dfrac{dy}{dx} = ky$.

19.   The rate of change of y with respect to x is three times the value of y. Write an equation for y given that $y = 9$ when $x = 0$.

A. $y = 3x + 9$

B. $y = 3x$

C. $y = 9e^{3x}$

D. $y = 9(3)^x$

E.   More information is needed.

Answer   C. The differential equation described in the question is $\dfrac{dy}{dx} = 3y$, which gives $y = Ae^{3x}$.
Since $y = 9$ when $x = 0$, the value of A is 9, so $y = 9e^{3x}$.

20.   The differential equation $\dfrac{dy}{dt} = \frac{1}{2}y$ for $y > 0$:

I. Has a whole family of solutions

II. Has one single solution

III. Represents an example of exponential decay

A.   I only

B.   I and III only

C.   II only

D.   II and III only

E.   III only

Answer   A. The solution $y = Ae^{\frac{1}{2}t}$ represents a whole family of functions and not just one solution (because the constant A can take on any value).

III is false because $\dfrac{dy}{dt} = \frac{1}{2}y$ for $y > 0$ represents growth, not decay.

21. Which of these graphs represents a solution to the differential equation $\dfrac{dy}{dx} = -2y$?

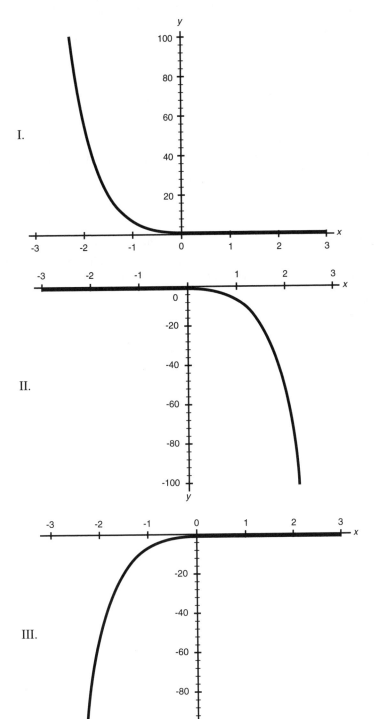

I.

II.

III.

A. I only

B. I and III only

C. II only

D. II and III only

E. III only

Answer   B.

I is straight-forward decay.

II is false because if y is negative, $\dfrac{dy}{dx} = -2y$ should give a positive slope.

III represents a solution where $y < 0$.

---

22. A radioactive element decays according to the differential equation $\dfrac{dN}{dt} = -.005N$, where N is the number of atoms and t is years. What is the half-life of this element? (The half-life is the time required for the material to be reduced by half.)

A. 73.303 years

B. 138.629 years

C. 219.722 years

D. -138.629 years

E. Not enough information is given.

Answer   B.

$\dfrac{dN}{dt} = -0.005N$ gives $N = N_0 e^{-0.005t}$, where N_0 represents the initial quantity of radioactive elements (which is unknown). Let $N = \dfrac{N_0}{2}$ when $t = t_{\text{half-life}}$. Solving for t_h:

$$\frac{N_0}{2} = N_0 e^{-0.005t_h}$$

$$\frac{1}{2} = e^{-0.005t_h}$$

$$\ln\left(\frac{1}{2}\right) = -0.005t_h$$

$$t_h = \frac{\ln 2}{0.005} \approx 138.629 \text{ years.}$$

---

23. Which of these graphs could **not** represent a possible solution to the differential equation $\dfrac{dy}{dt} = k(y - a)$?

A.

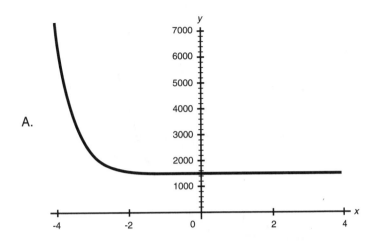

B.

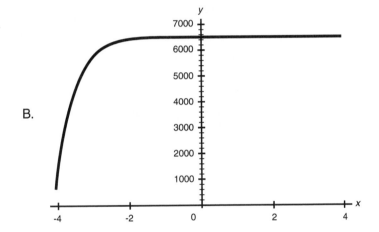

C.

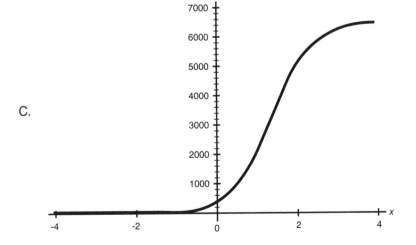

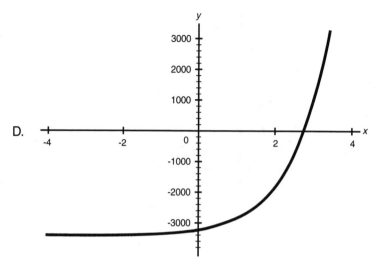

E.   All of these might be possible solutions.

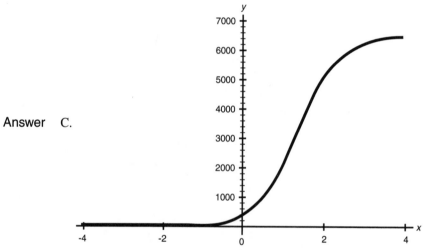

Answer   C.

Remember that $\dfrac{dy}{dt} = k(y-a)$ has a solution of $y = Ae^{kt} + a$, which is just the exponential solution, shifted up or down. The best clue on this one is the concavity.

$y = Ae^{kt} + a$ will not have any points of inflection for any values A, k, and a.

---

24.   The velocity of a ball $v(t)$ rolling down a hill changes according to the differential equation

$$\frac{dv}{dt} = -5 - v,$$

where the velocity is measured in $\dfrac{m}{s}$ and time is measured in seconds. If the ball starts from rest, then write an equation for $v(t)$.

(Note that this differential equation takes into account the friction acting on the ball, and we've defined velocity down the hill to be positive.)

A. $v(t) = -5e^{-t} - 5$

B. $v(t) = 5e^{-t} + 5$

C. $v(t) = -5e^{-t+5}$

D. $v(t) = 5e^{-t} - 5$

E. Not enough information is given.

Answer   D.

$$\frac{dv}{dt} = -5 - v$$

$$\frac{dv}{v+5} = (-1)dt$$

$$\int \frac{dv}{v+5} = \int (-1)dt$$

$$\ln|v+5| = -t + C$$

$$|v+5| = e^{(-t+C)} = Ae^{-t}$$

allow A to be negative:

$$v + 5 = Ae^{-t}$$

$$v = Ae^{-t} - 5,$$

Since $v = 0$ when $t = 0$

$$0 = A - 5$$

$$A = 5$$

$$v(t) = 5e^{-t} - 5.$$

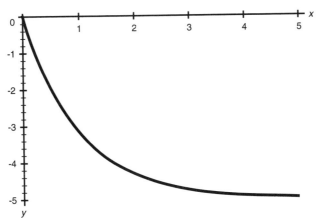

Notice that the velocity is always negative, but it approaches a limit toward a terminal velocity representing the speed where gravity and friction forces cancel each other.

---

25.   The rate of change of y with respect to x is one-half times the value of y.

Find an equation for y, given that $y = -7$ when $x = 0$.

A. $\dfrac{dy}{dx} = \frac{1}{2}y$

B. $y = -7(\frac{1}{2})e^{x}$

C. $y = e^{0.5x} - 7$

D. $y = -7e^{0.5x}$

E. $y = -7(\frac{1}{2})^{x}$

Answer   D. The differential equation that's described by the problem is $\dfrac{dy}{dx} = \dfrac{1}{2}y$.

Separating the variables gives you $\dfrac{1}{y}\,dy = \frac{1}{2}\,dx$. Integrating then gives you $\ln|y| = \frac{1}{2}x + C$.
The next step is exponentiating:

$$|y| = e^{0.5x+C} = e^{C}e^{0.5x}.$$

Finally, replacing e^{C} with A and dropping the absolute values gives you $y = Ae^{0.5x}$.
Now the initial condition is that $y = -7$ when $x = 0$. So $-7 = Ae^{0.5(0)} = A$, and the final solution is $-7e^{0.5x}$.

---

26.   The rate of change of y with respect to t is 3 times the value of the quantity 2 less than y. Find an equation for y, given that $y = 212$ when $t = 0$.

A. $y = 212e^{3t} + 2$

B. $y = 210e^{3t} - 2$

C. $y = 210e^{3t} + 2$

D. $y = 212e^{3t} - 2$

E. $y = 214e^{3t} - 2$

Answer   C. Since the rate of change of y with respect to t is 3 times the value of 2 less than y, the differential equation that represents this is $\dfrac{dy}{dt} = 3(y - 2)$.

If you remember the pattern, you know right away that $y = Ae^{3t} + 2$. If not,

$$\frac{dy}{dt} = 3(y-2) \Rightarrow \frac{1}{(y-2)}\,dy = 3\,dt \Rightarrow \ln|y-2| = 3t + C$$
$$\Rightarrow |y-2| = e^{C}e^{3t} \Rightarrow y - 2 = Ae^{3t} \Rightarrow y = Ae^{3t} + 2.$$

When $t = 0$, $y = 212$, so you have

$$212 = A + 2 \Rightarrow A = 210.$$

Therefore, the equation for y is $y = 210e^{3t} + 2$.

27. Which of the following curves exhibits modified exponential decay, has a y value of 110 when t is 0, and approaches the value $y = 30$ as t goes to infinity?

A. $y = 110e^{7t} - 80$

B. $y = 110 - 80e^{-5t}$

C. $y = 30 + 80e^{-2t}$

D. $y = 110 + 80e^{2t}$

E. $y = 30 - 80e^{-4t}$

Answer   C. This curve represents modified exponential decay, since it's just an exponential decay curve ($80e^{-2t}$) shifted up 30 units.

At $t = 0$, $y = 30 + 80 = 110$, so that condition is satisfied.

Finally, as $t \to \infty$ the term $80e^{-2t} \to 0$, and therefore $30 + 80e^{-2t} \to 30$ as $t \to \infty$.

So, $y = 30 + 80e^{-2t}$ is the curve you're looking for.

---

28. Which of the following statements best describes the solution curve to the differential equation $\dfrac{dy}{dx} = -1.233(y - 80)$, with initial condition $y(0) = -40$?

A. The solution curve $y = y(x)$ is modified exponential decay, and $y \to -40$ as $x \to \infty$.

B. The solution curve $y = y(x)$ is modified exponential decay, and $y \to 80$ as $x \to \infty$.

C. The solution curve $y = y(x)$ is modified exponential growth, and $y \to 80$ as $x \to \infty$

D. The solution curve $y = y(x)$ is modified exponential growth, and $y \to -\infty$ as $x \to \infty$.

E. The solution curve $y = y(x)$ is modified exponential growth, and $y \to \infty$ as $x \to \infty$.

Answer   B.

The solution is read off as $y = Ae^{-1.233x} + 80$.

Recall that the solution to the differential equation $\dfrac{dy}{dx} = k(y - a)$ is $y = Ae^{kx} + a$.

Next, the initial condition $y(0) = -40$ requires that

$$-40 = Ae^{-1.233(0)} + 80 = A + 80.$$

Thus, $A = -120$.

So, the solution to the differential equation with the given initial condition is

$$y = -120e^{-1.233x} + 80,$$

which is modified exponential decay (because of the -1.233 and the vertical shift).

Furthermore, $y \to 80$ as $x \to \infty$, since the term $-120e^{-1.233x} \to 0$ as $x \to \infty$.

29. Suppose that interest on money in the bank accumulates at an annual rate of 6% per year compounded continuously. If you deposit $3000 in the account today, how much will it be worth 20 years from now?

    (Hint: First find the equation that represents the balance $B = B(t)$ in the account at time t years.)

    A. $9,960.35

    B. $10,030.45

    C. $903.58

    D. $10,066.36

    E. $6,775.68

Answer   A. You may realize that the equation that gives you the account balance $B = B(t)$ where t is in years, is $B(t) = 3,000e^{0.06t}$, since the growth rate of the balance is 6% per year, which corresponds to the 0.06, and the initial investment at time $t = 0$ is $3,000.

You can always solve the problem by modeling the balance $B = B(t)$ with a differential equation and initial condition. You have that the balance is increasing at a continuous rate of 6% per year, which mathematically is expressed by the equation $\dfrac{dB}{dt} = 0.06B$.

Now you should realize that the solution to this differential equation is $B(t) = Ae^{0.06t}$.

Then, using the initial condition that $3000 is deposited at time $t = 0$, you get

$$B(0) = A = 3,000.$$

So the balance in the account is given by the function $B(t) = 3,000e^{0.06t}$.

Finally, at $t = 20$ years, the balance in the account is $B(20) = 3,000e^{0.06(20)} = \$9,960.35$.

30. Suppose that interest on money in the bank accumulates at an annual rate of 5% per year compounded continuously. How much money should be invested today, so that 20 years from now it will be worth $20,000?

    (Hint: If you're stuck, then model the account balance $B = B(t)$ with a differential equation and an initial condition, keeping in mind that the initial condition here is **not** at $t = 0$.)

    A. $5,498.23

    B. $5,909.04

    C. $6,766.49

    D. $7,357.59

    E. $7,982.22

Answer   D. Since the interest rate is 5% per year (compounded continuously), we have $B(t) = Ae^{0.05t}$, where A is yet undetermined.

To see this, you can set up the differential equation $\dfrac{dB}{dt} = 0.05B$, since the rate of change of the balance is equal to .05 of the current balance.

To solve for A, use the initial condition that $B(20) = 20,000$, since you want to enforce the fact that after 20 years, the balance in the account is $20,000.

Solving for A,

$$20,000 = B(20) = Ae^{0.05(20)} = Ae^1 \Rightarrow A = \frac{20,000}{e}.$$

Then, since you're investing at time $t = 0$ and $B(0) = Ae^{0.05(0)} = A$, the amount you should invest is $A = \dfrac{20,000}{e} = \$7,357.59$.

---

31.   In an experiment to study the growth of bacteria, a medical student measured $5,000$ bacteria at time 0 and $8,000$ at time 10 minutes. Assuming that the number of bacteria grows exponentially, how many bacteria will be present after 30 minutes?

A. $14,000$ bacteria

B. $20,480$ bacteria

C. $17,830$ bacteria

D. $24,332$ bacteria

E. $29,333$ bacteria

Answer   B. If $P = P(t)$ represents the population of the bacteria at time t minutes, then the assumption that the population grows exponentially means that $P(t) = Ae^{kt}$, where $k > 0$ is the proportional rate of increase (per minute) of the population, and $A > 0$ is the population at time $t = 0$.

Since $P(0) = 5,000$ (the population at time 0 is given as $5,000$), you know that

$$P(t) = 5,000e^{kt}.$$

Next, enforcing the condition that there are 8000 bacteria at 10 minutes gives you

$$8000 = P(10) = 5,000e^{10k}.$$

Then solving for k,

$$8,000 = 5,000e^{10k} \Rightarrow \frac{8,000}{5,000} = e^{10k} \Rightarrow \ln\frac{8}{5} = 10k \Rightarrow k = \frac{1}{10}\ln\frac{8}{5}.$$

Finally, the question is how many bacteria will be present at $t = 30$, and the answer is

$$P(30) = 5,000e^{30k} = 5,000e^{30\left(\frac{1}{10}\ln\frac{8}{5}\right)} = 20,480 \text{ bacteria}.$$

32. A certain radioactive material is known to decay at a rate proportional to the amount present. A block of this material originally having a mass of 100 grams is observed after 20 years to have a mass of only 80 grams. Find the half-life of this radioactive material.

(Recall that half-life is the time required for the material to be reduced by half.)

A. 54.343 years

B. 56.442 years

C. 59.030 years

D. 61.045 years

E. 62.126 years

Answer E. Let $M(t)$ represent the mass of the material at time t years. Since the radioactive material decays at a rate proportional to the amount present, the function $M(t)$ is exponential decay, and you have that $M(t) = Ae^{kt}$, where $k < 0$ and $A > 0$ is the mass of the material at time 0. Thus, you have

$$M(t) = 100e^{kt}.$$

To enforce the condition that after 20 years the mass of the material is 80 grams, set $M(20) = 80$, and you get $80 = 100e^{20k}$, then solve for k,

$$\Rightarrow \frac{8}{10} = e^{20k}$$

$$\Rightarrow \ln \frac{8}{10} = 20k$$

$$\Rightarrow k = \frac{1}{20} \ln \frac{8}{10}$$

Next, since the mass at time 0 is 100 grams, you can find the half-life of the material by finding the time t at which the mass of the material is 50 grams. Solving $M(t) = 50$ for t:

$$50 = 100e^{kt}$$

$$\Rightarrow \frac{1}{2} = e^{kt}$$

$$\Rightarrow \ln \frac{1}{2} = kt$$

$$\Rightarrow t = \frac{1}{k} \ln \frac{1}{2}$$

(In fact, for exponential decay, the half life is always given by $\frac{1}{k} \ln \frac{1}{2}$.)

$$\Rightarrow t \approx 62.126 \text{ years}.$$

You can always check your answer by seeing if $M(62.126) \approx 50$. It is.

33. Suppose you're going to sell a certain product, and your marketing team has determined that the maximum number of units of the product that can be sold (based on number of households, desirability, and so on) is given by a constant $M > 0$, and that the rate of increase in unit sales will be proportional to the difference between M and the number of units that have currently been sold. Then the cumulative number of units $U = U(t)$ sold for any time t will be given by a function of the form:

A. $U(t) = Ae^{kt}$, where A is some undetermined constant, and $k > 0$ is some proportionality constant.

B. $U(t) = Ae^{kt} - M$, where A is some undetermined constant, and $k > 0$ is some proportionality constant.

C. $U(t) = Ae^{-kt} - M$, where A is some undetermined constant, and $k > 0$ is some proportionality constant.

D. $U(t) = Ae^{-kt} + M$, where A is some undetermined constant, and $k > 0$ is some proportionality constant.

E. $U(t) = Ae^{kt} + M$, where A is some undetermined constant, and $k > 0$ is some proportionality constant.

Answer D.

Rewrite the equation as $\dfrac{dU}{dt} = -k(U - M)$, which has solutions of the form

$$U(t) = Ae^{-kt} + M,$$

where A is some undetermined constant, and $k > 0$ is some proportionality constant.

Notice that if you assume that 0 units are sold at time 0, that is, $U(0) = 0$, then $A = -M$, and you have the modified exponential growth curve

$$U(t) = -Me^{-kt} + M.$$

34. A dish of ice cream is left on the table in a room with a constant temperature of $70°F$.

At $t = 0$, the temperature of the ice cream is $20°F$. Five minutes later, its temperature is $25°F$. When will it melt? (The ice cream will melt at $35°F$.)

A. 15 minutes

B. 16.93 minutes

C. 11.477 minutes

D. 47.456 minutes

E. There's not enough information to tell.

Answer   B. The difference D in the temperatures decays exponentially, so $D(t) = Ce^{-kt}$.

The initial difference is $D(0) = 50 = C$ and $D(5) = 45 = 50e^{-5k}$, so $k = \frac{-\ln(.9)}{5}$.

Therefore, $D(t) = 50e^{\frac{t\ln(.9)}{5}}$, and using this equation to solve $D(t) = 35$ for t gives

$$t \approx 16.93 \text{ minutes.}$$

---

35. The pattern we use with air resistance works on objects that are traveling on level ground, too. For example, if you stop pedaling your bicycle and just coast on a straight, level road, eventually you'll coast to a stop. The resistance here is proportional to your velocity, just like it is for a falling body. The general differential equation for coasting to a stop is

$$\frac{dv}{dt} = -kv,$$

where k is a constant. What's the terminal velocity in this case?

A. 0

B. v

C. k

D. $-k$

E. $-\infty$

Answer   A. This is coasting to a stop, after all.

---

36. A tank originally contains 100 gallons of fresh, water. Water with $\frac{1}{2}$ pound per gallon of salt is poured into the tank at a rate of 2 gallons per minute, and the well-stirred mixture is allowed to leave the tank at the same rate (so the volume remains constant). After ten minutes, the process is stopped, and fresh water is poured into the tank at a rate of 2 gallons per minute (with the mixture again leaving at the same rate). Let $y(t)$ be the amount of salt in the tank at time t.

Which of the graphs below is the correct graph of $y(t)$ for $0 \le t \le 20$?

A.

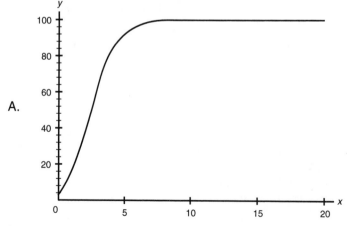

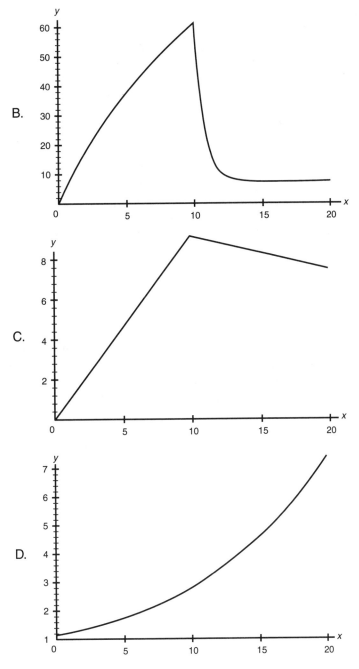

B.

C.

D.

E.   None of these

Answer   C.

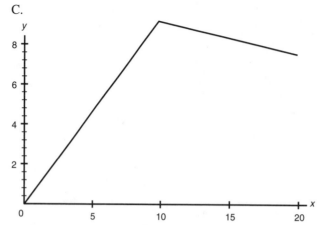

It's surprising how straight these curve segments look, isn't it? But this is the graph of the function in this example. The salt in the tank gradually increases to about 9 pounds, and then gradually decreases to about 7.5 pounds. The curves look linear because the interval we're looking at is small from the function's point of view. (It's that locally linear thing again!)

---

37.   A room containing $1,200$ cubic feet of air is originally free of carbon monoxide. Beginning at time $t = 0$, cigarette smoke containing 4% carbon monoxide is introduced into the room at a rate of 0.1 cubic feet per minute, and the well-circulated mixture is allowed to leave at the same rate. Which of these differential equations describes this situation?

A. $\dfrac{dy}{dt} = 0.1 - y$

B. $\dfrac{dy}{dt} = .004 - 0.1y$

C. $\dfrac{dy}{dt} = .004 + 0.1y$

D. $\dfrac{dy}{dt} = .004 - \dfrac{0.1y}{1200}$

E. $\dfrac{dy}{dt} = .004 - 120y$

Answer   D. The carbon monoxide is coming in at 0.004 cubic feet per minute, and it's leaving at 0.1 cubic feet per minute times the concentration of carbon monoxide in the room, which is $\dfrac{y}{1200}$. So, the correct differential equation is $\dfrac{dy}{dt} = .004 - \dfrac{0.1y}{1200}$.

38. A rabbit population is given by the formula $P(t) = \dfrac{500}{1 + e^{-2t}}$.

Which of these is the carrying capacity for this habitat?

A. 500 rabbits

B. Unlimited

C. 250 rabbits

D. 250 times as many as we started with

E. There's not enough information to tell

Answer A. The carrying capacity is the maximum number of rabbits that can be sustained here. You can find it by looking at $\lim\limits_{t\to\infty} P(t) = 500$.

---

39. At time $t = 0$, a country's population is 300 million, and its death rate exceeds its birth rate by k. There is also a constant immigration of I million people each year. Which of these differential equations describes this situation?

A. $\dfrac{dy}{dt} = -k + I$

B. $\dfrac{dy}{dt} = I - ky$.

C. $\dfrac{dy}{dt} = 300e^{-kt} + I$

D. $\dfrac{dy}{dt} = ky + I$

E. $\dfrac{dy}{dt} = ky - I + 300$

Answer B.

The rate at which the country's population is changing equals the rate at which people are entering the population minus the rate at which people are leaving the population.

The rate at which people are entering is I million people per year, and the rate at which they are leaving is ky (since the death rate exceeds the birth rate, so there is a net decrease).

The differential equation that describes this is $\dfrac{dy}{dt} = I - ky$.

40. The situation described in the last problem is mathematically similar to a(n):

A. compound interest problem.

B. logistic growth problem.

C. exponential decay problem.

D. Newton's Law of Cooling.

E. mixing problem.

Answer E. This situation is really the same as saying that I million people per year are coming into the tank, and the well-mixed population is allowed to leave at k times y people per year.

SECTION III:

Practice Tests

Practice Calculus AB Tests

In this final section are two complete practice examinations, each including 45 multiple-choice and 6 free response questions in a format similar to that found on the AB examination. Use the answer sheet provided to record your multiple-choice answers. At the end of each examination you will find an answer key. Solutions to the examinations include a discussion of each multiple-choice question and answers to the free-response questions. A scoring guide for free-response questions is included in the answers to Practice Test One.

Practice Calculus AB Test One

You'll get the most benefit from this practice examination if you take it under conditions exactly like those of the actual AP Exam.

Estimated Time to Complete: 200 minutes

Section I Part A, 55 minutes, 28 multiple-choice questions: NO calculator allowed.

Section I Part B, 50 minutes, 17 multiple-choice questions: Calculator allowed.

Short break, 5 minutes.

Section II Part A, 45 minutes, 3 free-response questions: Calculator allowed.

Section II Part B, 45 minutes, 3 free-response questions: NO calculator allowed.

Work on only one section at a time. You may not go back to a previous section. You may however, continue to work on part A of Section II after you have turned in your calculator.

Practice Test Section 1—Multiple-Choice

Time limit: 105 minutes

Part A consists of **28** questions. A calculator may **not** be used on this part of the examination. Part B consists of **17** questions. In the multiple-choice section of the examination, as a correction for guessing, one-fourth of the number of questions answered incorrectly will be subtracted from the number answered correctly.

Directions: Solve each of the following problems, using the available space for scratchwork. After examining the form of the choices, decide which is the best of the choices given and fill in the corresponding oval on the answer sheet. No credit will be given for any math work written on the test paper. Do not spend too much time on any one problem.

In this test: Unless otherwise specified, the domain of a function f is assumed to be the set of all real numbers x for which $f(x)$ is a real number.

Formulas

Here is a short list of area and volume formulas that you might need.

For a circle of radius r: Area $= \pi r^2$, Circumference $= 2\pi r$.

For a sphere of radius r: Volume $= \frac{4}{3}\pi r^3$, Surface Area $= 4\pi r^2$.

For a right circular cylinder with radius r and height h: Volume $= \pi r^2 h$, Surface Area $= 2\pi r h + 2\pi r^2$.

For a right circular cone with radius r and height h: Volume $= \frac{1}{3}\pi r^2 h$.

Answer Sheet—Multiple Choice

PART A

1 (A) (B) (C) (D) (E)	8 (A) (B) (C) (D) (E)	15 (A) (B) (C) (D) (E)	22 (A) (B) (C) (D) (E)
2 (A) (B) (C) (D) (E)	9 (A) (B) (C) (D) (E)	16 (A) (B) (C) (D) (E)	23 (A) (B) (C) (D) (E)
3 (A) (B) (C) (D) (E)	10 (A) (B) (C) (D) (E)	17 (A) (B) (C) (D) (E)	24 (A) (B) (C) (D) (E)
4 (A) (B) (C) (D) (E)	11 (A) (B) (C) (D) (E)	18 (A) (B) (C) (D) (E)	25 (A) (B) (C) (D) (E)
5 (A) (B) (C) (D) (E)	12 (A) (B) (C) (D) (E)	19 (A) (B) (C) (D) (E)	26 (A) (B) (C) (D) (E)
6 (A) (B) (C) (D) (E)	13 (A) (B) (C) (D) (E)	20 (A) (B) (C) (D) (E)	27 (A) (B) (C) (D) (E)
7 (A) (B) (C) (D) (E)	14 (A) (B) (C) (D) (E)	21 (A) (B) (C) (D) (E)	28 (A) (B) (C) (D) (E)

PART B

29 (A) (B) (C) (D) (E)	34 (A) (B) (C) (D) (E)	39 (A) (B) (C) (D) (E)	44 (A) (B) (C) (D) (E)
30 (A) (B) (C) (D) (E)	35 (A) (B) (C) (D) (E)	40 (A) (B) (C) (D) (E)	45 (A) (B) (C) (D) (E)
31 (A) (B) (C) (D) (E)	36 (A) (B) (C) (D) (E)	41 (A) (B) (C) (D) (E)	
32 (A) (B) (C) (D) (E)	37 (A) (B) (C) (D) (E)	42 (A) (B) (C) (D) (E)	
33 (A) (B) (C) (D) (E)	38 (A) (B) (C) (D) (E)	43 (A) (B) (C) (D) (E)	

Multiple-Choice, Section I, Part A (noncalculator): 55 minutes

You may NOT use a calculator on this section.

---

1. $\displaystyle\lim_{x\to\infty} \frac{6x^{7/3} - 2x^2 + 3x}{1 - x + 7x^2} =$

   A. 0

   B. ∞

   C. -6

   D. $\frac{6}{7}$

   E. $-\frac{2}{7}$

---

2. $\dfrac{d}{dx}(x + (x+1)^2)^3 =$

   A. $3(x + (x+1)^2)^2$

   B. $x^2 + 3(x+1)^2$

   C. $3x(x + (x+1)^2)^2$

   D. $(2x+1)(x^2 + 3x + 1)^2$

   E. $(6x+9)(x^2 + 3x + 1)^2$

---

3. $\displaystyle\int_0^1 (t^2 + 1)^2 \, dt =$

   A. $\frac{28}{15}$

   B. $\frac{4}{3}$

   C. $\frac{7}{3}$

   D. $\frac{23}{15}$

   E. $\frac{6}{5}$

4. If $y = 2x^{\frac{5}{2}} \tan x$, then $\dfrac{dy}{dx} =$

A. $x^{\frac{3}{2}}(5 \tan x + 2x \sec^2 x)$

B. $5x^{\frac{3}{2}} \tan x - x \csc^2 x$

C. $5x^{\frac{5}{2}} \tan x + 2x^{3/2} \sec^2 x$

D. $x^{\frac{3}{2}}(5 \tan x + x \sec^2 x)$

E. $x^2 \tan x - 5x \csc^2 x$

5. The acceleration of an object is given by $a(t) = t + \cos t$. If the velocity at $t = 0$ is 1 and the position of the object at $t = 0$ is 2, find the position of the object as a function of t.

A. $\dfrac{t^3}{6} - \cos t + t + 3$

B. $\dfrac{t^3}{3} + \sin t + 2$

C. $\dfrac{t^3}{6} + \cos t - t$

D. $t^3 - \cos t + 2$

E. $\dfrac{t^3}{6} - \sin t + t + 2$

6. The line tangent to the curve $x^4 + 3xy^2 + 5y^3 = 9$ at the point $(1, 1)$ is given by:

A. $y - 1 = 3(x - 1)$

B. $y - 1 = -\frac{1}{3}(x - 1)$

C. $y = -\frac{1}{3}(x - 1)$

D. $y = -3(x - 1) + 1$

E. $y = \frac{1}{3}(x - 1) + 1$

7. $\int \dfrac{x}{(x^2+1)^{\frac{1}{3}}}\,dx =$

A. $x(x^2+1)^{\frac{2}{3}} + C$

B. $\frac{3}{2}(x^2+1)^{\frac{2}{3}} + C$

C. $\dfrac{(x^2+1)^{\frac{2}{3}}}{x} + C$

D. $(x^2+1)^{\frac{3}{2}} + C$

E. $\frac{3}{4}(x^2+1)^{\frac{2}{3}} + C$

8. Let $g(x) = \dfrac{(nx+3)^2}{(nx)^2}$, where $n > 0$. Which of the following statements is **not** true?

A. $y = 1$ is a horizontal asymptote of g.

B. g is decreasing for $x > 0$.

C. g has a critical point at $x = -\frac{3}{n}$.

D. $x = 0$ is a vertical asymptote of g.

E. g has no relative maxima or minima for $x < 0$.

9. $f(x)$ is shown in the following curve:

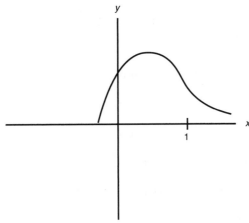

Which of the following is (are) true?

I. $f''(x) > 0$ for all $x > 1$.

II. $F(x) = \displaystyle\int_1^x f(y)\, dy > 0$ for all $x > 1$.

III. $f'(x) > 0$ for all $x > 1$.

A. I only

B. III only

C. I and II

D. II and III

E. I, II, and III

10. The height of a cylinder is increasing at a rate of 6 inches per minute. The radius of the cylinder is decreasing at a rate of 2 inches per minute. The volume is:

A. always increasing.

B. always decreasing.

C. increasing when $r < \frac{2}{3}h$.

D. decreasing when $r < \frac{2}{3}h$.

E. constant.

11. Given that $f(x)$ is continuous for all real numbers,

x	0	1	2	3	4
$f(x)$	1	-1	3	5	2

which of the following statements is necessarily true?

A. There is some c in the interval $(0, 2)$ such that $f(c) = 4$.

B. There is some c in the interval $(1, 2)$ such that $f(c) = -5$.

C. There is some c in the interval $(2, 4)$ such that $f(c) = 6$.

D. There is some c in the interval $(0, 3)$ such that $f(c) = 2$.

E. There is some c in the interval $(0, 4)$ such that $f(c) = -3$.

12. If $f(x) = \dfrac{\sqrt{x+1}}{x^3 + 2}$, then $f'(0) =$

A. $-\frac{1}{4}$

B. $\frac{1}{2}$

C. $\frac{3}{8}$

D. $\frac{1}{4}$

E. Undefined

13. $\displaystyle\lim_{x \to 4} \frac{2x^2 - 32}{x^2 + x - 20} =$

A. 0

B. $\frac{16}{9}$

C. $\frac{1}{3}$

D. $\frac{8}{9}$

E. Does not exist

14. If $g(x) = \displaystyle\int_{x}^{x^2} \cos^3 t \, dt$, then $g'(x) =$

A. $\dfrac{\sin^4 x^2}{4} - \dfrac{\sin^4 x}{4}$

B. $2x \cos^3 x$

C. $\cos^3 x^2 - \cos^3 x$

D. $2x \sin^4 x - \sin^4 x^2$

E. $2x \cos^3 x^2 - \cos^3 x$

---

15. $\displaystyle\lim_{x \to \frac{\pi}{2}} \dfrac{\cot\left(\frac{\pi}{2}\right) - \cot x}{\frac{\pi}{2} - x} =$

A. 0

B. -1

C. 1

D. $\frac{\sqrt{2}}{2}$

E. ∞

16. $\int \dfrac{t}{(5t^2 + 4)^4}\, dt =$

A. $\dfrac{-1}{30(5t^2 + 4)^3} + C$

B. $\dfrac{-1}{30} \ln(5t^2 + 4) + C$

C. $\dfrac{1}{10}(5t^2 + 4)^{-3} + C$

D. $\dfrac{-1}{3(5t^2 + 4)^3} + C$

E. $\dfrac{-t^2}{3(5t^2 + 4)^3} + C$

17. If $k(x) = 2(6^{3x})$, then $k'(x) =$

A. 6^{3x+1}

B. $2e^{3x}$

C. $\dfrac{6^{3x}}{2}$

D. $6^{3x+1}(\ln 6)$

E. $\dfrac{6^{3x}}{\ln 6}$

Questions 18 and 19 refer to the following graph of the velocity of an object as a function of time:

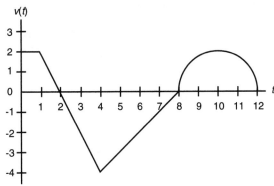

18. What is the total distance traveled by the object between time $t = 0$ and $t = 12$?

A. $-9 + 2\pi$

B. $15 + 4\pi$

C. $15 + 2\pi$

D. $12 + 2\pi$

E. 15

19. The time at which the object has its maximum speed is:

A. 1

B. 2

C. 4

D. 6

E. 10

20. $f(x)$ is a function that satisfies the following conditions:

	$x < 0$	$0 < x < 1$	$1 < x$
$f'(x)$	$+$	$-$	$-$
$f''(x)$	$+$	$+$	$-$

Which of the following could be the graph of $f(x)$?

A.

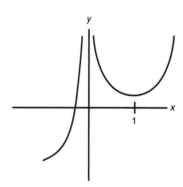

B.

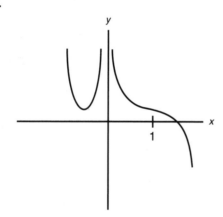

C.

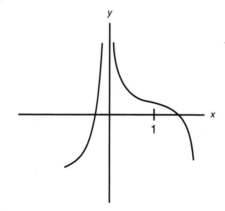

D.

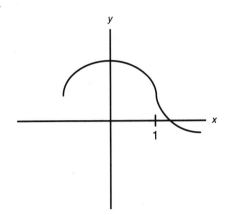

E.

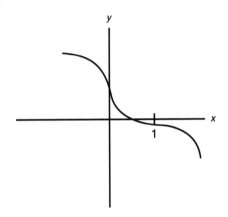

21. If $f(x) = \displaystyle\int_1^{x^2} \dfrac{t^3 + 1}{t^2 + 1}\, dt$, then $f'(x) =$

A. $2x(x^6 + 1)\arctan x$

B. $\dfrac{x^6 + 1}{x^4 + 1}$

C. $\dfrac{x^3 + 1}{x^2 + 1}$

D. $2x\left(\dfrac{x^6 + 1}{x^4 + 1}\right)$

E. $2x\left(\dfrac{x^3 + 1}{x^2 + 1}\right)$

22. An object has a velocity $v(t) = 3t^4 - 3t^2$, where $t \geq 0$. Find an expression for the total distance traveled by the object between time $t = 0$ and time $t = 4$.

A. $\displaystyle\int_0^1 (3t^4 - 3t^2)\,dt + \int_1^4 (3t^2 - 3t^4)\,dt$

B. $\displaystyle\int_0^2 (3t^2 - 3t^4)\,dt + \int_2^4 (3t^4 - 3t^2)\,dt$

C. $\displaystyle\int_0^1 (3t^2 - 3t^4)\,dt + \int_1^4 (3t^4 - 3t^2)\,dt$

D. $\displaystyle\int_0^4 (3t^4 - 3t^2)\,dt$

E. $\displaystyle\int_0^4 \left(\frac{3t^5}{5} - t^3\right)\,dt$

23. R, the base of a solid, is the region in the first quadrant bounded by the x-axis, the y-axis, and the circle $x^2 + y^2 = 4$. Each cross section of the solid, perpendicular to the x-axis, is an equilateral triangle. What is the volume of this solid? Recall that the area of an equilateral triangle with a side of length s is $\frac{\sqrt{3}}{4}s^2$.

A. $\frac{16\pi}{3}$

B. $\frac{4}{3}$

C. $\frac{16}{3}$

D. $\frac{4\sqrt{3}}{3}$

E. $2\sqrt{3}$

24. $\displaystyle\lim_{x \to 3} \frac{|x-3|}{x} =$

A. 1

B. 6

C. 0

D. -1

E. Does not exist

25. The function $f(x) = xe^x$ has inflection points at:

A. 0

B. -1

C. -2

D. 1

E. There are no inflection points of f.

26. If $f(2) = 3$, $f'(2) = 4$, and $g(x)$ is the inverse function to $f(x)$, the equation of the tangent line to $g(x)$ at $x = 3$ is:

A. $y - 2 = -\frac{1}{4}x - 3$

B. $y - 2 = 4(x - 3)$

C. $y - 3 = -\frac{1}{4}x - 2$

D. $y - 2 = \frac{1}{4}(x - 3)$

E. $y - 3 = \frac{1}{4}x - 2$

27. Let $h(x) = \begin{cases} \dfrac{\sin 2x}{x}, & \text{if } x \neq 0. \\ a, & \text{if } x = 0. \end{cases}$ Find a value of a so that h is continuous at $x = 0$.

A. $a = 1$

B. $a = 0$

C. $a = 2$

D. $a = \frac{1}{2}$

E. It is not possible to find a value of a so that $h(x)$ is continuous at $x = 0$.

28. Suppose $\dfrac{dy}{dt} = t(y + 1)$ and $y(0) = 0$. Find an explicit expression for $y(t)$ as a function of t.

A. $y(t) = e^{t^2} - 1$

B. $\ln|y(t) + 1| = \dfrac{t^2}{2}$

C. $y(t) = 1 - e^t$

D. $y(t) = 2e^{\frac{t^2}{2}}$

E. $y(t) = t^2 + e^t - 1$

Multiple-Choice, Section 1, Part B (with calculator): 50 minutes

You may use your calculator on any of these questions.

---

29. The rectangle shown below has vertices of $(0,0)$, $(4,0)$, $(0,2)$, and $(4,2)$. What percent of the rectangle is shaded?

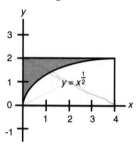

A. 25%

B. 33.3%

C. 50%

D. 66.7%

E. 75%

---

30. If $h(x)$ is an **odd** function and $\int_{3}^{5} h(x)\,dx = -2$, then $\int_{-3}^{5} h(x)\,dx =$

A. 0

B. 8

C. -2

D. -4

E. Not enough information to compute

31. The curve below is the graph of $y = g(x)$ on the domain $[0, 4]$.

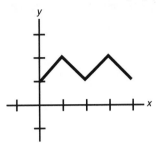

Which of the following statements about g are true?

I. There is some point c in the interval $[0, 4]$ such that $g'(c) = 0$.

II. g is continuous at $x = 2$.

III. $g'(\frac{3}{2}) < 0$.

A. I only

B. II only

C. I and II only

D. II and III only

E. I, II, and III

32. Suppose $f(x) = (x + 1)^{5/2}$. Use a tangent line approximation to approximate $f(3.1)$.

A. 34.038

B. 32

C. 34

D. 30.038

E. 28.434

33. Let $b(x) = \displaystyle\int_0^{2x^2 - x + 1} \sqrt{t^2 + 1}\, dt$. Which of the following statements is true?

A. $b(x)$ has a relative maximum at $x = \frac{1}{4}$.

B. $b(x)$ has no relative maximum or minimum, since $b'(x)$ is never 0.

C. $b(x)$ has a relative minimum at $x = 1$.

D. $b(x)$ has a relative minimum at $x = \frac{1}{4}$.

E. $b(x)$ has a relative maximum at $x = 1$.

34. If $\dfrac{dy}{dt} = \dfrac{3}{2y(t+1)}$ and $y(0) = 1$, find an explicit expression for $y(t)$.

A. $y(t) = \dfrac{3t}{2} \ln|t+1| + 1$

B. $y(t) = \sqrt{3 \ln|t+1| + 1}$

C. $y(t) = e^{3t/2}$

D. $y(t) = \frac{3}{2}t + 1 - \frac{1}{2}$

E. $y(t) = 6t + 1$

35. Find the area of the region bounded by the curves $y = \sqrt{x} + 2$ and $y = x$ and the lines $x = 0$ and $x = 4$.

A. $\frac{8}{3}$

B. $\frac{16}{3}$

C. 2

D. $\frac{11}{3}$

E. 5

36. What is the average value of $\dfrac{x+1}{x^3+2}$ over the interval $0 \le x \le 3$?

   A. 1.340

   B. 2.033

   C. 0.501

   D. 0.746

   E. 0.447

37. Let $f(x)$ be defined by $f(x) = \begin{cases} x^3 & \text{if } x \le 0 \\ x^2 & \text{if } x > 0 \end{cases}$. Then $f''(0) =$

   A. 0

   B. 1

   C. 2

   D. 6

   E. Does not exist

38. Let g and h be continuous on $[a, b]$ and differentiable on (a, b), where $a < b$. Assume that $g(x) \ge h(x)$ for all x in $[a, b]$. Which of the following are necessarily true?

   I. If $p(x) = g(x) - h(x)$, there exists a c between a and b such that $\dfrac{p(b) - p(a)}{b - a} = p'(c)$.

   II. $\displaystyle\int_a^b g(x)\,dx \ge \int_a^b h(x)\,dx$.

   III. $g'(x) \ge h'(x)$ for all x in (a, b).

   A. I only

   B. III only

   C. I and II only

   D. II and III only

   E. I, II, and III

39. If $\displaystyle\int_0^c (x^6 - e^x)\,dx = 0$ and $0 < c < 10$, then $c =$

A. 0

B. 2.557

C. 8.443

D. 1.610

E. 1.678

40.

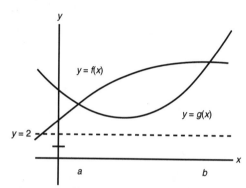

The region bounded by the graphs of $y = f(x)$ and $y = g(x)$ in the figure above is rotated around the line $y = 2$ to form a solid. Which of the following integrals represents the volume of the solid that is generated?

A. $\displaystyle\int_a^b (f(x) - g(x))\,dx$

B. $\displaystyle\pi\int_a^b (f(x) - 2)^2\,dx + \pi\int_a^b (g(x) - 2)^2\,dx$

C. $\displaystyle\pi\int_a^b (f(x))^2\,dx - \pi\int_a^b (g(x))^2\,dx$

D. $\displaystyle\pi\int_a^b (f(x) - 2)^2 - (g(x) - 2)^2)\,dx$

E. $\displaystyle\pi\int_a^b (f(x) - g(x) - 2)^2\,dx$

41. If $2xy^4 - x^3 \cos y + y = 2\pi^4 + 1 + \pi$, then in terms of x and y, $\dfrac{dy}{dx} =$

A. $-\dfrac{3x^2 \cos y + 2y^4}{8x^3 + x^3 \sin y + 1}$

B. $\dfrac{3x^2 \cos y + 2y^4}{8x^3 + x^3 \sin y}$

C. $\dfrac{-3x^2 \cos y + y^4}{8x^3 + x^3 \sin y}$

D. $\dfrac{-3x^2 \cos y + y^4 + 1}{x^3 \sin y + 1}$

E. $\dfrac{3x^2 \cos y - 2y^4}{8x^3 + x^3 \sin y + 1}$

42. Let $f(x) = 2x^5 + 2x^3 + 2x$. If g is the inverse function of f, then $g'(6) =$

A. 18

B. 1

C. $\frac{1}{13178}$

D. $\frac{1}{6}$

E. $\frac{1}{18}$

43. A bug walks along the circle $x^2 + y^2 = 4$. When it is at the point $(1, \sqrt{3})$, $\dfrac{dx}{dt} = 2$.

Find $\dfrac{dy}{dt}$ at this instant.

A. $\frac{2}{\sqrt{3}}$

B. $-\frac{1}{\sqrt{3}}$

C. 0

D. $-\frac{2}{\sqrt{3}}$

E. $\frac{1}{3}$

44. The left-hand Riemann sum approximation for $\int_0^{12} \sqrt{y}\, dy$ using 4 intervals of equal width is:

   A. 21.545

   B. 27.713

   C. $\left(\frac{2}{3}\right)12^{3/2}$

   D. 31.937

   E. 26.741

45. The velocity of an object moving along the x-axis is given by $v(t) = -2t^4 + 3t^2 + t + 1$ where $t > 0$. At what time $(t > 0)$ does the object's maximum acceleration occur?

   A. 0

   B. $\frac{1}{4}$

   C. $\frac{1}{2}$

   D. 1

   E. $\frac{5}{4}$

End of the multiple-choice portion of the practice examination

Section II (90 minutes) Directions for Free-Response Questions

NOTE to students: These are instructions for the actual AP Exam, where you have a test booklet and a separate bubble sheet for answering multiple-choice questions.

You may wish to look the problems over before starting to work on them, since it isn't expected that everyone will be able to complete all parts of all problems. All problems are given equal weight, but the parts of a particular problem aren't necessarily given equal weight. The problems are printed in the booklet and in the green insert. It may be easier for you to first look over all problems in the insert. When you're told to begin, open your booklet, carefully tear out the green insert, and start to work.

A GRAPHING CALCULATOR IS REQUIRED FOR SOME PROBLEMS OR PARTS OF PROBLEMS ON THIS SECTION OF THE EXAMINATION.

* You should write all work for each part of each problem in the space provided for that part in the booklet. Be sure to write clearly and legibly. If you make an error, you may save time by crossing it out rather than trying to erase it. Erased or crossed-out work will not be graded.

* Show all your work. You will be graded on the correctness and completeness of your methods as well as the accuracy of your final answers. Correct answers without supporting work may not receive credit.

* Justifications require that you give mathematical (noncalculator) reasons and that you clearly identify functions, graphs, tables, or other objects you use.

* You are permitted to use your calculator to solve an equation, find the derivative of a function at a point, or calculate the value of a definite integral. However, you must clearly indicate the setup of your problem, namely the equation, function, or integral you are using. If you use other built-in features or programs, you must show the mathematical steps necessary to produce your results.

* Your work must be expressed in standard mathematical notation rather than calculator syntax. For example, $\int_{1}^{5} x^2\,dx$ may not be written as fnInt (X ^ 2, X, 1, 5). Unless otherwise specified, answers (numeric or algebraic) need not be simplified. If your answer is given as a decimal approximation, it should be correct to three places after the decimal point.

* Unless otherwise specified, the domain of a function f is assumed to be the set of all real numbers x for which $f(x)$ is a real number.

Free-Response Questions 1–3 (with calculator): 45 minutes

You may use your calculator for these questions.

---

1. Let $f(x) = 1 + \dfrac{1}{x^4 + 1}$.

A. Sketch the graph of $f(x)$ from $x = 0$ to $x = 4$ and shade in the region M enclosed by the graph of f, the lines $x = 0$ and $x = 4$, and the x-axis.

B. Find the area of the region M.

C. Let $A(z)$ be the area of the region bounded by the graph of f, the x-axis, and the line $x = z$. Write an integral for $A(z)$.

D. Compute $\dfrac{dA}{dz}$ at $z = 2$.

2. An object moves along the x-axis with velocity $v(t) = t\sqrt{2t + 3} - 3t$, where $t \geq 0$.

A. Find the acceleration $a(t)$ of the object.

B. For which t is the object moving to the right?

C. If the particle is at $x = 2$ at time $t = 0$, what is its position at $t = 3$?

D. What is the total distance the particle traveled between time $t = 0$ and time $t = 6$?

3. Readings of the power being used at a given time are taken at hourly intervals at an electric substation. The units of power are in kilowatts (kW) and the hourly readings over a 12-hour period are shown below.

t (hours)	$f(t)$ (kW)
0	900
1	1,200
2	1,300
3	1,200
4	1,000
5	850
6	1,200
7	1,000
8	1,100
9	1,400
10	1,400
11	1,700
12	1,500

A. Find the average rate of change of power usage over the interval $3 \le t \le 9$. Give the correct units.

B. Using a midpoint approximation with 4 equally spaced subintervals, approximate the average power usage from time $t = 4$ to time $t = 12$.

C. Explain the meaning of the integral $\int_{0}^{12} f(t)\, dt$. Give the correct units.

D. Over which hour-long period was the power usage increase the most? Explain.

Free-Response Questions 4-6 (no calculator): 45 minutes

You may not use your calculator for questions 4-6

---

4. A function $y(t)$ satisfies the differential equation $\dfrac{dy}{dt} = y^2 \cos t$ and the initial condition $y(0) = 1$.

A. Find an explicit expression for $y(t)$.

B. For what value(s) of t between $-\pi$ and π does $y(t)$ have vertical asymptotes?

C. Find an expression for $\dfrac{d^2y}{dt^2}$ in terms of t only.

5. The graph of a function f is shown below.

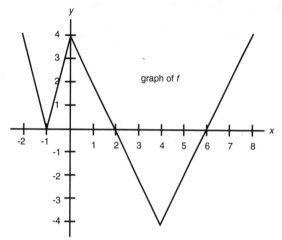

graph of f

The function q is defined by the formula $q(x) = \int_0^x f(t)\, dt$.

A. Compute $q(-1)$ and $q(4)$.

B. Find $q'(0)$.

C. Find the absolute maximum of $q(x)$ on the interval $[2, 8]$.

D. $q''(x)$ is not defined at $x = -1$ and at $x = 0$. At which of these points are there inflection points of q? Explain your answer.

6. Consider the curve defined by the equation $x^2 \cos(y) + y \sin(2\pi x) - x = -1$.

A. Compute $\dfrac{dy}{dx}$.

B. Write the equation of the tangent line to the curve at the point $(1, \frac{\pi}{2})$.

C. Use the tangent line approximation to approximate the y-coordinate of the point on the curve whose x-coordinate is 1.1.

Answer Key for Practice Test One—Multiple-Choice Questions

Part A	Part B
1. B	29. B
2. E	30. C
3. A	31. D
4. A	32. C
5. A	33. D
6. B	34. B
7. E	35. B
8. E	36. E
9. C	37. E
10. D	38. C
11. D	39. D
12. D	40. D
13. B	41. E
14. E	42. E
15. B	43. D
16. A	44. A
17. D	45. C
18. C	
19. C	
20. C	
21. D	
22. C	
23. D	
24. C	
25. C	
26. D	
27. C	
28. A	

Practice Calculus AB Test One—Multiple-Choice Answers and Explanations

1. B.

The only significant terms are the terms with $x^{\frac{7}{3}}$ (in the numerator) and x^2 (in the denominator), the terms with the largest exponents:

$$\lim_{x \to \infty} \frac{6x^{\frac{7}{3}} - 2x^2 + 3x}{1 - x + 7x^2} = \lim_{x \to \infty} \frac{6x^{\frac{7}{3}}}{7x^2} = \lim_{x \to \infty} \frac{6x^{\frac{1}{3}}}{7} = \infty.$$

2. E.

The Chain Rule gives $\dfrac{d}{dx}(x + x + 1^2)^3 = 3(x + (x + 1)^2)^2 \times \dfrac{d}{dx}(x + (x + 1)^2).$

This simplifies to $3(x + (x + 1)^2)^2 \times (2x + 3) = (6x + 9)(x^2 + 3x + 1)^2.$

3. A.

$$\int_0^1 (t^2 + 1)^2 \, dt = \int_0^1 (t^4 + 2t^2 + 1) \, dt = \left[\frac{t^5}{5} + \frac{2t^3}{3} + t \right]_0^1$$

$$= \frac{1}{5} + \frac{2}{3} + 1 = \frac{3}{15} + \frac{10}{15} + \frac{15}{15} = \frac{28}{15}.$$

4. A.

Using the product rule, you find that

$$\frac{dy}{dx} = 2(\frac{5}{2})x^{\frac{3}{2}} \tan x + 2x^{\frac{5}{2}} \sec^2 x = 5x^{\frac{3}{2}} \tan x + 2x^{\frac{5}{2}} \sec^2 x.$$

By factoring out an $x^{\frac{3}{2}}$, we have $\dfrac{dy}{dx} = x^{\frac{3}{2}}(5 \tan x + 2x \sec^2 x).$

5. A.

The velocity $v(t) = \displaystyle\int a(t) \, dt = \int (t + \cos t) \, dt = \frac{t^2}{2} + \sin t + C_1 = 1.$

Since $v(0) = 1$, $C_1 = 1$.

Then the position is $x(t) = \displaystyle\int v(t) \, dt = \int \left(\frac{t^2}{2} + \sin t + 1 \right) dt = \frac{t^3}{6} - \cos t + t + C_2.$

Because $x(0) = 2$, $C_2 = 3$. So, the position function is $x(t) = \dfrac{t^3}{6} - \cos t + t + 3.$

6. B.

Implicit differentiation gives

$$4x^3 + 3y^2 + 6xy\frac{dy}{dx} + 15y^2\frac{dy}{dx} = 0$$

$$4 + 3 + 6\frac{dy}{dx} + 15\frac{dy}{dx} = 0$$

$$21\frac{dy}{dx} = -7$$

$$\frac{dy}{dx} = -\frac{1}{3}.$$

The equation of the tangent line is $y = -\frac{1}{3}(x-1) + 1$.

7. E.

$\displaystyle\int \frac{x}{(x^2+1)^{\frac{1}{3}}} \, dx = \frac{3}{4}(x^2+1)^{\frac{2}{3}} + C$. Notice that you can either guess and check or just take the derivative of each possible answer to find the correct one.

8. E.

$$g(x) = \frac{(nx+3)^2}{(nx)^2} = \left(\frac{nx+3}{nx}\right)^2 = \left(1 + \frac{3}{nx}\right)^2$$

$$g'(x) = 2\left(1 + \frac{3}{nx}\right)\left(\frac{-3n}{n^2x^2}\right) = \frac{-6(1 + \frac{3}{nx})^2}{nx^2}.$$

$g'(x) = 0$ when $x = -\frac{3}{n}$. For $x < -\frac{3}{n}$, $g'(x) < 0$, and for $x > -\frac{3}{n}$, $g'(x) > 0$.
Thus, $g(x)$ has a relative minimum at $x = -\frac{3}{n}$.

9. C.

I is true because the graph is concave up for all $x > 1$.

II is true because $f(x) > 0$ for all $x > 1$, so the area function $F(x)$ is positive as well.

III is false because $f(x)$ is decreasing for all $x > 1$.

10. D.

Since $A = \pi r^2 h$, $\dfrac{dA}{dt} = \pi(2r\dfrac{dr}{dt}h + r^2\dfrac{dh}{dt})$, where $\dfrac{dr}{dt} = -2$ and $\dfrac{dh}{dt} = 6$.

Thus, $\dfrac{dA}{dt} = \pi(2rh \cdot -2 + r^2 6) = \pi(6r^2 - 4rh) = \pi r(6r - 4h)$.

When $r < \frac{2}{3}h$, $\dfrac{dA}{dt} < 0$.

11. D.

Since $f(x)$ is continuous, $f(0) = 1$, and $f(3) = 5$, there is some c in the interval $(0,3)$ such that $f(c) = 2$ (Intermediate Value Theorem).

12. D.

$$f'(x) = \frac{\frac{1}{2\sqrt{x+1}}(x^3 + 2) - (3x^2)\sqrt{x+1}}{(x^3 + 2)^2} \text{ from the quotient rule.}$$

Thus $f'(0) = \frac{\frac{1}{2}(2) - 0}{2^2} = \frac{1}{4}$.

13. B.

By factoring you see that $\dfrac{2x^2 - 32}{x^2 + x - 20} = \dfrac{2(x+4)(x-4)}{(x+5)(x-4)} = \dfrac{2(x+4)}{x+5}$.

Thus, $\displaystyle\lim_{x \to 4} \frac{2x^2 - 32}{x^2 + x - 20} = \lim_{x \to 4} \frac{2(x+4)}{x+5} = \frac{2 \cdot 8}{9} = \frac{16}{9}$.

14. E.

$$g'(x) = \cos^3(x^2) \cdot \frac{d}{dx}(x^2) - \cos^3 x \cdot \frac{d}{dx}(x) = 2x\cos^3 x^2 - \cos^3 x.$$

15. B.

If $f(x) = \cot x$, then $f'(x) = -\csc^2 x$. By the definition of the derivative,

$$f'\left(\frac{\pi}{2}\right) = \lim_{x \to \frac{\pi}{2}} \frac{\cot x - \cot\left(\frac{\pi}{2}\right)}{x - \frac{\pi}{2}}.$$

So $\displaystyle\lim_{x \to \frac{\pi}{2}} \frac{\cot x - \cot\left(\frac{\pi}{2}\right)}{x - \frac{\pi}{2}} = f'\left(\frac{\pi}{2}\right) = -\csc^2\left(\frac{\pi}{2}\right) = -1.$

16. A.

Use the substitution $u = 5t^2 + 4$, $du = 10t\, dt$.

So, $\displaystyle\int \frac{t}{(5t^2 + 4)^4}\, dt = \frac{1}{10}\int u^{-4}\, du = -\frac{1}{30}u^{-3} + C = -\frac{1}{30}(5t^2 + 4)^{-3} + C = \dfrac{-1}{30(5t^2 + 4)^3} + C.$

17. D.

$$k'(x) = 2(\ln 6)(6^{3x}) \cdot \frac{d}{dx}(3x) = 6(\ln 6)(6^{3x}) = 6^{3x+1}(\ln 6).$$

18.   C.

Moves to the right: $1(2) + \frac{1}{2}(2) + \frac{1}{2}\pi(2^2) = 3 + 2\pi$. Moves to the left: $\frac{1}{2}(6)(4) = 12$. Total distance traveled for $0 \leq t \leq 12$: $3 + 2\pi + 12 = 15 + 2\pi$.

19.   C.

The maximum speed is the largest numerical value of the velocity, regardless of sign. From the graph, the object has its maximum speed of 4 at $t = 4$.

20.   C.

The graph must increase for $x < 0$ and then decrease for $x > 0$, which eliminates A, B, and E. The curve is concave up when x is between 0 and 1, which eliminates D.

21.   D.

$$f'(x) = \frac{d}{dx}(x^2) \times \frac{(x^2)^3 + 1}{(x^2)^2 + 1} = 2x\left(\frac{x^6 + 1}{x^4 + 1}\right).$$

22.   C.

$v(t) = 0$ when $0 = 3t^4 - 3t^2 = 3t^2(t^2 - 1) = 3t^2(t + 1)(t - 1)$. This has solutions $t = -1$, $t = 0$, and $t = 1$. For $0 < t < 1$, $v(t) < 0$ and for $t > 1$, $v(t) > 0$. So, the total distance traveled between $t = 0$ and $t = 1$ is

$$\int_0^1 (-v(t))\, dt + \int_1^4 v(t)\, dt = \int_0^1 (3t^2 - 3t^4)\, dt + \int_1^4 (3t^4 - 3t^2)\, dt.$$

23.   D.

One cross-sectional area is given by $A = \frac{\sqrt{3}}{4}y^2 = \frac{\sqrt{3}}{4}(4 - x^2)$.

The volume is given by

$$\int_0^2 A(x)\, dx = \int_0^2 \frac{\sqrt{3}}{4}(4 - x^2)\, dx = \frac{\sqrt{3}}{4}\left[4x - \frac{x^3}{3}\right]_0^2 = \frac{4\sqrt{3}}{3}.$$

24.   C.

$\lim\limits_{x \to 3} |x - 3| = 0$ and $\lim\limits_{x \to 3} x = 3$. So, by the properties of limits,

$$\lim_{x \to 3} \frac{|x - 3|}{x} = \frac{\lim\limits_{x \to 3} |x - 3|}{\lim\limits_{x \to 3} x} = \frac{0}{3} = 0.$$

25. C.

$f(x) = xe^x \implies f'(x) = e^x + xe^x \implies f''(x) = e^x + e^x + xe^x = 2e^x + xe^x = e^x(2 + x).$

Thus, $f''(x) = 0$ when $x = -2$, and f'' changes sign at $x = -2$, so there is an inflection point of f at $x = -2$.

26. D. Using the fact that $g'(x) = \dfrac{1}{f'(g(x))}$, you should find that

$$g'(2) = \frac{1}{f'(g(2))} = \frac{1}{f'(2)} = \frac{1}{4}.$$

Since the point $(2, 3)$ is on the graph of $f(x)$, the point $(3, 2)$ is on the graph of $g(x)$. Therefore the equation of the tangent line to $g(x)$ at $x = 3$ is $y - 2 = \frac{1}{4}(x - 3)$.

27. C.

For h to be continuous at $x = 0$, it's necessary that $\lim\limits_{x \to 0} h(x) = h(0) = a$.

$$\lim_{x \to 0} h(x) = \lim_{x \to 0} \frac{\sin 2x}{x} = 2 \lim_{x \to 0} \frac{\sin 2x}{2x} = (2)(1) = 2,$$

so $a = 2$.

28. A.

Separating the variables gives $\dfrac{dy}{y + 1} = t\, dt$.

Integrating both sides gives $\ln|y + 1| = \dfrac{t^2}{2} + C_1$ and then $|y + 1| = C_2 e^{\frac{t^2}{2}}$.

Allowing C_2 to take both positive and negative values, we get $y + 1 = C_2 e^{\frac{t^2}{2}}$.

Thus, $y(t) = C_2 e^{\frac{t^2}{2}} - 1$, and since $y(0) = 0$, we have $y(t) = e^{t^2} - 1$.

29. B.

The rectangle has area $A = bh = 2 \times 4 = 8$. The white area is $A = \displaystyle\int_0^4 \sqrt{x}\, dx = \dfrac{16}{3}$.

The percent shaded is $\dfrac{8 - \frac{16}{3}}{8} = 1 - \frac{2}{3} = \frac{1}{3} = 0.333 = 33.3\%$.

30. C.

By the properties of definite integrals,

$$\int_{-3}^{5} h(x)\, dx = \int_{-3}^{3} h(x)\, dx + \int_{3}^{5} h(x)\, dx.$$

Because $h(x)$ is an odd function, $\displaystyle\int_{-3}^{3} h(x)\, dx = 0$. So $\displaystyle\int_{-3}^{5} h(x)\, dx = -2$.

31. D.

I. is false: $g'(x)$ does not exist, is greater than 0, or is less than 0 at each x in $[0, 4]$.

II. is true: There are no gaps in the graph of $g(x)$ at $x = 2$.

III. is true: $g'(\frac{3}{2})$ exists and is negative, since on the subinterval $[1, 2]$ the graph of g is a straight line with negative slope.

32. C.

By the tangent line approximation, $f(3.1) \approx f(3) + (0.1) \times f'(3)$.

$f'(x) = \frac{5}{2}(x + 1)^{3/2}$, so $f'(3) = 20$ and $f(3) = 32.0$.

Therefore $f(3.1) \approx 32 + (0.1) \times 20 = 32 + 2 = 34.0$.

33. D.

$$b'(x) = \frac{d}{dx}(2x^2 - x + 1) \times \sqrt{(2x^2 - x + 1)^2 + 1} = (4x - 1)\sqrt{(2x^2 - x + 1)^2 + 1}.$$

This is equal to 0 only when $x = \frac{1}{4}$. Finally, note that for $x < \frac{1}{4}$, $b'(x) < 0$, and for $x > \frac{1}{4}$, $b'(x) > 0$, so that $b(x)$ has a relative minimum at $x = \frac{1}{4}$.

34. B.

Separating variables gives us $2y\, dy = \frac{3}{t + 1}\, dt$.

Integrating both sides then gives $y^2 = 3\ln|t + 1| + C$. Since $y(0) = 1$, we have $C = 1$.

Also, since $y(0) > 0$, we take the positive square root so that $y(t) = \sqrt{3\ln|t + 1| + 1}$.

35. B.

On the interval $[0, 4]$, the curve $y = \sqrt{x} + 2$ is above the curve $y = x$. So, the area is

$$\int_0^4 (\sqrt{x} + 2 - x)\, dx = \left[\frac{2}{3}x^{3/2} + 2x - \frac{1}{2}x^2\right]_0^4 = \frac{2}{3}4^{3/2} + 2 \times 4 - \frac{4^2}{2} = \frac{16}{3}.$$

36. E.

You need a calculator for this one:

$$\frac{1}{b - a}\int_a^b f(x)\, dx = \frac{1}{3}\int_0^3 \left(\frac{x + 1}{x^3 + 2}\right) dx = 0.447.$$

37. E.

For $x \neq 0$, $f'(x) = \begin{cases} 3x^2 & \text{if } x < 0 \\ 2x & \text{if } x > 0 \end{cases}$.

Thus, $f'(0) = 0$, since $\lim\limits_{x \to 0^-} 3x^2 = \lim\limits_{x \to 0^+} 2x = 0$.

For $x \neq 0$, $f''(x) = \begin{cases} 6x & \text{if } x < 0 \\ 2 & \text{if } x > 0 \end{cases}$.

Thus, $f''(0)$ does not exist, since $\lim\limits_{x \to 0^-} 6x \neq \lim\limits_{x \to 0^+} 2$.

---

38. C.

I is true; it is a statement of the Mean Value Theorem.

II is true; it is a property of definite integrals.

III is false. To see this, consider $g(x) = \dfrac{1}{x}$ and $h(x) = 0$ on the interval $[1, 2]$.

Note that $g'(x) = -\dfrac{1}{x^2}$ and $h'(x) = 0$. Thus $g(x) \geq h(x)$ on $[1, 2]$, but $g'(x) < h'(x)$ on $(1, 2)$.

---

39. D.

$$\int_0^c (x^6 - e^x)\, dx = \left[\frac{x^7}{7} - e^x \right]_0^c = \frac{c^7}{7} - e^c + 1$$

$$\frac{c^7}{7} - e^c + 1 = 0$$

$$c = 1.610 \,(\text{using a calculator}).$$

---

40. D.

Since we're rotating about the line $y = 2$, and $f(x) \geq g(x)$ over the given domain, the volume of the solid of revolution is

$$\pi \int_a^b (f(x) - 2)^2\, dx - \pi \int_a^b (g(x) - 2)^2\, dx = \pi \int_a^b \left((f(x) - 2)^2 - (g(x) - 2)^2 \right) dx.$$

41. E.

Using implicit differentiation, we have

$$2y^4 + 8xy^3\frac{dy}{dx} - 3x^2\cos y + x^3\sin y\frac{dy}{dx} + \frac{dy}{dx} = 0.$$

Isolate the $\frac{dy}{dx}$ terms:

$$8xy^3\frac{dy}{dx} + x^3\sin y\frac{dy}{dx} + \frac{dy}{dx} = 3x^2\cos y - 2y^4$$

$$\frac{dy}{dx}(8xy^3 + x^3\sin y + 1) = 3x^2\cos y - 2y^4$$

$$\frac{dy}{dx} = \frac{3x^2\cos y - 2y^4}{8xy^3 + x^3\sin y + 1}.$$

42. E.

First recognize that $f(1) = 6$, so that $g(6) = 1$, and $f'(x) = 10x^4 + 6x^2 + 2$.

Then, using the fact that $g'(x) = \dfrac{1}{f'(g(x))}$, we have:

$$g'(6) = \frac{1}{f'(g(6))} = \frac{1}{f'(1)} = \frac{1}{10 + 6 + 2} = \frac{1}{18}.$$

43. D.

Differentiating $x^2 + y^2 = 4$ with respect to time t gives us $2x\dfrac{dx}{dt} + 2y\dfrac{dy}{dt} = 0.$

Dividing both sides by 2 gives us $x\dfrac{dx}{dt} + y\dfrac{dy}{dt} = 0.$

Substitute the values given to get $2 + \sqrt{3}\dfrac{dy}{dt} = 0$ and solve for $\dfrac{dy}{dt} = -\dfrac{2}{\sqrt{3}}.$

44. A.

$3(\sqrt{0} + \sqrt{3} + \sqrt{6} + \sqrt{9}) \approx 21.545.$

45. C.

$a(t) = v'(t) = -8t^3 + 6t + 1$ and $a'(t) = -24t^2 + 6.$

Set $a'(t) = -24t^2 + 6 = 0$ and solve for the positive value of t to get $t = \frac{1}{2}.$

Note that $a'(t)$ is positive for $0 < t < \frac{1}{2}$ and negative for $\frac{1}{2} < t$ so that the acceleration is a maximum at $t = \frac{1}{2}.$

Practice Calculus AB Test One: Free-Response Answers and Scoring Guide

1.

A.

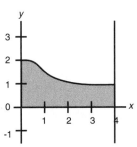

2 points
1: correct graph for $f(x)$
1: correct region M depicted

B. Area $= \displaystyle\int_0^4 \left(1 + \frac{1}{x^4 + 1}\right) dx \approx 5.106$ from the calculator.

2 points
1: correct integral
1: answer

C. $A(z) = \displaystyle\int_0^z \left(1 + \frac{1}{x^4 + 1}\right) dx$

2 points

2: correct integral

D. By the Fundamental Theorem of Calculus, $\dfrac{dA}{dz} = 1 + \dfrac{1}{z^4 + 1}$.

At $z = 2$, this is $1 + \dfrac{1}{2^4 + 1} = 1 + \dfrac{1}{17} \approx 1.059$.

3 points
2: correct derivative and use of FTC
1: evaluation at $z = 2$

2.

A. $a(t) = v'(t) = \sqrt{2t+3} + \dfrac{t}{\sqrt{2t+3}} - 3.$

1 point

1: correct acceleration

B. Object moves to the right when $v(t) > 0$.

$$v(t) = t\sqrt{2t+3} - 3t > 0$$
$$\sqrt{2t+3} - 3 > 0$$
$$\sqrt{2t+3} > 3$$
$$2t+3 > 9$$
$$t > 3.$$

3 points

1: correct inequality (or equation) for motion to the right

1: solve for t

1: answer

C. Change in position from $t = 0$ to $t = 3$ is given by

$$\int_0^3 v(t)\, dt = \int_0^3 (t\sqrt{2t+3} - 3t)\, dt = -1.661$$

(using the calculator). Since the particle was already at $x = 2$ when $t = 0$, at $t = 3$ it is at $x = 2 - 1.661 = .339$.

3 points

1: correct integral with limits

1: correct net change (value of integral)

1: answer

D. Total distance traveled from $t = 0$ to $t = 3$ is given by

$$= \int_0^6 |v(t)|\, dt = \int_0^6 |t\sqrt{2t+3} - 3t|\, dt = 8.456$$

(using the calculator).

2 points

1: correct integral with limits

1: answer

3.

A. Average rate of change of power usage:
$$= \frac{f(9) - f(3)}{9 - 3} = \frac{1400 - 1200}{6} = \frac{200}{6} = 33.333 \text{ kW per hour.}$$

2 points
1: correct expression
1: answer with units

B. Average $= \frac{1}{12-4} \int_4^{12} f(t)\, dt = \frac{1}{8} \int_4^{12} f(t)\, dt$. Using a midpoint approximation with 4 equally spaced subintervals, we get
$$\tfrac{1}{8}(2)(f(5) + f(7) + f(9) + f(11)) = \tfrac{1}{4}(850 + 1000 + 1400 + 1700) = 1237.5 \text{ kW.}$$

3 points
2: integral:

    1: limits and $\frac{1}{12-4}$

    1: integrand
1: answer

Note: 0/1 if integral not of the form $\frac{1}{b-a} \int_a^b F(t)\, dt$

C. $\int_0^{12} f(t)\, dt$ is the total number of kilowatt hours used over the period $0 \le t \le 12$.

Note: A kilowatt hour is one kilowatt being used for one hour.

2 points
1: correct meaning of integral
1: correct units

D. The power usage increased the most over the interval $5 \le t \le 6$ hours, since the difference $f(6) - f(5) = 1,200 - 850 = 350$ is the largest among all of the time intervals (and all of the intervals are 1 hour long).

2 points
1: correct answer
1: explanation

4.

A. Separating variables gives us $\dfrac{dy}{y^2} = \cos t\, dt$.

Integrating both sides now gives $\displaystyle\int \dfrac{dy}{y^2} = \int \cos t\, dt$, and then $-\dfrac{1}{y} = \sin t + C$.

Since $y(0) = 1$, $C = -1$. Solving for y: $\dfrac{1}{y} = 1 - \sin t \implies y(t) = \dfrac{1}{1 - \sin t}$.

4 points
1: separates variables
1: integrates both sides correctly
1: evaluates C
1: solves for y

B. $y(t)$ has a vertical asymptote at $t = a$ if $\displaystyle\lim_{t \to a^-} y(t) = -\infty$ or $\displaystyle\lim_{t \to a^+} y(t) = -\infty$.

This happens when the denominator approaches 0, or when $1 - \sin t$ approaches 0.

The only value of t between $-\pi$ and π for which this happens is at $t = \frac{\pi}{2}$.

3 points
1: recognizes condition for vertical asymptote
1: gets condition $1 - \sin t = 0$
1: solves for t in the stated interval

C. We know $\dfrac{dy}{dt} = y^2 \cos t = \dfrac{\cos t}{(1 - \sin t)^2}$.

Thus,

$$\frac{d^2 y}{dt^2} = \frac{-\sin t (1 - \sin t)^2 - 2(1 - \sin t)(-\cos t)(\cos t)}{(1 - \sin t)^4}$$

$$= \frac{2\cos^2 t (1 - \sin t) - \sin t (1 - \sin t)^2}{(1 - \sin t)^4}.$$

2 points
1: computes (or recognizes) $\dfrac{dy}{dt}$
1: explicit expression for $\dfrac{d^2 y}{dt^2}$

5.

A. $q(-1) = \int_0^{-1} f(t)\, dt = -\int_{-1}^0 f(t)\, dt = -\frac{1}{2}(1)4 = -2.$

$q(4) = \int_0^4 f(t)\, dt = \frac{1}{2}(2)4 - \frac{1}{2}(2)4 = 0.$

2 points

2: 1 point for each correct value of $q(x)$

B. $q'(x) = f(x)$ (Fundamental Theorem of Calculus). So, $q'(0) = f(0) = 4$.

1 point

1: correct value of $q'(0)$

C. Check for critical points in $[2, 8]$: $q'(x) = f(x)$, which equals 0 when $x = 6$.

We must also check the endpoints $x = 2$ and $x = 8$: $q(2) = 4$, $q(6) = -4$, $q(8) = 0$.

So, the absolute maximum of $q(x)$ on $[2, 8]$ is 4, which happens at $x = 2$.

3 points
1: finds critical point at $x = 6$
1: checks endpoints
1: answer

D. Since $q'(x) = f(x)$, we have $q''(x) = f'(x)$ wherever f' exists.

At $x = -1$, f' does change sign from positive to negative, so at $x = -1$ there is an inflection point of q.

At $x = 0$, f' does change sign from positive to negative, so at $x = 0$ there is an inflection point of q.

3 points
1: correct answer for $x = -1$
1: correct answer for $x = 0$
1: explanation

6.

A. $2x\cos(y) - x^2\sin(y)\dfrac{dy}{dx} + \dfrac{dy}{dx}\sin(2\pi x) + 2\pi y\cos(2\pi x) - 1 = 0$

$2x\cos(y) + 2\pi y\cos(2\pi x) - 1 = x^2\sin y\dfrac{dy}{dx} - \dfrac{dy}{dx}\sin(2\pi x)$

$2x\cos(y) + 2\pi y\cos(2\pi x) - 1 = \dfrac{dy}{dx}(x^2\sin(y) - \sin(2\pi x))$

$\dfrac{dy}{dx} = \dfrac{2x\cos(y) + 2\pi y\cos(2\pi x) - 1}{x^2\sin(y) - \sin(2\pi x)}.$

4 points
1: implicit differentiation

1: isolates terms with $\dfrac{dy}{dx}$ on one side

1: solves for $\dfrac{dy}{dx}$

1: answer

B. Using $\dfrac{dy}{dx}$ from part A, the slope of the curve at the point $\left(1, \dfrac{\pi}{2}\right)$ is $\dfrac{0 + \pi^2 - 1}{1} = \pi^2 - 1.$

So, the equation of the tangent line is $y - \dfrac{\pi}{2} = (\pi^2 - 1)(x - 1)$, or $y = (\pi^2 - 1)(x - 1) + \dfrac{\pi}{2}.$

2 points
1: correct slope at point $\left(1, \dfrac{\pi}{2}\right)$
1: equation of tangent line

C. Using the tangent line from part B to approximate $y(1.1)$ gives us

$$y(1.1) \approx (\pi^2 - 1)(1.1 - 1) + \dfrac{\pi}{2} = (.1)(\pi^2 - 1) + \dfrac{\pi}{2} = 2.458$$

3 points
1: uses tangent line from B
1: evaluates tangent line at $x = 1.1$
1: answer

Practice Calculus AB Test Two

You'll get the most benefit from this practice examination if you take it under conditions exactly like those of the actual AP Exam.

Estimated Time to Complete: 200 minutes

Section I Part A, 55 minutes, 28 multiple-choice questions: NO calculator allowed.

Section I Part B, 50 minutes, 17 multiple-choice questions: Calculator allowed.

Short break, 5 minutes.

Section II Part A, 45 minutes, 3 free-response questions: Calculator allowed.

Section II Part B, 45 minutes, 3 free-response questions: NO calculator allowed.

Work on only one section at a time. You may not go back to a previous section. You may, however, continue to work on part A of Section II after you have turned in your calculator.

Practice Exam Section 1—Multiple-Choice

Time limit: 105 minutes

Part A consists of **28** questions. A calculator may **not** be used on this part of the examination. Part B consist of **17** questions. In the multiple-choice of the examination, as a correction for guessing, one-fourth of the number of questions answered incorrectly will be subtracted from the number answered correctly.

Directions: Solve each of the following problems, using the available space for scratchwork. After examining the form of the choices, decide which is the best of the choices given and fill in the corresponding oval on the answer sheet. No credit will be given for any math work written on the test paper. Do not spend too much time on any one problem.

In this test: Unless otherwise specified, the domain of a function f is assumed to be the set of all real numbers x for which $f(x)$ is a real number.

Formulas

Here is a short list of area and volume formulas that you might need.

For a circle of radius r: Area $= \pi r^2$, Circumference $= 2\pi r$.

For a sphere of radius r: Volume $= \frac{4}{3}\pi r^3$, Surface Area $= 4\pi r^2$.

For a right circular cylinder with radius r and height h: Volume $= \pi r^2 h$, Surface Area $= 2\pi r h + 2\pi r^2$.

For a right circular cone with radius r and height h: Volume $= \frac{1}{3}\pi r^2 h$.

Answer Sheet—Multiple Choice

PART A

1 Ⓐ Ⓑ Ⓒ Ⓓ Ⓔ   8 Ⓐ Ⓑ Ⓒ Ⓓ Ⓔ   15 Ⓐ Ⓑ Ⓒ Ⓓ Ⓔ   22 Ⓐ Ⓑ Ⓒ Ⓓ Ⓔ
2 Ⓐ Ⓑ Ⓒ Ⓓ Ⓔ   9 Ⓐ Ⓑ Ⓒ Ⓓ Ⓔ   16 Ⓐ Ⓑ Ⓒ Ⓓ Ⓔ   23 Ⓐ Ⓑ Ⓒ Ⓓ Ⓔ
3 Ⓐ Ⓑ Ⓒ Ⓓ Ⓔ   10 Ⓐ Ⓑ Ⓒ Ⓓ Ⓔ   17 Ⓐ Ⓑ Ⓒ Ⓓ Ⓔ   24 Ⓐ Ⓑ Ⓒ Ⓓ Ⓔ
4 Ⓐ Ⓑ Ⓒ Ⓓ Ⓔ   11 Ⓐ Ⓑ Ⓒ Ⓓ Ⓔ   18 Ⓐ Ⓑ Ⓒ Ⓓ Ⓔ   25 Ⓐ Ⓑ Ⓒ Ⓓ Ⓔ
5 Ⓐ Ⓑ Ⓒ Ⓓ Ⓔ   12 Ⓐ Ⓑ Ⓒ Ⓓ Ⓔ   19 Ⓐ Ⓑ Ⓒ Ⓓ Ⓔ   26 Ⓐ Ⓑ Ⓒ Ⓓ Ⓔ
6 Ⓐ Ⓑ Ⓒ Ⓓ Ⓔ   13 Ⓐ Ⓑ Ⓒ Ⓓ Ⓔ   20 Ⓐ Ⓑ Ⓒ Ⓓ Ⓔ   27 Ⓐ Ⓑ Ⓒ Ⓓ Ⓔ
7 Ⓐ Ⓑ Ⓒ Ⓓ Ⓔ   14 Ⓐ Ⓑ Ⓒ Ⓓ Ⓔ   21 Ⓐ Ⓑ Ⓒ Ⓓ Ⓔ   28 Ⓐ Ⓑ Ⓒ Ⓓ Ⓔ

PART B

29 Ⓐ Ⓑ Ⓒ Ⓓ Ⓔ   34 Ⓐ Ⓑ Ⓒ Ⓓ Ⓔ   39 Ⓐ Ⓑ Ⓒ Ⓓ Ⓔ   44 Ⓐ Ⓑ Ⓒ Ⓓ Ⓔ
30 Ⓐ Ⓑ Ⓒ Ⓓ Ⓔ   35 Ⓐ Ⓑ Ⓒ Ⓓ Ⓔ   40 Ⓐ Ⓑ Ⓒ Ⓓ Ⓔ   45 Ⓐ Ⓑ Ⓒ Ⓓ Ⓔ
31 Ⓐ Ⓑ Ⓒ Ⓓ Ⓔ   36 Ⓐ Ⓑ Ⓒ Ⓓ Ⓔ   41 Ⓐ Ⓑ Ⓒ Ⓓ Ⓔ
32 Ⓐ Ⓑ Ⓒ Ⓓ Ⓔ   37 Ⓐ Ⓑ Ⓒ Ⓓ Ⓔ   42 Ⓐ Ⓑ Ⓒ Ⓓ Ⓔ
33 Ⓐ Ⓑ Ⓒ Ⓓ Ⓔ   38 Ⓐ Ⓑ Ⓒ Ⓓ Ⓔ   43 Ⓐ Ⓑ Ⓒ Ⓓ Ⓔ

Multiple-Choice, Section 1, Part A (noncalculator): 55 minutes

NOTE: You may NOT use your calculator on this section.

---

1. $\displaystyle \lim_{x \to \infty} \frac{2x - 4x^3 - 9}{6x^3 + 4x^2 + 5} =$

A. 0

B. ∞

C. 1

D. $-\frac{2}{3}$

E. $-\frac{9}{5}$

---

2. The radius of a circle is decreasing at 2 cm/sec. Find the rate of change of the area of the circle with respect to time (in cm^2/sec).

A. $-4\pi r$

B. $-2\pi r$

C. 2

D. $2\pi r$

E. $4\pi r$

---

3. $\displaystyle \int_0^1 (4y + 1)^2 \, dy =$

A. $\frac{1}{3}$

B. 0

C. $\frac{11}{3}$

D. $\frac{31}{3}$

E. 1

4. If $f(x) = \ln x^3$ and $g(x)$ is the inverse function of $f(x)$, then $g'(x) =$

A. $\ln y^3$

B. $\frac{1}{3}e^{\frac{x}{3}}$

C. $3x^2 e^{x^3}$

D. $\dfrac{3}{x}$

E. $3x^2 \ln x^3$

5. If the second derivative of f is given by $f''(x) = x^3 + \sin x$, which of the following could be $f(x)$?

A. $\dfrac{x^5}{20} - \sin x + 5x + 6$

B. $x^5 + \sin x - x - 3$

C. $\dfrac{x^5}{20} + \cos x + 5x + 2$

D. $6x - \sin x$

E. $\dfrac{x^4}{4} - \cos x - 1$

6. The line tangent to the curve $3x^2 - xy + y^2 = 5$ at point $(1, 2)$ is given by:

A. $y - 2 = -\frac{4}{3}(x - 1)$

B. $y - 2 = -\frac{3}{4}(x - 1)$

C. $y = 2$

D. $y = \frac{3}{4}(x - 1) + 2$

E. $y = \frac{4}{3}(x - 1) + 2$

7. A particle b is moving along the x-axis. Its velocity is given by $v(t) = t^3 - nt$, where $n > 0$. Which of the following are necessarily true?

   I. b has a velocity of 0 when $t = \sqrt{n}$.

   II. b moves to the right for all t.

   III. b has a minimum speed when $t = \sqrt{n}$.

   A. I only

   B. II only

   C. III only

   D. I and III only

   E. None

8. $g(x) = |x|$

   Which of the following are true statements?

   I. $g(x)$ is continuous at $x = 0$.

   II. $g(x)$ is differentiable at $x = 0$.

   III. $\lim\limits_{x \to 0} g(x) = 0$

   A. I only

   B. I and II

   C. III only

   D. I and III

   E. I, II, and III

9. $f(x)$ is shown in the curve below:

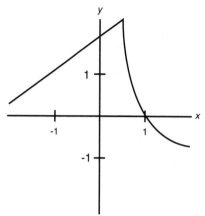

Which of the following is true?

A. $f''(1) < f'(1) < f(1)$

B. $f'(1) < f(1) < f''(1)$

C. $f(1) < f'(1) < f''(1)$

D. $f''(1) < f(1) < f'(1)$

E. $f(1) < f''(1) < f'(1)$

10. The height of a triangle is decreasing at 4 inches a minute. The base of the triangle is increasing at 4 inches a minute. The area is

A. always increasing.

B. always decreasing.

C. increasing when $h > b$.

D. decreasing when $h > b$.

E. constant.

11. Given that $f(x)$ is continuous for all real numbers,

x	0	1	2
$f(x)$	1	?	1

which of the following possible values of $f(1)$ would guarantee two solutions for $f(x) = \frac{1}{2}$ on the domain $0 \le x \le 2$?

A. $\frac{1}{2}$

B. 1

C. 2

D. 0

E. $f(1)$ could be any value.

12. If $f(x) = x^2 \cos^2 x$, then $f'(\pi) =$

A. -1

B. 2π

C. 0

D. π

E. -2π

13. $\displaystyle \lim_{x \to a} \frac{a^2 - x^2}{a^4 - x^4} =$

A. 0

B. 1

C. $\dfrac{1}{2a^2}$

D. a

E. Does not exist

14. Let $g(x) = \displaystyle\int_{-1-x}^{\frac{3x^2}{2}-6x} t^2\, dt$. Calculate $g'(x)$.

A. $g(x)$

B. $(3x-6)(\frac{3x^2}{2}-6x)^2 - (-1-x)^2$

C. $(\frac{3x^2}{2}-6x)^2 - (-1-x)^2$

D. $(3x-6)(\frac{3x^2}{2}-6x)^2 + (-1-x)^2$

E. $\frac{1}{3}(\frac{3x^2}{2}-6x)^3 - \frac{1}{3}(-1-x)^3$

15. $\displaystyle\lim_{x \to \pi} \frac{\sin \pi - \sin x}{\pi - x} =$

A. 0

B. -1

C. 1

D. ∞

E. does not exist

Questions 16–19 refer to the following graph of the velocity of an object over the time interval $0 \le t \le 9$.

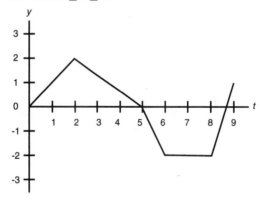

16. During which of these time intervals does the object have the greatest acceleration?

    A. $2 < t < 5$

    B. $5 < t < 8$

    C. $t = 6$

    D. $t = 8$

    E. $8 < t < 9$

17. The object is farthest from the starting point when $t =$

    A. 2

    B. 5

    C. 6

    D. 8

    E. 9

18. Given that at $t = 8$ the object was at position $x = 10$, at $t = 5$ the position was $x =$

    A. -5

    B. 5

    C. 7

    D. 13

    E. 15

19. The total distance traveled by the object for $2 \leq t \leq 7$ is:

    A. 0

    B. 2

    C. 4

    D. 6

    E. 10

20. $f(x)$ is a function that satisfies the following conditions:

	$x < 0$	$0 < x < 1$	$1 < x$
$f'(x)$	$-$	$+$	$+$
$f''(x)$	$+$	$+$	$-$

Which of the following could be $f(x)$?

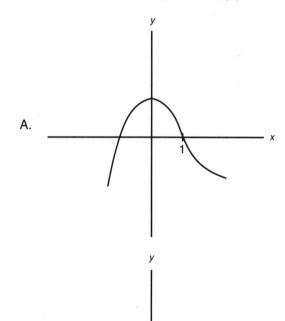

A.

B.

C.

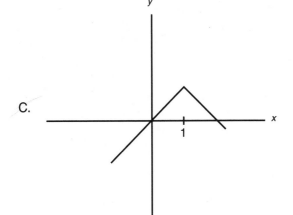

D.

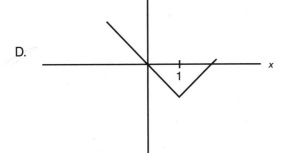

E.

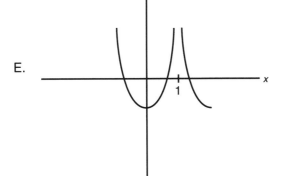

---

21. For all $x > 1$, if $f(x) = \int_1^x \left(t - \frac{2}{t} \right) dt$, then $f'(x) =$

   A. $x - \dfrac{2}{x}$

   B. $x - 2\ln x$

   C. $-2\ln x$

   D. $1 + \dfrac{2}{x^2}$

   E. $x - e^x$

---

22. $\displaystyle\int \left(\frac{x^2 + 1}{x} \right) dx =$

   A. $\frac{1}{2}\ln(x^2 + 1) + C$

   B. $\dfrac{2}{x} + C$

   C. $\dfrac{x^3}{3} + \ln|x| + C$

   D. $\dfrac{2x}{3} + \dfrac{2}{x} + C$

   E. $\dfrac{x^2}{2} + \ln|x| + C$

---

23. The acceleration of an object moving along the x-axis is given by $a(t) = 18t - 2$, where the velocity is 12 when $t = 2$ and the position is 2 when $t = 1$. The position $x(t) =$

   A. $9t^3 - t^2 - 20t + 14$

   B. $9t^3 - t^2 + 20t - 26$

   C. $3t^3 - t^2 - 20t + 20$

   D. $3t^3 - t^2 + 20t - 20$

   E. $t^3 - t^2 - 20t + 22$

24. Let $b > 0$ be a fixed constant. Then $\displaystyle\lim_{t \to b} \frac{\ln t - \ln b}{t - b} =$

    A. $e^{1/t}$

    B. e^b

    C. e^t

    D. $\dfrac{1}{b}$

    E. $\dfrac{1}{t}$

25. The number of inflection points of $f(x) = 3x^7 - 10x^5$ is:

    A. 0

    B. 1

    C. 2

    D. 3

    E. 5

26. If $y = e^x \cos x$, then $\dfrac{dy}{dx} =$

    A. $e^x \cos x$

    B. $-e^x \sin x$

    C. $e^x (\cos x - \sin x)$

    D. $e^x (\sin x - \cos x)$

    E. $e^x (\sin x + \cos x)$

27. Find where the absolute minimum of $g(x) = \int_0^x \cos(t-\pi)\, dt$ occurs on the interval $[0, 2\pi]$.

A. $x = 0$

B. $x = \frac{\pi}{2}$

C. $x = \pi$

D. $x = \frac{3\pi}{2}$

E. $x = 2\pi$

28. If $x\,\dfrac{dy}{dx} = 10$ and $y(1) = 0$, then $y =$

A. $5x^2 - 10$

B. $10x - 10$

C. $10\ln x$

D. $e^{10x} - e^{10}$

E. $10e^x - 10e$

Multiple-Choice, Section 1, Part B (with calculator): 50 minutes

You may use your calculator on any of these questions.

---

29. The following curve is the graph of $y = g(x)$. At which point are $g(x)$, $g'(x)$, and $g''(x)$ all less than zero?

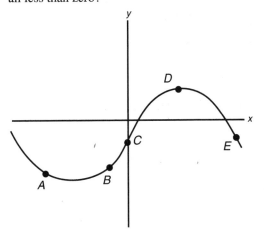

A. A

B. B

C. C

D. D

E. E

---

30. If $5x^2y^4 + 3y^3 + 9xy^2 = 17$, then in terms of x and y, we have $\dfrac{dy}{dx} =$

A. $-\dfrac{5y^2 + 8x}{20x^2y^3 + 9y^2 + 18xy}$

B. $-\dfrac{10xy^4 + 9y^2}{20xy^3 + 7y^2 + x^2y}$

C. $-\dfrac{xy^4 + 5y^2}{5x^2y^3 + 29y^2 + 18y}$

D. $-\dfrac{10xy^4 + 9y^2}{20x^2y^3 + 9y^2 + 18xy}$

E. $\dfrac{-5y^4 + 2xy^3}{10x^2y^3 + 10y^2 + 5xy^2}$

31. Let $f(x)$ be defined as follows, where $b \neq 0$:

$$f(x) = \begin{cases} \dfrac{b^2 - x^2}{b - x} & \text{for } x \neq b \\ 0 & \text{for } x = b. \end{cases}$$

Which of the following are true about f?

I. $\lim\limits_{x \to b} f(x)$ exists.

II. $f(b)$ exists.

III. $f(x)$ is continuous at $x = b$.

A. I only

B. II only

C. I and II only

D. All three are true

E. None are true

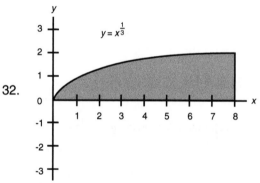

32.

The region on the graph above, from $x = 0$ to $x = 8$, is the base of a solid. Each cross section of the solid, perpendicular to the x-axis, is a semicircle. What is the volume of this solid?

A. 7.54

B. 12

C. 15.08

D. 16

E. 60.32

33. You are given the following table of values for $f(x)$:

x	2	5	7	8
$f(x)$	10	30	40	20

Using the intervals $[2,5]$, $[5,7]$, $[7,8]$, approximate the integral $\int_2^8 f(x)\,dx$ using the trapezoidal method.

A. 130

B. 190

C. 160

D. 150

E. 170

34. If p is positive and increasing, for what value of p is the instantaneous rate of increase of $\sqrt{p}$ with respect to the time t one-half times the instantaneous rate of increase of p with respect to the time t?

A. $\sqrt{2}$

B. $\dfrac{1}{\sqrt{2}}$

C. $\dfrac{1}{2}$

D. 1

E. 4

35. Calculate the area of the region bounded by $y = \sqrt{x}$, $y = -x$, and the line $x = 4$.

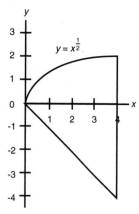

A. 6

B. 10

C. $13\frac{1}{3}$

D. $-2\frac{2}{3}$

E. 12

36. What is the average value of $\frac{1}{2}t^2 - \frac{1}{3}t^3$ over the interval $-2 \le t \le 1$?

A. $\frac{1}{36}$

B. $\frac{1}{12}$

C. $\frac{11}{12}$

D. $\frac{8}{9}$

E. $\frac{33}{12}$

37. Let $f(x)$ be defined by $f(x) = \begin{cases} x^2 & \text{for } x \leq 0 \\ x^3 & \text{for } x > 0 \end{cases}$. The value of $\int_{-2}^{2} f(x)\, dx =$

A. $\frac{20}{3}$

B. 8

C. $\frac{3}{4}$

D. $\frac{7}{2}$

E. 24

38. If f and g are continuous functions, $f(x) \geq 0$ for all real numbers x, and $a < b$, which of the following must be true?

I. $\int_{a}^{b} f(x)\, dx \geq 0$

II. $\int_{a}^{b} (f(x) - g(x))\, dx = \int_{a}^{b} f(x)\, dx + \int_{b}^{a} g(x)\, dx$

III. $\int_{a}^{b} (f(x))^3\, dx = \left(\int_{a}^{b} f(x)\, dx \right)^3$

A. I only

B. II only

C. I and II only

D. II and III only

E. I, II, and III

39. If $\int_{0}^{c} (cx^3 - x)\, dx = 0$ and $c > 0$, then $c =$

A. 1.26

B. 0

C. 1.48

D. 2.44

E. 0.66

40.

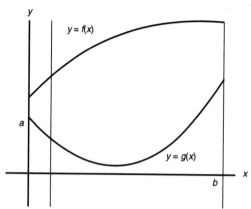

The region in the figure above is bounded by $y = f(x)$, $y = g(x)$, $x = a$, and $x = b$. The region is rotated around the x-axis to form a solid. Which of the following integrals represents the volume of the solid that is generated?

A. $\displaystyle\int_a^b (f(x) - g(x))\, dx$

B. $\displaystyle\int_a^b (f(x))^2\, dx + \int_a^b (g(x))^2\, dx$

C. $\displaystyle\pi\int_a^b f((x))^2\, dx - \pi\int_a^b g((x))^2\, dx$

D. $\displaystyle\pi\int_a^b (f(x) + g(x))\, dx$

E. $\displaystyle\int_a^b (f(x))^2\, dx - \int_a^b (g(x))^2\, dx$

---

41. If $\displaystyle\int_{-3}^{2} f(x)\,dx = 6$, and $\displaystyle\int_{2}^{1} f(x)\,dx = 11$, then $\displaystyle\int_{-3}^{1} f(x)\, dx =$

A. 5

B. 17

C. 3

D. −5

E. 9

42. Let $g(x) = x^3 + x + 1$. If h is the inverse function of g, then $h'(3) =$

    A. 28

    B. 4

    C. 1

    D. $\frac{1}{4}$

    E. $\frac{1}{28}$

43. A particle travels along the curve $y = -2x^2 + 8$ where $x > 0$. At the instant the particle crosses the x-axis, $\dfrac{dy}{dt} = 3$. Find $\dfrac{dx}{dt}$ at this instant.

    A. $-\frac{3}{8}$

    B. $-\frac{3}{4}$

    C. $-\frac{1}{3}$

    D. $\frac{1}{8}$

    E. $\frac{1}{28}$

44. If $\dfrac{dy}{dx} = \dfrac{4}{2x - 3}$ and $y(2) = 1$, then $y =$

    A. $2\ln|2x - 3| + 1$

    B. $\dfrac{4x}{x^2 - 3x + 1} + 1$

    C. $2\ln|2x - 3|$

    D. $e^{2x-3} + 1$

    E. $\frac{1}{2}\ln|2x - 3| - 1$

45. Various values for $f(x)$ are given in the table below. $f(x)$ is differentiable on the domain $-4 \le x \le 4$ and $f\prime(x) < 0$ on the domain $-4 \le x \le 4$.

x	-4	-3	-2	-1	0	1	2	3	4
$f(x)$	43	38	31	21	10	-1	-11	-18	-23

Which of the following statements about $f(x)$ are necessarily true?

I. $f(x)$ is monotonic on the domain $-4 \le x \le 4$.

II. There is some c between 0 and 1 such that $f'(c) = -11$.

III. The average value of $f(x)$ on the domain $-4 \le x \le 4$ is greater than zero.

A. None

B. I only

C. III only

D. I and III only

E. I, II, and III

End of the multiple-choice portion of the practice examination

Section II (90 minutes) Directions for Free-Response Questions

NOTE to students: These are instructions for the actual AP Exam, where you have a test booklet and a separate bubble sheet for answering multiple-choice questions.

You may wish to look the problems over before starting to work on them, since it isn't expected that everyone will be able to complete all parts of all problems. All problems are given equal weight, but the parts of a particular problem aren't necessarily given equal weight. The problems are printed in the booklet and in the green insert. It may be easier for you to first look over all problems in the insert. When you're told to begin, open your booklet, carefully tear out the green insert, and start to work.

A GRAPHING CALCULATOR IS REQUIRED FOR SOME PROBLEMS OR PARTS OF PROBLEMS ON THIS SECTION OF THE EXAMINATION.

* You should write all work for each part of each problem in the space provided for that part in the booklet. Be sure to write clearly and legibly. If you make an error, you may save time by crossing it out rather than trying to erase it. Erased or crossed-out work will not be graded.

* Show all your work. You will be graded on the correctness and completeness of your methods as well as the accuracy of your final answers. Correct answers without supporting work may not receive credit.

* Justifications require that you give mathematical (noncalculator) reasons and that you clearly identify functions, graphs, tables, or other objects you use.

* You are permitted to use your calculator to solve an equation, find the derivative of a function at a point, or calculate the value of a definite integral. However, you must clearly indicate the setup of your problem, namely the equation, function, or integral you are using. If you use other built-in features or programs, you must show the mathematical steps necessary to produce your results.

* Your work must be expressed in standard mathematical notation rather than calculator syntax. For example, $\int_{1}^{5} x^2 \, dx$ may not be written as fnInt (X ˆ 2, X, 1, 5). Unless otherwise specified, answers (numeric or algebraic) need not be simplified. If your answer is given as a decimal approximation, it should be correct to three places after the decimal point.

* Unless otherwise specified, the domain of a function f is assumed to be the set of all real numbers x for which $f(x)$ is a real number.

Free-Response Questions 1–3 (with calculator): 45 minutes

You may use your calculator for these questions.

---

1. The region R is bounded by the graphs of $f(x) = x^2$ and $g(x) = 3^{x-x^2}$.

A. Find the area of R.

B. Find the volume of the solid generated by revolving R around the x-axis.

C. Find the volume of the solid whose base is the region R and whose cross sections cut by planes perpendicular to the x-axis are equilateral triangles.

   (The area A of an equilateral triangle with side length s is given by $A = \dfrac{s^2\sqrt{3}}{4}$.)

---

2. A particle moves along the x-axis so that at any time t, $0 \le t \le 5$, its velocity is given by $v(t) = \ln(1.5 + \sin t)$. At $t = 0$, the position of the particle is $x(0) = 2$.

A. Find the acceleration of the particle at $t = 3$.

B. Find $x(3)$, the position of the particle at $t = 3$.

C. Find the total distance traveled by the particle from $t = 0$ to $t = 5$.

D. At what time t is the particle farthest to the right? Justify your answer.

3. The rate R at which a solar panel delivers electricity is a differentiable function of time t. The table below shows a sample of these rates over an 18 hour period.

t (hours)	4	6	8	10	12	14	16	18	20	22
$R(t)$ amps/hour	36	78	160	240	320	350	360	320	240	160

A. Use the table values to approximate $R'(10)$. Show your computations.

B. Use the trapezoidal rule with 9 equal subdivisions to approximate the average number of amps delivered by the panel from $t = 4$ to $t = 22$.

C. $R(t)$ is closely approximated by the function

$$A(t) = 170 \cos(0.26(t - 15)) + 200.$$

Find $A'(10)$ and find the average value of $A(t)$ over the interval $4 \le t \le 22$.

Free-Response Questions 4–6 (without calculator): 45 minutes

You may NOT use your calculator for these questions.

4. The function g is defined by $g(x) = \dfrac{3}{e^{-x} + 2}$.

A. Find $\lim\limits_{x \to \infty} g(x)$.

B. Find $\lim\limits_{x \to -\infty} g(x)$.

C. What are all values of x where g is increasing? Justify your answer.

D. What are all values of x where the graph of g is concave down? Justify your answer.

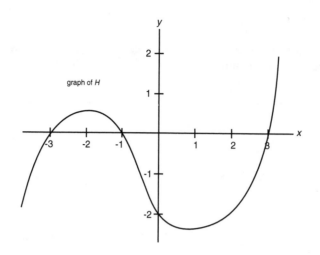

graph of H

---

5.   The function f is continuous for all real numbers. The graph of the function H, defined by $H(x) = \int_{-3}^{x} f(t)\, dt$ is given above. The graph of H has x-intercepts at -3, -1, and 3, and horizontal tangents at $x = -2$ and $x = 1$. Its only point of inflection is at $x = 0$.

A.   What are all values of x where $f(x) < 0$? Justify your answer.

B.   What are all values of x where $f'(x) < 0$? Justify your answer.

C.   Let G be the function defined by $G(x) = \int_{-2}^{x} f(t)\, dt$.

     Sketch a possible graph of $y = H(x) - G(x)$ on the axes provided.

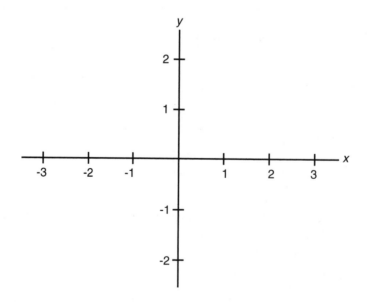

6. At all points (x, y) on the graph of a function $y = f(x)$, the slope is given by $\dfrac{dy}{dx} = \dfrac{1 - 3x^2}{e^{2y}}$. The point $(2, 0)$ is on the graph of f.

A. Find the x-coordinate of each point where the graph of f has a horizontal tangent.

B. Find the x-coordinate of a relative minimum point on the graph of f. Justify your answer.

C. Find $f(x)$.

Answer Key for Practice Test Two—Multiple-Choice Questions

Part A	Part B
1. D	29. E
2. A	30. D
3. D	31. C
4. B	32. A
5. A	33. C
6. A	34. D
7. D	35. C
8. D	36. C
9. B	37. A
10. C	38. C
11. D	39. A
12. B	40. C
13. C	41. B
14. D	42. D
15. B	43. A
16. E	44. A
17. B	45. E
18. E	
19. D	
20. B	
21. A	
22. E	
23. C	
24. D	
25. D	
26. C	
27. B	
28. C	

Practice Calculus AB Test Two—Multiple Choice Answers and Explanations

1. D.

   As $x \to \infty$, the behavior of the polynomials are dominated by the term of highest degree. Hence,
   $$\lim_{x \to \infty} \frac{2x - 4x^3 - 9}{6x^3 + 4x^2 + 5} = \lim_{x \to \infty} \frac{-4x^3}{6x^3} = -\frac{4}{6} = -\frac{2}{3}.$$

2. A.

   The radius of a circle is decreasing at 2 cm/sec, which means $\frac{dr}{dt} = -2$ (in cm/sec). Using the chain rule, the rate of change of the area $A = \pi r^2$ of the circle with respect to time (in cm^2/sec) is
   $$\frac{dA}{dt} = 2\pi r \frac{dr}{dt} = 2\pi r(-2) = -4\pi r.$$

3. D.

   $$\int_0^1 (4y + 1)^2 \, dy = \int_0^1 (16y^2 + 8y + 1) \, dy = \left(\frac{16y^3}{3} + 4y^2 + y \right) \Big|_0^1 = \frac{16}{3} + 4 + 1 = \frac{31}{3}.$$

4. B.

   If $f(x) = \ln x^3$ and $y = g(x)$ is the inverse function of $f(x)$, then
   $$x = \ln y^3 \implies e^x = y^3 \implies (e^x)^{\frac{1}{3}} = y \implies y = e^{\frac{x}{3}}.$$

   Thus, $g'(x) = \frac{1}{3} e^{\frac{x}{3}}$ by the chain rule.

5. A.

   You need to antidifferentiate twice here. If $f''(x) = x^3 + \sin x$, then $f'(x) = \frac{x^4}{4} - \cos x + C_1$ for some constant C_1. Now antidifferentiate again and you have
   $$f(x) = \frac{x^5}{20} - \sin x + C_1 x + C_2$$

   where C_2 is another constant. The only choice that fits this form is $\frac{x^5}{20} - \sin x + 5x + 6$ (with $C_1 = 5$ and $C_2 = 6$).

6. A.

Using implicit differentiation on the curve $3x^2 - xy + y^2 = 5$, you get

$$6x - y - x\frac{dy}{dx} + 2y\frac{dy}{dx} = 0.$$

At the point $(1, 2)$ you substitute $x = 1$ and $y = 2$ and solve for $\frac{dy}{dx}$:

$$6 \cdot 1 - 2 - 1 \cdot \frac{dy}{dx} + 2 \cdot 2 \cdot \frac{dy}{dx} = 0 \quad \Longrightarrow \quad 4 + 3\frac{dy}{dx} = 0 \quad \Longrightarrow \quad \frac{dy}{dx} = -\frac{4}{3}.$$

Finally, substitute this value for m in the point-slope form for a line to get

$$y - 2 = -\frac{4}{3}(x - 1).$$

7. D.

I is true since $v(\sqrt{n}) = (\sqrt{n})^3 - n\sqrt{n} = n^{\frac{3}{2}} - n^{\frac{3}{2}} = 0$.

II is false since $v(t) < 0$ when $0 < t < \sqrt{n}$. For example,

$$v\left(\frac{\sqrt{n}}{2}\right) = \left(\frac{\sqrt{n}}{2}\right)^3 - n\left(\frac{\sqrt{n}}{2}\right) = \frac{n^{\frac{3}{2}}}{8} - \frac{n^{\frac{3}{2}}}{2} = -\frac{3n^{\frac{3}{2}}}{8} < 0.$$

III is true since the *speed* of the particle is given by the absolute value of the velocity $|v(t)|$ and this is a minimum when $v(t) = 0$.

8. D.

The graph of g is unbroken but has a sharp corner at $x = 0$ (graph it to see). Thus, I is true, which means $\lim_{x \to 0} g(x) = g(0) = |0| = 0$ so III is also true. However, g is not differentiable at $x = 0$ so II is false.

9. B.

You can see that $f(1) = 0$ and the graph of f is decreasing and concave up at $x = 1$ so $f'(1) < 0$ and $f''(1) > 0$.

10. C.

The area A of a triangle is given by the formula $A = \frac{1}{2}bh$ where b is the base and h is the height. By the product rule, you have

$$\frac{dA}{dt} = \frac{1}{2}\left(\frac{db}{dt}h + b\frac{dh}{dt}\right)$$

and after substituting the given rates $\frac{dh}{dt} = -4$ inches per minute and $\frac{db}{dt} = 4$ inches per minute, you have

$$\frac{dA}{dt} = \frac{1}{2}(4h - 4b) = 2(h - b)$$

which is positive when $h > b$. So A is increasing when $h > b$.

11. D.

The Intermediate Value Theorem guarantees a solution to the equation $f(x) = c$ on the interval $[a, b]$ if f is continuous and c is between $f(a)$ and $f(b)$. If $f(1) = 0$, then there must be a solution to the equation $f(x) = \frac{1}{2}$ on the interval $[0, 1]$ and another solution on the interval $[1, 2]$.

12. B.

If $f(x) = x^2 \cos^2 x$, then by the Product Rule and Chain Rule you have

$$f'(x) = (2x) \cos^2 x + x^2 (2(\cos x)(-\sin x)),$$

so $f'(\pi) = (2\pi) \cos^2 \pi + \pi^2(2(\cos \pi)(-\sin \pi)) = 2\pi \cdot (-1)^2 + \pi^2(2(-1)(0)) = 2\pi$.

13. C.

$$\lim_{x \to a} \frac{a^2 - x^2}{a^4 - x^4} = \lim_{x \to a} \frac{a^2 - x^2}{(a^2 - x^2)(a^2 + x^2)} = \lim_{x \to a} \frac{1}{a^2 + x^2} = \frac{1}{a^2 + a^2} = \frac{1}{2a^2}.$$

14. D.

In general, if $g(x) = \displaystyle\int_{h(x)}^{k(x)} f(t)\, dt$, then by the Fundamental Theorem of Calculus,

$$g'(x) = f(k(x))k'(x) - f(h(x))h'(x).$$

In this case, you have $f(t) = t^2$, $k(x) = \frac{3x^2}{2} - 6x$, and $h(x) = -1 - x$, so $k'(x) = 3x - 6$, $h'(x) = -1$, and

$$g'(x) = (\tfrac{3x^2}{2} - 6x)^2(3x - 6) - (-1 - x)^2(-1) = (3x - 6)(\tfrac{3x^2}{2} - 6x)^2 + (-1 - x)^2.$$

15. B.

$$\lim_{x \to \pi} \frac{\sin \pi - \sin x}{\pi - x} = \lim_{x \to \pi} \frac{\sin x - \sin \pi}{x - \pi} = f'(\pi)$$

where $f(x) = \sin x$. Since $\dfrac{d}{dx}(\sin x) = \cos x$, you have $f'(\pi) = \cos \pi = -1$.

16. E.

The acceleration of the object is represented by the slope of the velocity graph. The slope of the graph has its greatest slope on the interval $8 < t < 9$.

17. B.

The signed distance of the object from its starting point at time $t = x$ is given by

$$\int_0^x v(t)\,dt$$

which is represented by the accumulated signed area under the velocity graph from $t = 0$ to $t = x$. Checking each choice, you have

$$\int_0^2 v(t)\,dt = \frac{1}{2}(2)(2) = 2 \qquad \int_0^5 v(t)\,dt = \frac{1}{2}(5)(2) = 5$$

$$\int_0^6 v(t)\,dt = 5 - \frac{1}{2}(1)(2) = 5 - 1 = 4 \qquad \int_0^8 v(t)\,dt = 5 - \frac{1}{2}(1)(2) - 2(2) = 5 - 1 - 4 = 0$$

and $\int_0^9 v(t)\,dt = 5 - \frac{1}{2}(1)(2) - 2(2) - \frac{1}{2}(\frac{2}{3})(2) + \frac{1}{2}(\frac{1}{3})(1) = 5 - 1 - 4 - \frac{1}{2} = -\frac{1}{2}.$

The greatest value is given by $t = 5$.

18. E.

If the position of the object at time t is written as $x(t)$, then the change in the object's position between time $t = 5$ and time $t = 8$ is given by

$$\int_5^8 v(t)\,dt = x(8) - x(5).$$

You are given that at time $t = 8$ the object was at position $x = 10$, so $x(8) = 10$. The signed area under the velocity graph between time $t = 5$ and $t = 8$ is

$$\int_5^8 v(t)\,dt = -\frac{1}{2}(1)(2) - 2(2) = -5.$$

Thus, $-5 = x(8) - x(5) = 10 - x(5)$ and $x(5) = 15$.

19. D.

The total distance traveled by the object for $2 \le t \le 7$ is given by

$$\int_2^7 |v(t)|\,dt$$

so you must measure the absolute value of the area between the velocity graph and the t-axis. This is given by the areas of two triangles and a rectangle:

$$\frac{1}{2}(3)(2) + \frac{1}{2}(1)(2) + (1)(2) = 3 + 1 + 2 = 6.$$

---

20. B.

    To fit the data in the table, the graph of f needs to have negative slope for $x < 0$ and positive slope for $0 < x < 1$ and $1 < x$. The graph of f must also be concave up for $x < 0$ and $0 < x < 1$ and concave down for $1 < x$. Thus, the graph will have a minimum at $x = 0$ and an inflection point at $x = 1$. The only graph that satisfies these conditions is B.

---

21. A.

    For all $x > 1$, if $f(x) = \int_1^x \left(t - \dfrac{2}{t} \right) dt$, then by the Fundamental Theorem of Calculus you have $f'(x) = \left(x - \dfrac{2}{x} \right)$.

---

22. E.

    $$\int \left(\frac{x^2 + 1}{x} \right) dx = \int \left(x + \frac{1}{x} \right) dx = \frac{x^2}{2} + \ln |x| + C.$$

---

23. C.

    Since the acceleration $a(t) = 18t - 2$, and $v'(t) = a(t)$, you have

    $$v(t) = 9t^2 - 2t + C.$$

    Since the velocity is 12 when $t = 2$, you have $12 = v(2) = 9(2)^2 - 2(2) + C = 36 - 4 + C = 32 + C$. Thus, $C = -20$ and $v(t) = 9t^2 - 2t - 20$. Since $x'(t) = v(t)$, you have

    $$x(t) = 3t^3 - t^2 - 20t + K.$$

    Since the position is 2 when $t = 1$, you have $2 = x(1) = 3(1)^3 - (1)^2 - 20(1) + K = 3 - 1 - 20 + K = -18 + K$. Thus, $K = 20$ and

    $$x(t) = 3t^3 - t^2 - 20t + 20.$$

---

24. D.

    $\displaystyle \lim_{t \to b} \frac{\ln t - \ln b}{t - b} = f'(b)$, where $f(t) = \ln t$. Since $\dfrac{d}{dt}(\ln t) = \dfrac{1}{t}$, you have $f'(b) = \dfrac{1}{b}$.

---

25. D.

    If $f(x) = 3x^7 - 10x^5$, then $f'(x) = 21x^6 - 50x^4$ and $f''(x) = 126x^5 - 200x^3 = x^3(126x^2 - 200)$. You can see that $f''(x)$ changes sign at $x = 0$ and at $x = \pm\sqrt{\frac{200}{126}}$, so the number of inflection points of $f(x) = 3x^7 - 10x^5$ is three.

26. C.

If $y = e^x \cos x$, then $\dfrac{dy}{dx} = e^x \cos x + e^x(-\sin x) = e^x(\cos x - \sin x)$.

27. B.

If $g(x) = \displaystyle\int_0^x \cos(t - \pi)\,dt$, then then by the Fundamental Theorem of Calculus, you have $g'(x) = \cos(x - \pi)$ which changes sign from negative to positive at $x = \dfrac{\pi}{2}$. Thus, the minimum of $g(x)$ on the interval $[0, 2\pi]$ occurs at $x = \dfrac{\pi}{2}$.

28. C.

Separating variables in the differential equation $x\dfrac{dy}{dx} = 10$ gives you

$$dy = \frac{10}{x}\,dx \implies \int dy = \int \frac{10}{x}\,dx \implies y = 10\ln|x| + C.$$

Substituting $x = 1$ and $y = 0$, you have

$$0 = 10\ln(1) + C \implies 0 = 0 + C \implies C = 0$$

so the solution is $y = 10\ln x$ since $x > 0$.

29. E.

For $g(x)$, $g'(x)$, and $g''(x)$ to all be less than zero, the point on the graph must be below the x-axis and the graph must be decreasing and concave down. This occurs only at point E.

30. D.

Using implicit differentiation on the equation $5x^2y^4 + 3y^3 + 9xy^2 = 17$, you have

$$10xy^4 + 5x^2(4y^3)\frac{dy}{dx} + 9y^2\frac{dy}{dx} + 9y^2 + 9x(2y)\frac{dy}{dx} = 0.$$

Solving for $\dfrac{dy}{dx}$, you get

$$\frac{dy}{dx} = -\frac{10xy^4 + 9y^2}{20x^2y^3 + 9y^2 + 18xy}.$$

31. C.

$$\lim_{x \to b} f(x) = \lim_{x \to b} \frac{b^2 - x^2}{b - x} = \lim_{x \to b} \frac{(b + x)(b - x)}{b - x} = \lim_{x \to b}(b + x) = 2b \neq 0 = f(b).$$

Thus, I and II are true since $\lim\limits_{x \to b} f(x)$ exists and $f(b)$ exists, but III is false since they are not equal and $f(x)$ is not continuous at $x = b$.

32. A.

The diameter d of each semicircular cross-section from $x = 0$ to $x = 8$ is the height of the graph $y = x^{\frac{1}{3}}$ at that point. A semicircle has area $\dfrac{\pi r^2}{2}$ where $r = \dfrac{d}{2}$. Thus, the volume of the solid is given by the definite integral

$$\int_0^8 \frac{\pi(\frac{d}{2})^2}{2}\,dx = \int_0^8 \frac{\pi\left(\dfrac{x^{\frac{1}{3}}}{2}\right)^2}{2}\,dx = \pi\int_0^8 \frac{x^{\frac{2}{3}}}{8}\,dx$$

$$= \frac{\pi}{8}\frac{3}{5}x^{\frac{5}{3}}\Big|_0^8 = \frac{3\pi}{40}(8^{\frac{5}{3}} - 0) = \frac{96\pi}{40} \approx 7.54$$

33. C.

Using the intervals $[2, 5]$, $[5, 7]$, $[7, 8]$, the trapezoidal method approximates the integral

$$\int_2^8 f(x)\,dx \approx \frac{f(5) + f(2)}{2}(5 - 2) + \frac{f(7) + f(5)}{2}(7 - 5) + \frac{f(8) + f(7)}{2}(8 - 7)$$

$$= \frac{40}{2}(3) + \frac{70}{2}(2) + \frac{60}{2}(1) = 60 + 70 + 30 = 160.$$

34. D.

$$\frac{d}{dt}(\sqrt{p}) = \frac{d}{dt}(p^{\frac{1}{2}}) = \frac{1}{2}p^{-\frac{1}{2}}\frac{dp}{dt}.$$

For this rate of increase to be one-half of $\dfrac{dp}{dt}$ you must have $p^{-\frac{1}{2}} = 1$ or $p = 1$.

35. C.

The area of the region bounded by $y = \sqrt{x}$, $y = -x$, and the line $x = 4$ is given by

$$\int_0^4 (x^{\frac{1}{2}} - (-x))\,dx = \int_0^4 (x^{\frac{1}{2}} + x)\,dx$$

$$= \left(\frac{2}{3}x^{\frac{3}{2}} + \frac{x^2}{2}\right)\Big|_0^4 = \frac{2}{3}4^{\frac{3}{2}} + \frac{4^2}{2} = \frac{16}{3} + 8 = \frac{40}{3} = 13\frac{1}{3}.$$

36. C.

The average value of $\frac{1}{2}t^2 - \frac{1}{3}t^3$ over the interval $-2 \le t \le 1$ is given by

$$\frac{1}{1 - (-2)}\int_{-2}^1 \left(\frac{1}{2}t^2 - \frac{1}{3}t^3\right)dt = \frac{1}{3}\left(\frac{1}{6}t^3 - \frac{1}{12}t^4\right)\Big|_{-2}^1$$

$$= \frac{1}{3}\left[\left(\frac{1}{6} - \frac{1}{12}\right) - \left(\frac{1}{6}(-8) - \frac{1}{12}(16)\right)\right] = \frac{1}{3}\left[\left(\frac{1}{12}\right) + \left(\frac{8}{3}\right)\right] = \frac{1}{3}\frac{33}{12} = \frac{11}{12}.$$

---

37. A.

$$\int_{-2}^{2} f(x)\, dx = \int_{-2}^{0} x^2\, dx + \int_{0}^{2} x^3\, dx = \frac{x^3}{3}\Big|_{-2}^{0} + \frac{x^4}{4}\Big|_{0}^{2} = -\frac{(-2)^3}{3} + \frac{2^4}{4} = \frac{8}{3} + \frac{16}{4} = \frac{20}{3}.$$

---

38. C.

I is true since $f(x) \geq 0$ over $[a, b] \implies \int_{a}^{b} f(x)\, dx \geq 0$.

II is true since $\int_{b}^{a} g(x)\, dx = -\int_{a}^{b} g(x)\, dx$, and hence

$$\int_{a}^{b} (f(x) - g(x))\, dx = \int_{a}^{b} f(x)\, dx - \int_{a}^{b} g(x)\, dx = \int_{a}^{b} f(x)\, dx + \int_{b}^{a} g(x)\, dx.$$

III is false. For example, let $f(x) = x$ with $a = 0$ and $b = 1$. Note that

$$\int_{0}^{1} x^3\, dx = \frac{x^4}{4}\Big|_{0}^{1} = \frac{1}{4}$$

but, in contrast,

$$\left(\int_{0}^{1} x\, dx\right)^3 = \left(\frac{x^2}{2}\Big|_{0}^{1}\right)^3 = \left(\frac{1}{2}\right)^3 = \frac{1}{8}.$$

---

39. A.

If $\int_{0}^{c} (cx^3 - x)\, dx = 0$, then

$$\left(\frac{cx^4}{4} - \frac{x^2}{2}\right)\Big|_{0}^{c} = 0 \implies \frac{c^5}{4} - \frac{c^2}{2} = 0 \implies c^2\left(\frac{c^3}{4} - \frac{1}{2}\right) = 0.$$

Since $c > 0$, then $\frac{c^3}{4} - \frac{1}{2} = 0 \implies c^3 = 2 \implies c \approx 1.26$.

---

40. C.

Since $f(x) > g(x)$, for $a < x < b$, when the region is rotated around the x-axis to form a solid, the area of each cross-section from $x = a$ to $x = b$ will be $\pi(f(x))^2 - \pi(g(x))^2$. Thus, the integral that represents the volume of the solid that is generated is

$$\pi \int_{a}^{b} (f(x))^2\, dx - \pi \int_{a}^{b} (g(x))^2\, dx.$$

41. B.

If $\displaystyle\int_{-3}^{2} f(x)\,dx = 6$, and $\displaystyle\int_{2}^{1} f(x)\,dx = 11$, then

$$\int_{-3}^{1} f(x)\,dx = \int_{-3}^{2} f(x)\,dx + \int_{2}^{1} f(x)\,dx = 6 + 11 = 17.$$

42. D.

If $g(x) = x^3 + x + 1$, then $g'(x) = 3x^2 + 1$. If h is the inverse function of g, then $h(3) = g^{-1}(3) = 1$, since $g(1) = 1^3 + 1 + 1 = 3$. Thus,

$$h'(3) = \frac{1}{g'(1)} = \frac{1}{3(1^2) + 1} = \frac{1}{4}.$$

43. A.

The particle travels along the curve $y = -2x^2 + 8$ where $x > 0$, so $\dfrac{dy}{dx} = -4x$. At the instant the particle crosses the x-axis, you must have $y = -2x^2 + 8 = 0 \implies x^2 = 4 \implies x = 2$, since $x > 0$. By the chain rule,

$$\frac{dy}{dt} = \frac{dy}{dx} \cdot \frac{dx}{dt} \quad \implies \quad \frac{dx}{dt} = \frac{dy/dt}{dy/dx}.$$

Substitute $-4(2) = -8$ for $\dfrac{dy}{dx}$ and $\dfrac{dy}{dt} = 3$, and you find $\dfrac{dx}{dt} = -\dfrac{3}{8}$.

44. A.

If $\dfrac{dy}{dx} = \dfrac{4}{2x - 3}$, then $y = \displaystyle\int \dfrac{4}{2x - 3}\,dx$. Let $u = 2x - 3$ so that $du = 2\,dx$ and

$$y = \int \frac{4}{2x - 3}\,dx = \int \frac{2}{u}\,du = 2\ln|u| + C = 2\ln|2x - 3| + C$$

for some constant C. You also know $y(2) = 1$, so you can substitute $x = 2$ and $y = 1$ and solve for C:

$$1 = 2\ln|2(2) - 3| + C \implies 1 = 2\ln(1) + C \implies 1 = 0 + C \implies C = 1.$$

Thus, $y = 2\ln|2x - 3| + 1$.

45.   E.

You are given that $f(x)$ is differentiable on the domain $-4 \leq x \leq 4$, and all critical points for the curve $y = f(x)$ are included in this table:

x	-4	-3	-2	-1	0	1	2	3	4
$f(x)$	43	38	31	21	10	-1	-11	-18	-23

I. $f(x)$ is monotonic on the domain $-4 \leq x \leq 4$. This statement is true. You can see the values in the table are strictly decreasing, but you have to consider the behavior of the function between these values. However, if $f(x)$ changed from decreasing to increasing at some value x, then $f'(x) = 0$ there and x would be a critical point. Since you know that there are no other critical points than those included in the table, you can rule out this possibility.

II. There is some c between 0 and 1 such that $f'(c) = -11$. This statement is true. Using the Mean Value Theorem, there must exist some c between 0 and 1 such that

$$f'(c) = \frac{f(1) - f(0)}{1 - 0} = \frac{(-1) - 10}{1} = -11.$$

III. The average value of $f(x)$ on the domain $-4 \leq x \leq 4$ is greater than zero. This statement is true. Since you already have determined that $f(x)$ is strictly decreasing, using a right-hand Riemann sum will guarantee you an underestimate of the integral $\int_{-4}^{4} f(x)\, dx$. Thus, the average value of $f(x)$ over the domain $-4 \leq x \leq 4$ is

$$\frac{1}{4 - (-4)} \int_{-4}^{4} f(x)\, dx$$

$$> \frac{1}{8}(38 + 31 + 21 + 10 + (-1) + (-11) + (-18) + (-23)) = \frac{47}{8} > 0.$$

So, all three statements are true!

Answers to Free-Response Questions

1.

A. $a = -.594268$, $b = 1$, $Area = \int_{a}^{b} (g(x) - f(x))\, dx \approx 1.195$

B. $V_x = \pi \int_{a}^{b} (g(x)^2 - f(x)^2)\, dx \approx 4.802$

C. $V = \frac{\sqrt{3}}{4} \int_{a}^{b} (g(x) - f(x))^2\, dx \approx 0.476$

2.

A. $a(t) = v'(t) = \dfrac{1}{1.5 + \sin(t)}\cos(t)$ $a(3) = v'(3) = \dfrac{\cos(3)}{1.5 + \sin(3)} \approx -0.603.$

B. $x(t) = 2 + \displaystyle\int_0^t v(u)\, du$ $x(3) = 2 + \displaystyle\int_0^3 (\ln(1.5 + \sin(u)))\, du \approx 4.286.$

C. $v(t) > 0$ for $0 < t < 3.665191$ $v(t) < 0$ for $3.665191 < t < 5.$

$$\text{Total distance} = \int_0^{3.665191} (\ln(1.5 + \sin(t)))\, dt + \int_{3.665191}^5 (-\ln(1.5 + \sin(t)))\, dt$$

$$\approx 2.46105 + 0.63161 \approx 3.093.$$

D. At both endpoints, $x(t)$ has a minimum. As shown in part C,

$v(t) = x'(t) > 0$ for $0 < t < 3.665191$ and $v(t) < 0$ for $3.665191 < t < 5.$

So, $x(t)$ has its absolute maximum at $t = 3.665191.$

3.

A. $R'(10) \approx \dfrac{R(12) - R(8)}{12 - 8} = \dfrac{320 - 160}{4} = 40.$

B. total $\approx 2 \cdot \left(\dfrac{36+78}{2} + \dfrac{78+160}{2} + \dfrac{160+240}{2} + \dfrac{240+320}{2} + \dfrac{320+350}{2} \right.$

$\left. + \dfrac{350+360}{2} + \dfrac{360+320}{2} + \dfrac{320+240}{2} + \dfrac{240+160}{2} \right) = 4332$ amps

C.) $A'(t) = -170(0.26) \sin(0.26(t - 15))$

$A'(10) = -170(0.26) \sin(-1.3) \approx 42.589$

$\dfrac{1}{18} \displaystyle\int_{4}^{22} A(t)\, dt \approx 245.297$ amps

4.

A. $\displaystyle\lim_{x \to \infty} \dfrac{3}{e^{-x} + 2} = \dfrac{3}{2}.$

B. $\displaystyle\lim_{x \to -\infty} \dfrac{3}{e^{-x} + 2} = 0.$

C.) $g'(x) = -3(e^{-x} + 2)^{-2} \cdot (-e^{-x}) = \dfrac{3e^{-x}}{(e^{-x} + 2)^2} > 0.$

Since $g'(x) > 0$ for all real numbers x, $g(x)$ is increasing for all x.

D. $g''(x) = \dfrac{(e^{-x} + 2)^2(-3e^{-x}) - 3e^{-x}(2)(e^{-x} + 2)(-e^{-x})}{(e^{-x} + 2)^4} = \dfrac{3e^{-x}(e^{-x} + 2)(e^{-x} - 2)}{(e^{-x} + 2)^4}.$

Thus, $g''(x) = 0$ for $x = -\ln(2)$. Since the graph of g is concave down where $g''(x) < 0$, this will occur for $x > -\ln(2)$.